Design, Mediation, and the Posthuman

Postphenomenology and the Philosophy of Technology

Technological advances affect everything from our understandings of ethics, politics, and communication, to gender, science, and selfhood. Philosophical reflection on technology helps draw out and analyze the nature of these changes, and helps us understand both the broad patterns and the concrete details of technological effects. This book series provides a publication outlet for the field of the philosophy of technology in general, and the school of thought called "postphenomenology" in particular. Philosophy of technology applies insights from the history of philosophy to current issues in technology, and reflects on how technological developments change our understanding of philosophical issues. In response, postphenomenology analyzes human relationships with technologies, while integrating philosophical commitments of the American pragmatist tradition of thought.

Design, Meditation, and the Posthuman, Edited by Dennis M. Weiss, Amy D. Propen, and Colbey Emmerson Reid

Design, Mediation, and the Posthuman

Edited by
Dennis M. Weiss, Amy D. Propen,
and Colbey Emmerson Reid

LEXINGTON BOOKS
Lanham • Boulder • New York • London

Published by Lexington Books
An imprint of The Rowman & Littlefield Publishing Group, Inc.
4501 Forbes Boulevard, Suite 200, Lanham, Maryland 20706
www.rowman.com

16 Carlisle Street, London W1D 3BT, United Kingdom

British Library Cataloguing in Publication Information Available

Library of Congress Cataloging-in-Publication Data

Library of Congress Cataloging-in-Publication Data Available
ISBN 978-0-7391-9177-4 (cloth : alk. paper)
ISBN 978-0-7391-9178-1 (electronic)
ISBN 978-1-4985-0115-6 (pbk : alk. paper)

∞™ The paper used in this publication meets the minimum requirements of American National Standard for Information Sciences Permanence of Paper for Printed Library Materials, ANSI/NISO Z39.48-1992.

Printed in the United States of America

Contents

Acknowledgments

This book is about how complex the interface between technology, design, and human users has become—and always has been. It's also about how hard it is to claim to think about anything so complicated without the host of perspectives facilitated by the kind of specialized knowledge it is impossible for any one person to hold. But the book is not only about this condition; it enacts it. Fourteen authors from twelve institutions, six disciplines, and three continents have contributed to the ideas voiced here. All have been generous in sharing their original work, working over the course of two years to respond to several rounds of multiple editors' commentaries, questions, and probes. Our contributors have not only been patient through the process—they have been enthusiastically responsive, and have made the volume better than the high hopes we had for it.

Design, Mediation, and the Posthuman began as a conference panel in 2011, and we would like to thank the intellectual generosity of the Humanities and Technology Association and the audience who attended our panel for sparking an ongoing, multi-year conversation between the editors of this book. All of our thinking, on this project and others, has been enriched by it. The Society for the Philosophy of Technology in Lisbon, Portugal, in 2013 served as a valuable opportunity to check in face-to-face as the project reached its final stages. We are grateful to the Faculty Development Committee at York College of Pennsylvania for funding both of these trips (also supported by the business management department at North Carolina State University), as well as grants to support the authoring of two of this book's chapters.

Thanks to Lexington Books, Jana Hodges-Kluck, and Natalie Mandziuk for their support of interdisciplinary, multi-authored inquiries into postphenomenology and philosophy of technology. We believe that the kind of scholarly interchange that takes place through projects like this is indispensable to the advancement of thought, and are glad that such opportunities are afforded by the visionary editors and presses who materially support them.

Finally, thanks to Norma, Karen, and Todd for supporting at home what happens on the page (and sometimes, Portugal).

Introduction

Mind vs. Thing, and Other Central Events of the Twenty-First Century

This introductory chapter is about a face-off that is taking place right now, between two widespread assumptions about the present era. The assumptions are about the character of our age and the outcome of their contest, a conflict of ideas that bears material repercussions for us all, may determine the direction of our culture's design: its environment, technology, economy, domesticities, and sociality. Like any good ideological face-off, both the coming together and the outcome of the contest represent "planetary dilemmas" that defy any kind of "final solution, since any resolution generates further issues, and . . . solutions are not true or false or good or bad, but the best that can be done at the time." Such dilemmas might be called "wicked problems," not in the moral sense but in the sense that they are "diabolical" in their resistance to "all the usual attempts to resolve them" (Brown, Harris, and Russell 2010, 4). They call, as such, for a new kind of imagination, a "transdisciplinary imagination" that draws on "all our intellectual resources, valuing the contributions of all the academic disciplines as well as other ways in which we construct our knowledge." The present collection of chapters enacts just such an instance of transdisciplinary imagination. It exists because we believe that only the kind of coalition that transgresses the conventional siloing of knowledge by academic discipline and professional specialization can—not *solve*, but even begin to *think* about what is happening now.

And what is *happening now*? Specialists and academics alike tend to agree that a technological revolution has reshaped the very fiber of civilization, reforming characteristics of cultures and human beings that have long been considered foundational to those entities, even displacing culture and human beings with a computer-centric network the likes of which has never been seen, in precisely this way, before. Even if everyone seems to disagree about the outcome, value, extent, origin, and degree of uniqueness of the change, most agree that how we live now is not quite like how we lived before, and that the distance between now and before seems to be getting wider as the speed of computational advancement increases, exponentially, with every new incarnation of it.

In the wake of our technological revolution, which has either, depending on whom we ask about it, just begun or simply accelerated something begun over two hundred and fifty years ago, and will either be our ruin or our salvation but rarely anything in between, two primary assumptions have begun to constitute our civilization's understanding of its ever-increasingly techno-centric self. The first, which emerged in the 1990s, proclaims that computers have introduced us into a virtual era in which the material world is becoming irrelevant. The second emerged just a few years after the turn of the millennium, and declares a "comeback" for matter, which hasn't become irrelevant after all. Two quintessential examples: social media on the one hand, and digital devices on the other. The former have brought about a much decried but nevertheless ubiquitously enjoyed dissolution of materiality's centrality to social exchanges, which now centers around the interaction of disembodied minds on a screen and does not require a human body or a natural or architectural space in which to occur. Facebook literally incarnates virtual culture's mind-centrality as paradoxically face*less.* On the other hand, the rise of hand-held digital devices—tablets, cell phones, MP3 players, eBooks, and other "smart objects" have made us increasingly aware of our bodies' interactions with machines. Thus the line is drawn in the sand, and the competitors emerge: software vs. hardware; production vs. consumption; Microsoft vs. Apple; MIND vs. THING.

As we discuss below, the ascent of mind over materiality facilitated by virtual culture has repeatedly been declared, and by various sources, the "central event of the twentieth century." The emergence of matter as a trenchant counter-presence in spite of this might be called the central event of the twenty-first. Moreover, the editors of this volume don't think the mind vs. thing event is, at present, an equal contest. We believe that currently, Thing is edging ahead—not, perhaps, fully and straightforwardly enough to declare a champion—the game is still on—but sufficiently to make us all doubt the triumph of the virtual that was all but assumed an inevitability merely fifteen years ago. Even Facebook *aficionados* react passionately to (and often against) the software's design, forced to realize that their virtual world is, actually, *something*, and that its shape shapes virtual culture. Things are going in a different direction than we all expected when we started along this path, not so much dissolving and retreating as reasserting themselves. That alone is a surprise worth noting. The present volume does so, but also reflects on what it means, in a variety of contexts, when Thing edges ahead.

We begin our collection with an overview: this chapter, which is designed to accomplish three primary goals. First, we spell out the origins of mind vs. thing: where (and who) did it come from? Without lingering too long on mind, which as the reigning discourse of humanism can claim most of that period as its *homage*, we also define the terms of Thing: what, exactly, is meant by it? Second, we survey and synthesize the diverse forms taken across contemporary academic culture of what the Center for Twenty-first-Century Studies at the University of Wisconsin-Milwaukee has called "the nonhuman turn." A central purpose of this volume is not so much to declare the Rise of Thing, a declaration that is already emerging across the academy and that constitutes more of a chorus than an argument at this point, but rather to point out the occurrence of the chorus as delineating the appearance of a new academic meta-trend. Finally, we acknowledge in the third and last section of this chapter some points of contest and divergence in the different incarnations of what we argue ought to be a common, transdisciplinary conversation about what is, perhaps, our era's quintessential "wicked problem." The differences in approach are important to maintain, because they delineate certain overarching questions that might otherwise remain invisible—that have, in any case, remained unarticulated by individual versions of the discussion. It is here, furthermore, in the spaces of contest and irresolution, where most of the chapters collected in our volume reside. Assuming Thing is on the rise, the face-off is still interesting when we query the stakes of a win, what we're rooting for when we root for a winner, or where the possibilities of a rematch, or a third competitor, might emerge. The collected chapters, which follow in three sections labeled "interface," "artifact," and "user," contemplate precisely these issues according to their implications for our understanding of mediation, the things themselves, and the new kind of human being constructed when we consider and accelerate our interface with nonhumanity in all its varied forms.

THE TURN FROM MIND TO MATTER

There's always a certain amount of risk entailed in looking from one's limited perspective toward the near past or close future and making pronouncements regarding the ways things have been or might yet be. But the risk doesn't often prevent scholars, journalists, commentators, and others from making such pronouncements, especially regarding the opening or closing of a new age or the historical turn from one period to another. Indeed, in his 1986 study *The Control Revolution: Technological and Economic Origins of the Information Society*, James Beniger notes that we have apparently lived through more than sixty societal transformations between 1950 and 1984. According to Beniger, more than half of these transformations were driven by computers and information technology. The succession of transformations identified have included the computer revolution, the electronics revolution, the information age, the age of the smart machine, and the micro revolution.

Beniger also notes the chronic inability of past generations to appreciate the major societal transformations of their own eras. And yet, despite this myopia, there has been no lessening of pronouncements concerning this transformation of society, that entry into a new age, or the dawning of another new era. Risk, it seems, we must, as we struggle to make sense of the transformations we collectively live through. So while recognizing the risk of oversimplifying and flirting with the possibility of proving Beniger's generalization true, consider, if for no other purpose than an analytical one, what has happened since 1984 and why, despite the risks of historical myopia, we might claim to have witnessed something of a turn, a turn we might denominate as moving us from the virtual age to a new age of materiality in which things matter.

It seems fitting to begin with 1984, a year noteworthy for perhaps failing to usher in Orwell's own future-oriented pronouncements, but which did witness the emergence of twin texts mediating our vision of the future: William Gibson's *Neuromancer* and Apple's Super Bowl commercial for the Macintosh. Both texts seemingly presaged the opening of a virtual age in which our technology would liberate us from the chains of the material world. They promised an age of freedom, direct democracy, and identity play in the virtual realm of cyberspace. These themes were almost immediately taken up by fanboys and cybertheorists alike, who proclaimed that in the movement from the offline world to the online world we were witnessing the demassification of culture.

Consider in this vein two founding political documents from the age of cyberspace. In "Cyberspace and the American Dream: A Magna Carta for the Knowledge Age" (1994), Esther Dyson, Alvin Toffler, George Keyworth and George Gilder argued that the central event of the twentieth century is the overthrow of matter. As they put it, "In technology, economics, and the politics of nations, wealth—in the form of physical resources—has been losing value and significance. The powers of mind are everywhere ascendant over the brute force of things." We were witnessing, they argued, a shift from a mass-production, mass media, mass-culture civilization to a demassified civilization. Matter no longer mattered. We were moving, the authors suggested, from the world of atoms to the world of bits. Power in the knowledge age would be about mind and consciousness and information, not material wealth. John Perry Barlow's 1996 cyberspace manifesto "Declaration of Independence for Cyberspace" concurred. Barlow, a former lyricist for The Grateful Dead and one of the founders of the Electronic Frontier Foundation, declared in that influential and widely shared document: "Governments of the Industrial World, you weary giants of flesh and steel, I come from Cyberspace, the new home of mind. On behalf of the future, I ask you of the past to leave us alone. You are not welcome among us." Barlow's Declaration and the Magna Carta set up an opposition between the weary world of flesh and matter, the rusted world of the past and the future world of cyberspace and mind. As Barlow noted, "This governance will arise according to the conditions of our world, not yours. Our world is different. Cyberspace consists of transactions, relationships, and thought itself, arrayed like a standing wave in the web of our communications. Ours is a world that is both everywhere and nowhere, but it is not where bodies live." Two years earlier, in an equally influential early cyberspace manifesto, Barlow closed his essay on "the galloping digitization of everything not obstinately physical," called "The Economy of Ideas," with the following age-defining pronouncement: "And finally, in the years to come, most human exchange will be virtual rather than physical, consisting not of stuff but the stuff of which dreams are made. Our future business will be conducted in a world made more of verbs than nouns."

Life, we were being told again and again in the early and mid-1990s, would become increasingly virtual, and then the world of things wouldn't matter much. In *The War of Desire and Technology at the Close of the Mechanical Age* (1995) Allucquere Rosanne Stone argued that we were entering a new historical situation in which communications technology challenged and transformed existing definitions of the self. Stone suggested that the virtual age was giving birth to a new social space, which she referred to as the "technosocial," in which technology is viewed as nature and has become invisible. This technosocial space gives rise to new collective structures and gives new meanings to presence, agency, identity, and the body: "The identities that emerge from these interactions—fragmented, complex, diffracted through the lenses of technology, culture, and new technocultural formations—seem to me to be, for better or worse, more visible as the critters we ourselves are in the process of becoming, here at the close of the mechanical age" (1995, 36). Such assertions assumed that digital subjectivity was becoming dislocated from the body and other anchoring materialities, inducting the self into a world of imaginative play and performative possibility ascendant over "the brute force of things" (Dyson et al., 1994). Stable identity, it was forecast, would become with material fixity a vestige of bygone eras.

But something unexpected happened on the way to the electronic frontier. Again, for analytical purposes, let's call it Y2K and date it around the turn of the century. It is perhaps hard for today's denizens of the Facebook, Instagram, Snap Chat generation to recall the near hysteria that greeted the emergence of the so-called millennium bug in the mid-1990s, in large part no doubt because so little actually happened, as the title of a 2013 documentary from Retro Report makes clear: "Y2K: Much Ado About Nothing?" Retro Report is itself in the business of re-mediation, as the mission statement posted to their website makes clear: "Combining documentary techniques with shoe-leather reporting, we peel back the layers of some of the most perplexing news stories of our past with the goal of encouraging the public to think more critically about current events and the media."

Retro Report's documentary re-examining the events surrounding the Y2K millennium bug begins with the statement, "At the close of the twentieth century, a tiny computer glitch threatened to disrupt nearly every aspect of modern life." And it's with this opening statement that we get a sense that this was not in fact a matter of "much ado about nothing," for Retro Report's account makes clear just how inextricably intertwined were the virtual and the material worlds. Y2K was perhaps the first major inkling the world had of the extent to which computer chips were now embedded in everything. Now, a mere software kluge, a matter of abbreviating the year in terms of two digits rather than four, could lead to international disaster in which planes fell from the sky, elevators ceased to work, and nuclear power plants went offline. As Senator Christopher Dowd warned, in a televised segment quoted in the report, "you wouldn't want to be in an airplane, you wouldn't want to be on an elevator, you wouldn't want to be in a hospital." Or consider another take on this immanent apocalypse included in Retro Report's documentary. A reporter is standing in a neonatal unit, his camera fixed on a young life enveloped by technology, as he intones: "The life of this baby depends on computer chips, they're embedded in the machines that monitor his heart rate and the question is will they work on the night of January 1, 2000." Here we witness an early nod to our increasingly cyborg existence in which computer chips were mediating the very processes sustaining infant life and human beings were increasingly and inextricably enrolled in complex networks of computers, software, and things. Technology design, humans' mediation by it, and our imminent posthuman existence suddenly come to the forefront.

Far from being a matter of nothing, then, we could read Y2K as being about nothing more than matter. As reporters tried to make the dimensions of the problem understandable for a skeptical public, reporter Gina Smith is shown comparing the digital world of zeros and ones to the bolts holding together the Brooklyn Bridge. "Imagine someone told you that you had to find and replace every bolt in the Brooklyn Bridge. That's daunting but doable. Now imagine you have to find and replace every bolt in every bridge in America and then the world. And you only have twenty-one months to do it."

Drawing on the very material bolts holding together the very thingly structure of the Brooklyn Bridge in order to illustrate the problems with virtual software code could stand as an apt metaphor for the resurgence of matter in the new century. Things, objects, and pure materiality were making a comeback. Kevin Ashton claims to have coined the phrase "The Internet of Things" during this period, in 1999, and in a 2009 piece looking back on that coinage, reminds us: "We're physical, and so is our environment. Our economy, society and survival aren't based on ideas or information—they're based on things. You can't eat bits, burn them to stay warm or put them in your gas tank. Ideas and information are important, but things matter much more." Graham Harman officially coined the term "object-oriented philosophy" in his 1999 doctoral dissertation "Tool-Being: Elements in a Theory of Objects." The year 2001 saw the publication of Bill Brown's special issue of *Critical Inquiry* devoted to Thing theory. Reality even made a comeback on television, with reality TV exploding as a phenomenon in the late 1990s and early 2000s. Even cyberspace guru John Perry Barlow was said to be getting in on the act. In an article titled "John Perry Barlow 2.0: The Thomas Jefferson of cyberspace reinvents his body—and his politics," Brian Doherty, writing for Reason Magazine, reported that Barlow was set to become the star of a new reality TV show, tentatively titled *Walking Time Bomb,* that would chronicle his extensive attempts to improve his health after many years of self-abuse. The virtual was out and authenticity and the material were back in. Consumer research reports that authenticity is trending as an overriding important value sought in products and services by consumers today, even in those that could ostensibly be categorized as part of the virtual revolution (Rose and Wood 2005; Beverland and Farrelly 2010, etc). It would seem, then, that over the course of the past twenty years we have witnessed a turn from the virtual age to the age of the thing, or at least to a recognition that in spite of our many technological advancements—and often through these advancements—matter continues to matter, and things still do things.

Michael Hardt and Antoni Negri have called the realization a "corporeal transformation," defined as a growing awareness that "human nature is in no way separate from nature as a whole," "that nature itself is an artificial terrain open to ever new mutations, mixtures, and hybridizations" (Hardt and Negri 2000, 215). The perception of nature as a recombinatory territory has decreased the value of metaphysical essences, often an immiscible bundle, and emphasized the tangible material. Though the word "corporeal" can carry specific associations with the human body, the OED emphasizes the term's historical association with matter and materiality, its reference to the "tangible, associated with material objects." It is no accident that contemporary usage divested corporeality of its association with things. In the original usage, a reference to the corporeal did not mandate a body, but when it did refer to one it positioned even the body as an object, divested of qualities that enliven it. A corporeal transformation describes a shift to the world of matter, emphasizing the thingness of and in a body. Hardt and Negri's adoption of the term enacts a useful anachronism, because it reminds us that the turn to matter doesn't simply abandon humanistic inquiry but repositions it. The corporeal transformation demarcates a new kind of fascination with the human as constituted by the world: the materiality that impacts us, and that makes us up. This is an important distinction from the version of interest in the philosophy of embodiment that was prominent in the academy in the 1990s, which led scholars to investigate corporeality only in the sense of human embodiment. Now, not only human bodies but also the forms of inorganic matter into which humans project and extend themselves (or are in ourselves—our bones, neurons, and hormones, not to mention our iPads, Kindles, and Androids) receive attention. And ultimately, theorists have begun to investigate the forms and lives of things distinct from human being in ways that emphasize the ultimate difference of objects from human phenomenology: their "alien phenomenology," in Ian Bogost's terms (2012). In the wake of the corporeal transformation, not only do things come into focus but also non-things (humans and non-human animals) come into a renewed and different kind of focus as we begin to see what they are through an alternate framework that reconsiders them through their affinity with inorganic matter rather than always underscoring and fetishizing the space of their distinction from it.

COMMON GROUND

Suddenly, then, the world has become very complicated. Complicated in the sense that a fundamental boundary, the one between people and things, has become a moving target. The concept for the present collaboration is a product of that complexity, in the context of the editors' ongoing conversations about how technology design, very broadly defined, has become a point of mediation—of "radical interface"—between humans and things. The conversation has progressed among the editors from the vantage points of their respective home disciplines of literature, rhetoric, and philosophy, and has reached across the aisle of some of the most trenchant academic divisions: two of the authors work in humanities departments, and the third works in a business school. All of us found ourselves talking, furthermore, about the form and implications of specific hardware builds: neither a traditional humanistic nor managerial territory.

During these dialogues, we recognized the value of bringing together the conceptual lexicons and approaches through which various disciplines understand the role of technology design in facilitating the kinds of extreme mediation that can lead to subjectivities and embodiments that are neither fully human nor fully mechanical in a world where technology has begun to render the human-object dichotomy obsolete. That is, as we've discussed technology design, we've realized that specific artifacts are, in a Latourian sense, intricately "networked"—and so must be the scholarship that reflects upon them. Our questions about matter are not merely philosophical, aesthetic, rhetorical, sociological, archaeological, technological, or artisanal. They are all of these things at once. We thus contend that a rich discussion of things and of the thingness intrinsic to humanity cannot happen effectively within traditional academic disciplinary or conventional professional specialty silos, and herein construct an occasion for a more fluid approach.

Design, Mediation, and the Posthuman is not the first medium to attempt a more fluid approach. Numerous "subdisciplines" and specializations have recently emerged as dedicated to understanding artifacts, things, objects, and technologies in relation to human beings, and they will receive more detailed treatment below. For now, consider the common characteristics that reach across the varied incarnations of a corporeal transformation across the academic disciplines. They include:

- The assumption of a paradigm shift
- The rejection of anthropocentrism and belief in the moral value of removing humans from their privileged status in the ecosystem
- The rejection of a centered subjectivity, which is revised to account for the radical displacement of the human by non-human others, and the consequent sense of ontological destabilization
- The reconceptualization of central categories for thinking about the human condition: self, subject, agency, intentionality, responsibility, freedom
- The assumption of horizontal rather than hierarchical ontologies, and a consequential investment in network ontologies and hybrids
- The perception of humans as embedded in a natural-cultural-technological assemblage
- Boundary dissolution and transgression
- The rejection of modernist binaries in favor of fusions: nature with culture, human with animal, organic with inorganic
- Belief in the emancipatory effects of the paradigm shift
- The emergence of "monstrous subjectivity" (Soper 2003, 100) as a result of binary dissolution, and as a necessary vehicle to experiencing its emancipatory effects

As the editors have endeavored to synthesize the many corporeal turns taking place across discourses, we have identified what we might call a corporeal canon: a set of seminal chapters, each deriving from different home disciplines, but with implications and applications beyond their space of disciplinary origin. We offer brief summaries of these chapters below. Though all of their authors have published more recent book-length versions of their work that tend to be well known within their disciplines, we submit the chapter versions as more easily fungible (and accessible) across disciplinary lines. Indeed, and as an aside, we might pose the question of whether the chapter is emerging as the genre of transdisciplinary inquiry, a form that moves across boundaries more readily than book-length work because of its LEGO-like atomism: chapters are short enough to be endlessly paired and re-paired with alternate others. Something different is suggested by every fusion. The following chapters, at any rate, combine productively with each other, but also invite the creation of more blocks using other kits.

Jane Bennett's "The Force of Things: Steps Toward an Ecology of Matter" (2004) is a political science chapter that endeavors to sketch out what it might mean for public policy makers to see non-human entities as alive, and therefore as demanding human attention and even responsibility. Is "life," Bennett wonders, necessarily an exclusively humanistic term? Bennett contends that it is not: matter also has life, if in a form dramatically differently than ours. Moreover, this retooled "vibrant matter" is not inert; it exerts its force on humans every day, affecting us and even manipulating actions we mistakenly attribute to our own agency with its "thing-power." Matter thus belongs within our public contracts. Short of an attempt (however anthropomorphic) to depict the alien life forms of the nonhumanity that is everywhere in and around humans, Bennett believes it is impossible to understand the human position in relation to our environments, and thus impossible accurately to comprehend our political responsibilities to the living nonhumans with which we share our planet. It is Bennett's environmental politics that lead her to investigate a new kind of vital materialism, one that views other existing materialisms (Marxism's cultural materialism, for instance, or feminism's materialism of the body) as overly centered around human beings and our systems of exchange.

From the field of English literature, Bill Brown's "Thing Theory" (2001) questions the centrality of language to world-making that is assumed within his discipline, and in so doing hypothesizes the necessity of fostering a resurgence of interest in the nonlinguistic exertions of the material world. For Brown, things offer occasions for observing the contingency of human experiences on the material world, and for staging interruptions to the traditional mindset of anthropocentrism. For Brown, as for Bennett, there is an important difference between "objects" and "things," the former eternally positioning things in relation to their human users and the latter signifying the latent remainder of human instrumentation within every thing. By proposing a more horizontal form of relationality between people and things, removing the vertical hierarchy whereby things are made by and for humans, Brown is able to recast people as the occasional objects, or instruments, of things—themselves fundamentally evasive of instrumentation and therein possessing a "subjectivity" or "agency" in their own right. Brown emphasizes the liminality of thingness: outside of human instrumentation, on the margins of utility, matter returns to a state of non-identity. An effect of Brown's recovery of the thingness of the non-human world is an interrogation of Western philosophy's fetishization of subjectivity, as well as its fetishization of characteristics fundamental to subjectivity: language, representation, agency, and consciousness. What other modes of communication, being, and action exist? Like Bennet, Brown's "thing theory" is ethically oriented: he argues human identification with things is what enables the acceptance of radical difference, a form of empathy without which it is impossible to effect substantive social transformations. Brown's theory also is designed, like Bennett's, to disrupt fundamental disciplinary tenets; his work demonstrates the prevalence of things in literature, challenging the notion that only words, characters, intentions—all human-centered constructs—communicate, drive plots, structure relationships.

In media studies, Mark B. N. Hansen's "Media Theory" (2006) has offered an especially productive version of the corporeal transformation to those who seek to understand the impact and purpose of the contemporary media renaissance on things' resurgence. Whereas Bennett and Brown orient their analyses toward things broadly construed, even if they acknowledge as Brown does that contemporary digital culture peculiarly positions things to become preeminent, Hansen focuses on the computational revolution as giving birth to a particular category of objects. That is, he examines what kind of thing technology is. Focusing specifically on media technology, Hansen's claim is that humans and technology have evolved and still are evolving together, humans by reaching out to the *techne* that embody the wild reaches of their imagination, and technology by drawing people into every remediation and reshaping the structure of their lives by doing so. Media fundamentally displace the division between empirical and transcendental thinking, repeatedly demonstrating the empirical element of thought and the transcendental element of new media incarnations. Because media are phenomenological but also able to embody thought's infrastructure, they perform a hermeneutic function in every new instantiation. Each remediation, that is, dramatizes the conditions of its own existence. Ultimately, Hansen thus conceptualizes the medium as an environment for life, not mere representation of life: by giving concrete form to the evolution of human being, media are not just artifacts of but participants in "technogenesis," humans' co-evolution with technology. Hansen pushes the issue of mediation more than Bennett and Brown, asking us to consider the profound relationality of things and human beings to the extent of embodying a shared life form.

Likewise, Peter-Paul Verbeek's "Materializing Morality: Design Ethics and Technological Mediation" (2006) hails from philosophy to address the "script" concept in technical mediation. The script concept asserts that technologies prescribe human actions, reorienting human choice, action, and experience around the technologies they design rather than assuming that we control what we make. Verbeek, like Hansen, shows that technologies co-shape human actions, but unlike Hansen, Bennett, and Brown, for whom the question of ethics remains implicit, Verbeek asserts that technologies are explicit ethical instruments. In co-shaping human actions, technologies give material answers to ethical questions about the direction human actions *should* take. Engineers, in other words, are ethicists with a material bent: they "materialize morality." The problem, in Verbeek's view, is that engineers have become the moral philosophers of our era without self-awareness of this element to their work, and the human beings whose actions are everyday determined by the designed objects we use have unwittingly submitted ourselves to the particular moralities of—not just those things, but—their (usually invisible) engineers. Verbeek's argument implies the necessity of taking on the questions of both design and ethics—design ethics, without which humans are at the mercy of thoughtless mediations that seem to occur inevitably but in fact are only coincidental (and need not be). Verbeek thus considers design as behavior influence, addresses the morality of designing behavior-influencing technologies, and identifies two ways designers and philosophers of technology design can anticipate and evaluate the types of mediation that emerge from the design process: mediation analysis, and constructive technology assessment. Each method is construed to bring reflection back into the process of matter construction. Verbeek strives thusly to bring more human intentionality back into the construction and use of technology, which he positions, like Hansen, as a class of things that emphasizes human-thing interface more than, as in Bennett and Brown, thing ascendancy.

The essays by Bennett, Brown, Hansen, and Verbeek—authors who are seminal theorists within their own disciplines, but rarely cited or discussed each by the other or their disciplinary heirs—serve as a fruitful starting place from which to trace resonances between various accounts of the corporeal transformation, flesh out a robust transdisciplinary version of it, and identify open spaces that invite further query and research. Their combination fosters the sense of a corporeal transformation as stimulating a paradigm shift, restructuring the hierarchy of human-thing relations, invigorating attention to technology, emphasizing the status of mediation and relationality, drawing questions of design into theoretical prominence, and contrasting the emancipatory and dystopian potential implicit in variations on this process.

THING-TENSION

Of course, a good sign of a productive confluence of theoretical perspectives is that it leaves us with many worthwhile questions. As we continue to complicate issues of selves, things, and the Internet, we present here some of the primary questions that we consider in the following section of this chapter and in this volume as a whole. These questions, and others, emerge from a comparative analysis across multiple disciplinary and sub-disciplinary incarnations of the corporeal transformation.

Consider, for instance, the new approach to things that has emerged within the discipline of archaeology, generally known as a discipline devoted to studying the past and its people through its relics and remnants. For some twenty-five years archaeology has been reframed as an interpretive mechanism, less about things than generating sets of thing-meanings. "Symmetrical Archaeology," as defined on the Stanford Humanities Lab webpage, however, is a recent counter-movement founded on the premise that archaeology has drifted too far from materiality. Symmetrical archaeologists contend, following Latour, that meaning must be "reconfigured within heterogeneous networks comprised of collectivities of humans, materials, media and other companion species" to reacquire the lost symmetry between human beings, objects, and non-human others. By avoiding divisive in favor of symmetrical thinking, symmetrical archaeologists are able to study "complex socio-technical assemblages," as noted on the Symmetrical Archaeology website. The assemblages that serve as the reconstituted object of study drive Symmetrists to articulate "mixtures, imbroglios, hybrids of humans and things" rather than interpretations in which things themselves endlessly recede into the background—while still acknowledging that no object can be considered alone, "out of network." The premise is more that things and humans equally exert force on one another (a premise that physics might drive us to critique in certain circumstances, while interestingly reconfiguring the question around the size of the exerting object rather than the category or species). It is that they comingle and fuse, and that it makes little sense to prioritize one over the other since we must always come to know one through the other.

In their examination of hybrids the Symmetrical Archaeologists share a common object of study with Cyborg Studies, a field pioneered by Donna Haraway, and one that variously emphasizes the element of visibility in human-machine prostheticizations. Whereas Haraway and her immediate progenitors conceived of certain identity categories (primarily women, but also addressing people of color or the victims of class marginalization) as intrinsic hybrids of physical body parts and cultural constructs, Andy Clark has theorized a more radical cyborg in which all humans inevitably combine themselves with technologies in order to function in the world—that we all fundamentally (not occasionally) exceed our "skin bags." In this model, things are not diametrically opposed to organic entities, but are their continuations in the world, their prosthetic extensions. In this sense, cyborg studies has been plagued by many of the same theoretical difficulties as race in identity politics, relying on purist, or what Elaine Graham has called "ontologically hygienic," notions of human nature. In the real world, human-thing boundaries may be no more visually identifiable than racial differences are;

that is, as history and modern life have shown us, identifications of both of these forms of difference, if indeed they do exist as anything other than problematic metaphysical categories, are prone to error. Consider the extent to which humans with heart valves or hearing augmentations can "pass" as fully organic—or, for that matter, genetically engineered babies, or simply babies who exist because a prenatal screening submitted them to a form of passive selection. We can't see these cyborgs' seams, and in some cases there are no seams. Yet each sutures human to thing and/or cultural construct in a way that Cyborg Studies claims and recognizes. Note, however, that while the cyborg construct assumes the notion of assemblage prominent in Symmetrical Archaeology, the two fields tend to ignore each other and to forge radically different lexicons.

Our emerging fascination with non-human others has created a new and bigger ecological sensibility, one that incorporates not only animals, plants, minerals, chemicals, and computers to urge alternate approaches to environmental study, but also to religion. Herein, the divine is a biotechnical superhuman, achieved in an optimistic ethos of augmentation to the point of immortal omniscience. Such configurations mark the present era as, in some scholarly configurations, "post-secular" in the sense that they indicate the deeply religious impulse that often informs technological innovation:

> Artificial Intelligence advocates wax eloquent about the possibilities of machine-based immortality and resurrection, and their disciples, the architects of virtual reality and cyberspace, exult in their expectation of God-like omnipresence and disembodied perfection. Genetic engineers imagine themselves divinely inspired participants in a new creation. All of these technological pioneers harbor deep-seated beliefs which are variations upon familiar religious themes. (Noble 1999, 5)

The rhetoric of both apocalyptic and messianic change that often accompanies technological innovation underscores the extent to which technology has not taken society farther from religion but is one of the most persistent incarnations of religious desire:

> There has never been a fundamental contradiction between the two disciplines. . . . The current fascination with new technologies as capable of synthesizing new worlds . . . is akin to more ancient modes of spirituality, for both seek to "transcend" the contingencies of the profane and material world in search of the more enduring realms of heavenly perfection. Signs of a preoccupation with omnipotence and transcendence are still apparent . . . in the psychological investment humanity places in technologies not merely as items of convenience and utility but as instruments of "deliverance." (Graham 2002, 166)

The so-called "transhumanist" movement's relationship to matter is fundamentally paradoxical: it fetishizes things as a vehicle to circumventing their thingness, looks within the properties of matter for evidence that humanity and its limited properties are not all there is. Though transhumanism seems to bear more connection to virtual culture's mind-ascendancy than other incarnations of the corporeal transformation, the movement could be described as seeing human beings as more deeply embedded in object-relations than any other. Here, they are connected in spirit: the most imaginative and limit-transcending aspects of the human are not ethereal but embedded in extraordinary thing-phenomenologies. Nevertheless, this group remains even more sequestered from Cyborg Studies and Symmetrical Archaeology than they are from each other.

The newly mesmerizing objects buoyed by the corporeal transformation include extraordinary and re-enchanted ordinary matter, but also monsters, aliens and other "others" created in the body-thing nexus. In yet a fourth incarnation of the corporeal transformation, the whole group has been dubbed "posthuman," a species of beings that some scholars believe represents a co-evolution of humans and environment so codependent as to be formally inextricable for anyone unblinded by categorical distinctions that preemptively prohibit the connection. "In other words," writes Elaine Graham, what we're calling the corporeal transformation is "less a matter of the empirical impact of technologies than their significance for human ontology. The impossibility of isolating 'human nature' from its refracted others suggests a model of post/humanity as inextricably bound up in relationality, affinity and contingency" (223). Graham uses her characteristic forward-slash in the word post/human to indicate that for her, the very term "posthuman" is a denial—a disavowal that the so-called monstrosities that inhabit the category are not our others but are the definition of human. As though we could (though we have tried) banish the messiness of relationality, a physiological axiom as much as a social action, from our midst.

Nevertheless, the corporeal transformation has profoundly impacted our conceptualization of evolutionary biology in ways that the term posthuman has only begun to reflect. Mark Poster argues, for instance, that a "symbiotic merger between human and machine might literally be occurring, one that threatens the stability of our sense of the boundary of the human body in the world. What may be happening is that human beings create computers and then computers create a new species of humans" (Poster 1990, 4). As such, Ihab Hassan contends, "the human form—including human desire and all its external representations—may be changing radically. . . . [F]ive hundred years of humanism may be coming to an end" (Hassan 1977, 205). As humanism comes to an end (if the corporeal transformation does, in fact, indicate such an event) we all have to change—and here, by "we," we mean not just human biology or popular culture but also academic culture. What sense does "the humanities" make as a construct when the human is an entire ecology, and that ecology includes a hybridized host of flesh, buttons, blood, wires, skin, enamel, organs, flashes, beeps, and breath? What sense "physics" or "computer science" when the matter or computers we're talking about are as much themselves as they are . . . us?

Posthumanism is not merely an identity (or anti-identity) category, or even a hip new disciplinary amalgam, however; it is also, as Cary Wolfe notes, a "historical moment in which the decentering of the human by its imbrication in technical, medical, informatics, and economic networks is increasingly impossible to ignore." It is "a historical development that points toward the necessity of new theoretical paradigms (but also thrusts them on us), a new mode of thought that comes after the cultural repressions and fantasies, the philosophical protocols and evasions, of humanism as a historically specific phenomenon" (Wolfe 2009, xvi). Philosophically and historically speaking, the age of matter seems to suggest a paradigm shift in which non-anthropocentric thinking becomes not only permissible but assumed. The corporeal transformation is not merely conceptual in this configuration, but defines a kind of moral imperative. In Bill Brown's words, echoing Bruno Latour, "only by overcoming the opposition between 'mute entities and speaking subjects, only through 'a wild extension of speech to things,' will we begin to imagine a political ecology for our time" (Brown 2001, 120–121). In his association of thing studies with ecological reform, Brown joins Bennett, for whom the "intricate dance" (Bennett 2010, 31) of humanity with non-humanity has become harder to ignore in the present because of the environmental impact of humans' relative failure to pay attention to it since the Industrial Revolution. In the context of environmental crisis, the specific non-human pressures exerted on humans and other non-humans alike by electricity and stem cells, fish oils and sugar molecules, metal and trash have taken on the likeness of speech—and they will make themselves heard, whether we like it or not. Posthumanism, Wolfe explains, does not simply describe the concept of the shift, but names the historical moment of its occurrence.

Symmetrical Archaeology, Cyborg Studies, Transhumanism, the new ecologies and biologies, and Posthumanism are certainly not an exhaustive list of all the incarnations the corporeal transformation has taken on across various disciplines. Complicating things even further is the extent to which each of the incarnations bear surface similarities that often mask important tensions, tensions that only emerge when we cease to respect the boundaries between them and try to synthesize diverse variations on the corporeal transformation.

A few examples: nature holds a contested place within thing theories, sometimes indicating very specifically the otherness of non-human animals, others expanding to include non-human bio-matter, and others extending even further to include naturally-occurring inorganic materials—and then others to technology and various crafted or machine-made artifacts. Theorists like Kate Soper have suggested that nature-endorsement and the natural are pivotal standards in moral evaluations of technological advancement while others take a more constructionist approach. The status of consumption and use is likewise controversial, as some scholars view consumer culture as retaining emancipatory spaces for both people and things while others view it as fundamentally and ruinously objectifying for both people and things. Not all participants in the corporeal transformation consider technology as central to it, and the definition of technology is sometimes so broadly construed as to include the most basic, even prehistoric tools employed by humans all the way to the most tenuous and sophisticated advancements of contemporary technoscientific laboratories. The temporality of the corporeal transformation is also a point of conflict: did it begin in Y2K? Some scholars—including several in this volume—reach back to the beginnings of the Industrial Revolution. Others find evidence of posthuman subjectivities in medieval "mind upgrades" through mnemonic tools. Orientation toward other related philosophical concepts is also often contradictory. For instance, relationality and "networking" are recurring values among relational ontologists like Verbeek and other postphenomenologists inspired by Latour. But object-oriented philosophers, on the other hand, espouse a kind of naïve realism intent on studying objects independent of their relations, especially their human relations. Anthropomorphism's value is also unstable among the discourses. Many concur that anthropocentric thinking must be overcome in order for people to comprehend and respect radically different ontological forms, and value variations of life and aspects of the environment that aren't, or can't be easily integrated into an analogy with, humans. But for some theorists the whole purpose of non-anthropocentric thinking is to circumvent human modes of perception, cognition, affect, communication, and even sociality to explore and try to imagine, however imperfectly, radical difference. For others, the anthropomorphism inherent in redefining life, language, aesthetic taste, even spirituality in terms broad enough to apply—in however halting a form—to orchids, earthworms, or machines is a key cognition in the nonhuman turn.

The jumble of tensions emerging here might be organized into three core questions: (1) Is the focus of the corporeal transformation on relations and networks, or things and matter (objects)? (2) Precisely what things ought we to be focusing on? (3) What are the implications for rethinking subjectivity, agency, intentionality, and human exceptionalism? What follows fleshes out the terms of these questions, and outlines divisions in the way various figures and approaches might address them.

1. Is the focus of the corporeal transformation on relations and networks, or things and matter?

Many of the diverse theoretical perspectives brought together throughout this volume freely employ metaphors of networks and assemblages to describe the shift we have been noting in this introductory discussion. From Bruno Latour's actor-network theory to Verbeek's human-technology relations and Bennett's assemblages formed by human and non-human bodies, the emphasis is seemingly on the connections and relations existing between human and nonhuman, disavowing the movement of purification by which subject and object were sundered apart and consigned to distinct ontological registers. Mediation and relationality are now key. Simultaneously, though, we are apparently witnessing a turn to things and matter, to materiality and objects; we must re-member things. This final exhortation comes from Norwegian archaeologist Bjornar Olsen in his essay "Material Culture after Text: Re-Membering Things" (2003). There, Olsen argues that archaeology, as well as the other social and human sciences, have largely ignored material culture. Rejecting social constructivist and phenomenological approaches, Olsen argues that the hard physicality of the world has often been reduced to little more than discursive objects or "phenomena" of the subjects' cognitive experience. But while rejecting what he sees as a negative attitude toward matter and technology, and encouraging humanists and social scientists to re-member things, Olsen also adopts a network or relational ontology in which one might expect such re-membered matter to be simply swallowed up. As he notes:

> According to John Law, network theory may be understood as "a semiotics of materiality. It takes the semiotic insight, that of the relationality of entities, the notion that they are produced in relations, and applies this ruthlessly to all materials—and not simply to those that are linguistic" (Law 1999, 4). Instead of reducing the world to the regime of two opposing ontological realms, culture-nature, this approach claims that nearly everything happens between the two extremes, happens by way of mediation and translation, by heterogeneous networks linking all kinds of materials and entities. Reality is not to be found in essences, but in imbroglios and mixtures, the seamless and rhizome-like fabrics of culture and nature that link humans and non-humans in intimate relationships. (Olsen 2003, 98)

Apparent here is a tension between mediation and materiality, network relations and realism, rhizomes and things. On the one hand, the title alone of Olsen's essay and its manifesto-like nature speaks to an interest in focusing squarely on things, material culture, objects. On the other hand, in many other ways he emphasizes networks or relations rather than the focus on things itself, agreeing with Heidegger, for instance, that our being-in-the-world is enmeshed in networks of things (96).

The tension noted in Olsen's defense of symmetrical archaeology is pervasive in the theoretical work that forms the backdrop to this volume. On the one hand, figures such as Verbeek and Bennett emphasize the relationality of the paradigm shift currently underway. On the other hand, object-oriented philosophers and to some extent Bennett as well argue for a kind of naïve realism which rejects approaching objects through their relation with human beings and eschews mediation for a direct account of materiality. The tension here could be described as the tension between a relational ontology and an object-oriented ontology. In the turn we have been pursuing, then, what is more fundamental, the thing or the network? Do we have unmediated access to materiality or is our relation to the world always mediated? What is the relationship between technology and reality?

In his emphasis on technological mediation, Verbeek is clear that our relations to the world are mediated in a variety of ways by diverse technologies. As he notes, "for humans, reality can only be what it is when humans disclose it—by experiencing and acting in it—as a 'world.' Humans and world co-constitute each other. Humans are what they are on the basis of their interactions with and experiences of their world, and their world is 'disclosed reality.' Apart from each other, they remain undisclosed and isolated" (2002). From Verbeek's phenomenological perspective, the human-world relationship is mediated by technological artifacts. "When fulfilling their function, technological artifacts always are a medium between humans and reality" (2002). Our access to the real world, Verbeek suggests, is always mediated.

It is precisely this emphasis on mediation and human-technology relations that more object-oriented figures reject. As the object-oriented philosopher Graham Harman notes in an essay on literary criticism: "The central problem at stake is none other than realism: does a real world exist independently of human access or not?" (2012, 184). Harman suggests that his own non-relational conception of the reality of things will strike many as reactionary.

> After all, most recent advances in the humanities pride themselves on having abandoned the notion of stale autonomous substances or individual human subjects in favor of networks, negotiations, relations, interactions, and dynamic fluctuations. This has been the guiding theme of our time. But the wager of object-oriented philosophy is that this programmatic movement towards holistic interaction is an idea once but no longer liberating, and that the real discoveries now lie on the other side of the yard. . . . This is not the oft-lamented "naïve realism" of oppressive and benighted patriarchs, but a *weird* realism in which real individual objects resist all forms of causal or cognitive mastery. (2012, 187–88)

Where Verbeek and Harman are particularly clear in regard to where they stand in this debate, Bennett seemingly attempts to adopt a both/and strategy.

Bennett emphasizes that all bodies are kin and are "inextricably enmeshed in a dense network of relations" (2010, 13). She regularly focuses her analysis on assemblages and federations of actants (2010, 28). Simultaneously, though, Bennett emphasizes thing-power and the need to give voice to a "vitality intrinsic to materiality" (2010, 3). In "The Force of Things," Bennett suggests that in the background of this notion of thing-power materialism is Thoreau's notion of the Wild: "that is, his idea that there is an existence peculiar to a thing that is irreducible to the thing's imbrication with human subjectivity. It is due to this otherness or wildness, says Thoreau, that things have the power to addle and rearrange thoughts and perceptions" (2004, 348). Here, perhaps our experience of the Wild may be understood as mediated by something other than technology—a non-technological kind of seeing, such that we may understand Bennett's efforts to reveal vibrant matter as perhaps "differently mediated," especially given that so much of the current discussion of mediation is situated in terms of technological mediation. Bennett's remarks about play provide further support for these ideas; that is, she suggests that a version of the idea of vibrant matter can also be found in "childhood experiences of a world populated by animate things rather than passive objects" (2010, vii). (This focus on play will also resurface in later comments by Rey Lee about the object-oriented sensory experiences enabled by Lego, which he feels reflect our own ideas about the constructedness of the world.) Thus, we might interpret Bennett as trying to "see" differently, not in terms of unmediated access to the world, but rather a sort of access that is not technological, scientific, or modern.

In responding to Harman's emphasis on a non-relational conception of objects, Bennett more explicitly adopts a both/and strategy and speaks of doing justice both to systems and things (2012, 228). Such a strategy is suggested as well in "The Force of Things":

> My view is that while humans do indeed encounter things only in a mediated way, there nonetheless remains something to be said for the naivete of naive realism. A moment of naivete is, I think, indispensable for any discernment of thing-power, if there is to be any chance of acknowledging the force of matter. A naive realism (which, in my case functions as an onto-story rather than an apodictic account) allows nonhumanity to appear on the ethical radar screen. Yes, there is a sense in which any thing-power discerned is an effect of culture, and this insight is a valuable counter to moralistic appeals to "nature." But concentration on this insight alone also diminishes any potential we might possess to render more manifest the world of nonhuman vitality. (2004, 357)

At this point, though, we may return to a challenge underscored by Verbeek's emphasis on technological mediation and a relational ontology and suggest that Bennett must still come to terms with the role of technology in shaping the very vision of vibrant matter that she offers as an alternative to our more "human-centric" perceptual powers. When juxtaposed with Verbeek, it seems that Bennett could further address the technoscientific apparatus that makes possible many of her claims. Bennett attempts a sort of realism in an effort to hold at bay the human and discern instead a nonhuman vitality; however, many of her examples presuppose the "human" technological apparatus that makes such perception possible. We might consider, for example, the following claim: "My flesh is populated and constituted by different swarms of foreigners. . . . The *its* outnumber the *mes*. . . . We are . . . *an array of bodies*" (2010, 112). Such a perception arguably presupposes a complicated technological relation to biological science and a view of nature mediated by technological apparati.

The authors collected in *Design, Mediation, and the Posthuman* tend to emphasize a much more relational approach to nature-cultures and human/nonhuman interactions than that favored by either Bennett or Harman. Tinnell's account of seawater, drawing on Bernard Stiegler, provides an interesting anecdote in this regard, given the manner in which, as he notes, "nature becomes a function of the technical system" (72). Takehana too seems to favor a more relational approach, emphasizing in her analysis of the music video for "Pomme C," the hybrid connection of human and environment; in this way, she seems to favor Bennett's account of a shared dependence and shared agency. In her analysis of GPS, Propen also tends toward a much more relational approach to nature-cultures.

2. *Precisely what things ought we to be focusing on?*

The turn to things that we have been examining in this introductory chapter is often situated directly in the context of the developments of twenty-first century technoscience. The hybrids and networks that are the centerpiece of such analyses are products both of our increasing familiarity with networks of human beings and digital technologies and the products of such networks, including genetically engineered organisms, artificially intelligent computer systems, and cybernetic technologies. As British philosopher Kate Soper notes,

> Advances in genetics have prompted some of this kind of thinking, given the new questions they pose about where to draw the divide between the artificially contrived and the naturally given. So, too, have developments in IT, and there are so-called "connectionists" and advocates of "Emergent Artificial Intelligence" who emphasize the unpredictable, non-rule governed and non-determined qualities of the most sophisticated computers, and view these as "psychological machines." (2012, 367–68)

It is likely for similar reasons that while Verbeek maintains that our reality is a web of relations between human and nonhuman entities, his focus remains largely on contemporary technologies and he argues that such technological artifacts are fundamental to our relation to the world. Indeed, Verbeek's account of mediation focuses almost exclusively on technological mediation and he understands the relations that co-constitute the human being as largely technological. Technology fundamentally mediates what kind of humans we are, and serves as the starting point for moral self-practices (2011, 82). "Technology," Verbeek notes, "forms the tissue of meaning within which our existence takes shape" and technological mediations are a starting point for the moral subject (Verbeek 2011, 73). The object and power relations in which the subject must articulate itself are, he suggests, primarily technological (Verbeek 2011, 81). Given this emphasis on the role of technical artifacts and technology design in approaching mediation and the posthuman, it is fitting that a number of the analyses included in this anthology focus on our relations with and the mediation of advanced, digital technologies, including GPS (Propen), mobile augmented reality (Tinnell), Apple's SIRI (McArthur), and videogames (Keogh).

And yet, a cursory examination of the variety of disciplinary interests and foci revolving around the turn to things, as well of course as the very generalness of that term "thing," perhaps suggests that it would be overly quick to conclude that technoscience and contemporary digital artifacts ought to be our primary focus. Ranging over the variety of proposals and theoretical perspectives brought together under the umbrella notion of the turn to things, one is struck by the range of objects aligned on the side of the nonhuman, which, depending on one's theoretical interests, has included everything from digital technologies and other technical artifacts, media objects, the human microbiome, animals, plants, and hyperobjects. In "Media Theory," for instance, Mark Hansen traces the coupling of human beings and technologies much further back than our current fascination with digital technologies, arguing for the "co-originality of technics and the human" and suggesting that the human being is originally a prosthetic being (299). Hansen raises worthwhile questions about the emphasis often placed on new media and how precisely we ought to historically periodize our relationship to media and technology. Several of the contributors to *Design, Mediation, and the Posthuman* similarly contest the emphasis placed on our current experiences with advanced technologies and look to older forms of mediated posthuman experiences. Reid turns to Victorian decorative arts and argues that artisans habitually compromised the distinction between humans and things in normative bourgeois domestic settings, treating objects as prosthetics of the human body and insisting that subjectivities utilize decorative artifacts not simply as forms of self-expression but as actual components of the self without which the subject could not exist. Fleckenstein and Mehler similarly turn to two nineteenth-century technologies, the tourist guidebook and the bicycle, for a more expansive notion of the posthuman. As they note in their chapter:

> Our analyses of the "sporting gentleman" and the "New Woman" demonstrate that these two nineteenth-century figures rely on the distributed cognitive system comprising technology, semiosis, and enaction in much the same way as we see with posthuman figures in the twenty-first. The two artifacts that we offer as examples, in particular, highlight an expanded notion of technology that moves beyond the contemporary digital bias implicit in posthumanism while emphasizing the posthuman body as mobile. (192)

We see an even more expansive reading of the category of thing in Bennett, who implicitly challenges the idea that we ought to focus on technology as fundamental and seemingly more primordial than other forms of human-thing relations. Bennett expands beyond technical artifacts to include a larger range of objects and things in her vital materialism. Her focus on food, trash, and microbial life extends Verbeek's network out beyond "designed" things such as tools, technologies, and the technical artifacts of contemporary life in an industrialized environment. She seeks to cultivate our embeddedness in a natural-cultural-technological assemblage and to enhance our awareness of the vitality of the world in which we are enmeshed: "Thing-power materialism offers a contestable but, I think, auspicious account of how it is that things have the power to move humans. . . . It emphasizes the shared material basis, the kinship, of all things, regardless of their status as human, animal, vegetable, or mineral" (2004, 359). Here, Bennett opens her analysis not only to designed, technical artifacts but also to minerals and metals, trash, nature, food, and microbes. In the chapters included in this anthology, analysis ranges equally far and wide and includes trash (Miccoli), LEGOs (Rey Lee), bicycles and tour books (Fleckenstein and Mehler), video games (Keogh), and film and planets (Merola). Anthony Miccoli's focus on hoarders and trash, for instance, challenges questions about what counts as "natural" technical objects. By drawing on the illustrative example of hoarders, Miccoli extends Verbeek beyond thinking about technology and well-designed technical products, to thinking about the "natural" world of trash.

It is also worth noting here, as evidenced by juxtaposing the views of Bennett and Verbeek, that discussions about design and mediation have tended to focus predominantly on human relationships with technical objects and digital technologies, while obscuring perhaps our relationships with nonhuman animals and the natural environment. Animal-studies scholars such as Cary Wolfe have explored the communicative and philosophical implications of the ways in which nonhuman animals are portrayed within various social contexts (2009, 129). A posthumanist approach, says Wolfe, ought to "take[s] seriously the ethical and even political challenges of the existence of nonhuman animals" (2009, 148). Within this collection, we see contributors like Nicole Merola take up such issues when she explores, through her discussion of the film *Melancholia*, whether "designed things" might replicate a kind of anthropocentrism. Here, Merola understands animal life and the planet itself as things, thus challenging the notion of "thing power," and how Verbeek, and perhaps Bennett, conceptualize "what things do."

The broad range of objects included in the nonhuman raises difficult and potentially troubling questions. How ought we to distinguish between the symbolic and the material? Is there a meaningful distinction between natural organisms and designed artifacts? What counts as the natural world, especially once we recognize the implications of technological mediation? At what point does the category of "thing" risk an indiscriminate ontology lacking any conceptual borders or boundaries? Is such a mongrel metaphysics conducive to understanding? Without boundaries and borders, without some guidelines for how to carve up the world, and with a principle that treats media, animals, artifacts, things, and matter symmetrically, we may lack the conceptual guidelines and boundaries adequate for making reasonable judgments. Conceptual and boundary work may be needed for clarifying more precisely how to more adequately understand these calls for symmetry and more precisely delineating the boundaries of things.

3. *What are the implications for rethinking subjectivity, agency, intentionality, and human exceptionality?*

Finally, taking perhaps one final turn, we might look back at the human after the nonhuman turn and ask what becomes of us? What becomes of the user of technology? How do we understand the figure of the posthuman? In what ways are subjectivity and agency reconfigured after the turn to things, and how radically displaced is the human?

Returning one last time to that fateful shift from mind to thing, it is worth noting that it is in 1999 that N. Katherine Hayles publishes her influential *How We Became Posthuman* in which she observes that the key battleground isn't over human versus posthuman, it's over what kind of posthuman future we will have. Hayles points out that the posthuman presupposed in much of the early work on the cyberculture recuperated the worst aspects of liberal humanist subjectivity, especially the erasure of human embodiment. By separating information and materiality, marginalizing materiality, and making information the essential characteristic of information processing devices, mechanical and human alike, the cybernetic view of the posthuman perpetuated a long tradition in Western thought of downplaying human embodiment and valorizing a coherent, centered, autonomous, and rational self (1999, 85–86). Rejecting this liberal humanist version of the self, Hayles prefers to portray herself as a distributed "posthuman collectivity": "Speaking for myself, I now find myself saying things like, 'Well, my sleep agent wants to rest, but my food agent says I should go to the store.' Each person who thinks this way begins to envision herself or himself as a posthuman collectivity, an I transformed into the we of autonomous agents operating together to make a self" (1999, 6).

Hayles argues that we should think of the self as a multiagent system whose complexity has evolved from highly recursive processes being applied to rather simple rules. There is no stable, coherent self but only autonomous agents running programs. The self is a small subsystem running its program of self-construction and self-assurance while remaining ignorant of the actual dynamics of complex system. We don't have unambiguous boundaries and shouldn't think of the subject as an autonomous self, independent of the environment. Hayles suggests that conscious agency and control are mere illusions, really only a luxury of those few individuals who had the wealth, power, and leisure to conceptualize themselves that way (286). Media theorist Mark Poster would seemingly agree, for he observes, "Everyday life emerging in information society is a battleground over the nature of human identity" (2006, 230). Bennett too takes up this issue in *Vibrant Matter*, arguing for a new conception of the self that begins with the recognition that "The human is not exclusively human. We are made up of its. But I think this truism, and the cultivated talent for remembering it, forms a key part of the newish self that needs to emerge, the self of a new self-interest" (2010, 139).

In gesturing toward technology, design, and the posthuman, this volume seeks to further illuminate this model of subjectivity while simultaneously questioning and complicating it. As the reader will note, many of the chapters included in this anthology seek to disrupt the subject-object divide and with it the presumption that intentionality is a feature of the human being that serves to distinguish us from objects in the world. As Miccoli notes in his contribution to this volume, a focus on mediation and the posthuman requires us to rethink the interior/exterior binary. Miccoli's analysis of hoarding leads him to conclude that the hoarder's hoard is part of his or her "aggregate mind." As he writes, "When hoarders feel that they are losing their minds because the hoards are being altered or removed, they are actually onto something. Parts of their minds are being removed. For the hoard(er), the cognitive process is intrinsically affected when part of the hoard is altered or removed. . . . The idea that the 'outside' is separate from the 'inside' is illusory" (18). In Propen's analysis of GPS, Tinnell's account of mobile augmented reality, Keogh's reflections on videogames and memory, and Levy's account of the Cynics and their relationships with things, we see similar calls for rethinking agency, intentionality, and responsibility.

Despite the widespread nature of the claim that the posthuman and the turn to things requires that we rethink the liberal humanist subject and the related concepts of agency and intentionality, a number of questions arise. As we noted earlier, there is some ambiguity in the work of our seminal figures regarding whether the focus of analysis should be on relations and networks or things and matter. A similar question arises when contemplating the agency or intentionality of things. Is agency located in things themselves, that is, do things do things, or is the agency located in the assemblages and networks? When we speak of murmuring matter or what things do or objects that hail us, it seems that matter itself, independent of its enrollment in a network is active, intentional, agentic. But other times the focus seems to be on the network and agency is attributed to the relational network. As we have seen, Bennett's flattened horizontal ontology leads her to embrace a nonlinear, nonhierarchical, non-subject-centered mode of agency in which assemblages exhibit agency and we are moved beyond the life-matter binary (2010, 20) and which moves beyond formulations of autonomy and strong responsibility (2010, 37). But Bennett's formulation of thing-power seems to locate agency not in assemblages but in things themselves. As she notes in her account of trash, for instance, our trash is not "away" in landfills but generating lively streams of chemicals and volatile winds of methane as we speak (2010, vii); trash become a matter of lively and potentially dangerous matter (2010, viii) and thing-power can rise from trash. Coming out of a phenomenological tradition, Verbeek develops a more complicated and nuanced approach to intentionality and yet he too is ultimately ambiguous on this matter. While the title of one of his books suggests that things do in fact do things, he pulls back from embracing some of the stronger implications of even Bennett's position, suggesting that, "things do not have intentions and cannot be held responsible for what they do" (2011, 216). Collectively, the chapters brought together in this volume together with the seminal accounts that form their backdrop, suggest that more work is needed in articulating the nature of mediation and posthuman accounts of subjectivity and intentionality.

Consistent with the turn to things and the advocacy of an amodern and impure horizontal ontology in which agency and intentionality is distributed throughout a network is a critique of the privilege enjoyed by human beings and a call to more radically think beyond the distinction between human being and the nonhuman. Human beings, it is urged, have no special ontological privilege and humanism, anthropocentrism, and the notion of human exceptionality obscures a recognition that human beings are always ever embedded in natural-cultural-technological assemblages as well as hinders ecological sensibilities that follow from an acceptance of culture-nature hybrids. As Bill Brown notes, for instance, we must move beyond "modernity's unbridled fascination with the subject, which left the (muted) object far behind." In the context of a reflection on Harriet Beecher Stowe's *Uncle Tom's Cabin*, which had as a subtitle in its first edition, "The Man Who Was a Thing," Brown observes that perhaps it is time to leave behind the distinction between person and thing. "Necessary as it was to distinguish persons from things in the nineteenth century, the future of the twenty-first century may well lie in thinking beyond that distinction" (2006, 121). And indeed, the posthuman often signals leaving behind any vestige of human exceptionalism or distinctiveness. As more and more of our lives are mediated by artifacts, things, and technologies, as we become aware of the multiple ways in which human beings are incorporated into networks or complex assemblages, and as our things take on agency and intentionality, perhaps it is time to take a final turn on a path that began with Copernicus, Darwin, and Freud and leave behind all presumption of human privilege, autonomy, and distinctiveness.

As we have seen in regard to subjectivity and agency, though, not everyone is content to embrace these more radical implications of mediation and the posthuman. Verbeek distinguishes his postphenomenological approach from Latour's nonmodernism on this point. "I do not want to give up the distinction between humans and nonhumans. Human beings have the ability to experience a world, and to act intentionally in it; things don't" (2009, 255). Despite her calls for a "newish self," Bennett too pulls back from the more radical implications of a horizontal ontology: "To put it bluntly, my conatus will not let me 'horizontalize' the world completely. I also identify with members of my species, insofar as they are bodies most similar to mine. I so identify even as I seek to extend awareness of our interinvolvements and interdependencies. The political goal of a vital materialism is not the perfect equality of actants, but a polity with more channels of communication between members" (2010, 104). To what extent then do reflections on mediation and the posthuman entail collapsing the distinction between person and thing? Do human beings have unique or distinctive traits that ought to be recognized? What are the ethical implications that follow from denying any such human distinctiveness? In this volume, both Van Den Eede and Weiss grapple with questions about where and how to draw the line between human beings and nonhuman things and artifacts, drawing on frameworks suggested by Marshall McLuhan and feminist philosophy to complicate how we negotiate these difficult boundaries. Kate Soper has observed that a world bereft of our organic-inorganic and human-animal distinctions is a world bereft of the grounding conditions for the moral, political, and scientific critique (1999) and American philosopher of technologist Andrew Feenberg has similarly argued that the antiessentialist demand for permanent contestation, for dispersion and difference cannot provide the basis for a positive approach to technological reform (2002, 32). Contestations of the boundaries of the human and its implications for our understanding of mediation and the posthuman clearly remain central to continued work in this area.

REFERENCES

Ashton, Kevin. 2009. "That 'Internet of Things' Things: In the real world, things matter more than ideas." *RFID Journal*, June 22. Accessed February 8, 2014. www.rfidjournal.com/articles/view?4986.

Barlow, John Perry. 1994. "The Economy of Ideas." *Wired*, March 1994. Accessed January 8, 2014. www.wired.com/wired/archive/2.03/economy.ideas.html.

———. 1996. "Declaration of Independence for Cyberspace." Accessed January 8, 2014. projects.eff.org/~barlow/Declaration-Final.html.

Bennett, Jane. 2004. "The Force of Things: Steps toward an Ecology of Matter," *Political Theory* 32.3: 347–372.

———. 2010. *Vibrant Matter*. Durham: Duke University Press.

———. 2012. "Systems and Things: A Reply to Graham Harman and Timothy Morton," *New Literary History* 43.2: 225–233.

Beverland, M.B. and Farrelly, F.J. 2010. "The Quest for Authenticity in Consumption: Consumers' Purposive Choice of Authentic Cues to Shape Experienced Outcomes." *Journal of Consumer Research* 36.5: 838–856.

Bogost, Ian. 2012. *Alien Phenomenology, or What It's Like to Be a Thing*. Minneapolis: University of Minnesota Press.

Brown, Bill. 2001. "Thing Theory." *Critical Inquiry* 28.1: 1–22.

———. 2006. "Murmuring Matter." *Metascience* 15: 117–121.

Brown, Valerie, John Harris and Jacqueline Russell. 2010. *Tackling Wicked Problems Through the Transdisciplinary Imagination.* Washington, DC: Earthscan.

Clark, Andy. 2003. *Natural-Born Cyborgs: Minds, Technologies, and the Future of Human Intelligence.* New York: Oxford University Press.

Doherty, Brian. 2004. "John Perry Barlow 2.0." *Reason Magazine*. August/September. Accessed February 8, 2014. reason.com/archives/2004/08/01/john-perry-barlow-20.

Dyson, Esther, et al. 1994. "Cyberspace and the American Dream: A Magna Carta for the Knowledge Age." Accessed January 8, 2014. www.pff.org/issues-pubs/futureinsights/fi1.2magnacarta.html.

Feenberg, Andrew. 2002. *Transforming Technology: A Critical Theory Revisited.* Oxford: Oxford University Press.

Gibson, William. 1986. *Neuromancer.* New York: Ace.

Graham, Elaine. 2002. *Representations of the Post/Human: Monsters, Aliens, and Others in Popular Culture*. New Brunswick: Rutgers University Press.

Hansen. Mark. 2006. "Media Theory." *Theory, Culture and Society* 23.2–3: 297–306.

Haraway, Donna. 1991. "A Cyborg Manifesto: Science, Technology, and Socialist-Feminism in the Late Twentieth-Century." *Simians, Cyborgs, and Women: The Reinvention of Nature.* New York: Routledge. 149–181.

Hardt, Michael and Antonio Negri. 2001. *Empire*. Cambridge: Harvard University Press.

Harman, Graham. 2002. *Tool-Being: Heidegger and the Metaphysics of Objects*. Peru, IL: Open Court.

———. 2012. "The Well-Wrought Hammer: Object-Oriented Literary Criticism." *New Literary History* 43: 183–203.

Hassan, Ihab. 1977. "Prometheus as Performer: Toward a Posthumanist Culture?" *The Georgia Review* 31.4: 830–850.

Hayles, N. Katherine. 1999. *How We Became Posthuman.* Chicago: University of Chicago Press.

Noble, David. 1999. *The Religion of Technology*. New York: Penguin Books.

Olsen, Bjornar. 2003. "Material Culture After Text: Re-Membering Things." *Norwegian Archaeological Review* 36.2: 87–104.

Poster, Mark. 2006. *Information Please*. Durham: Duke University Press.

———. 1990. *The Mode of Information: Poststructuralism and Social Context*. Chicago: The University of Chicago Press.

Retro Report. 2013. "Y2K: Much Ado About Nothing?" Accessed January 8, 2014. retroreport.org/y2k-much-ado-about-nothing/.

Rose, Randall L. and Stacy Wood. 2005. "Paradox and the Consumption of Authenticity through Reality Television. *Journal of Consumer Research* 32.2: 284–296.

Soper, Kate. 1999. "Of OncoMice and Female/men: Donna Haraway on Cyborg Ontology. *Capitalism Nature Socialism* 10.3: 73–80.

———. 2003. "Humans, Animals, Machines." *New Formations*. 49: 99–109.

———. 2012. "The Humanism in Posthumanism." *Comparative Critical Studies* 9.3: 365–378.

Stone, Allucquere Rosanne. 1995. *The War of Desire and Technology at the Close of the Mechanical Age*. Cambridge: The MIT Press.

Symmetrical Archaeology. 2013. Last modified November 21, 2013. Accessed February 10, 2014. humanitieslab.stanford.edu/23/Home.

Verbeek, Peter-Paul. 2002. "Devices of Engagement: On Borgmann's Philosophy of Information and Technology." *Techne* 6.1. Accessed February 10, 2014. scholar.lib.vt.edu/ejournals/SPT/v6n1/verbeek.html.

———. 2005. *What Things Do*. University Park: Penn State University Press.

———. 2006a. "The Technological Mediation of Morality." [online] Paper for session on Postphenomenology and the Contemporary Lifeworld, Society for Phenomenology and the Human Sciences, SPEP 2006, Philadelphia PA. Available at: www.academia.edu/694646/The_Technological_Mediation_of_Morality. Accessed May 8, 2013.

———. 2006b. "Materializing Morality: Design Ethics and Technological Mediation." *Science Technology Human Values* 31.3: 361–380.

———. 2009. "Let's Make Things Better: A reply to my readers." *Human Studies* 32.2: 251–261.

———. 2011. *Moralizing Technology*. Chicago: University of Chicago Press.

Wolfe, Cary. 2009. *What Is Posthumanism?* Minneapolis: University of Minnesota Press.

I

Interface

In this section, we offer interventions into the effects of subject-object adjacency. That is, authors consider here the shared boundary spaces that compromise the dialectic of subject and object, that complicate the distinction between organic and inorganic matter, or that construct a dialogue between kinds of corporealities—or with non-corporealities—such that we can no longer find these categories straightforwardly distinct. Readers will also find common themes related to questions of agency, the transgression of boundaries, and issues of control, relative to human-artifact relations.

In "Posthuman Topologies: Rethinking the Interface," for example, Anthony Miccoli argues that no matter how deeply we push into posthuman explanations of technology as an underlying epistemology or ontology, the concept of "interface" remains a fundamental difficulty. It represents a seemingly insurmountable topological space which exists between the human, and the technological artifact that the human manipulates. Miccoli contends that posthumanism's tendency to subsume the technological into the self as an epistemological or ontological modality reveals its "vestigial humanist conceits: (I know) through technology; or (I am) technologically" (6). To fully emerge from the humanist shadow, he argues, we must rethink "the human" as a function which occurs across substrates, and which non-anthropocentrically distributes cognition, selfhood, and being through our topological environments. This radical re-imagining of being, says Miccoli, requires us to start with a posthuman perspective and move on, rather than characterize the posthuman as the destination. To achieve such a re-imagining, we must, perhaps counterintuitively, re-emphasize technology, and the material "exterior," as on equal footing with the ontological self/mind. "Such a starting point, he argues, will allow us to see that what we traditionally know as "the human" is a function of a distributed—or aggregate—cognition, whose logic is particular to the temporal and spatial placement of that entity" (6). To make more salient the material "exterior," Miccoli takes as his example the intriguing notion of hoarding. This, he feels, offers a broader perspective on the cognitive and ontological processes, or the sort of aggregate cognition, which he explores in this chapter.

In "The Rhetorical Work of the GPS: Geographic Knowledge-Making and the Technologically-Mediated Body," Amy Propen likewise interrogates the relationship between the human and the technological artifact. Like Miccoli, Propen understands technology as an artifact on equal discursive footing with the ontological self or "mind." To test this footing, so to speak, she explores the textual and discursive work of the GPS device, and its influence on, or relationship with, the technologically-mediated, posthuman body. Drawing on interviews with GPS users, Propen explores how the GPS helps mediate geographic navigation, and how GPS users work with and against the technology to make purposeful decisions that both foster and constrain spatial understanding. Drawing on Hayles's notion that the posthuman body is always already linked to the material world, and on Latour's questioning of what counts as mediation with regard to the interactions of human beings and technological objects, Propen argues that a material rhetorical approach can help illuminate both the processes that shape users' interactions with GPS technology and the material, corporeal implications of those interactions. She takes the GPS as one example of how technologies can muddy and call into question the boundaries between human and nonhuman, and subsequently understands GPS users as engaged in an ongoing, interactive exchange with the technology, whereby agency is co-constructed through multiple rhetorical practices that include using the GPS as a supplement to paper maps; negotiating, challenging, and subverting suggested routes; storing and retrieving information; and engaging with symbolic representations while remaining attentive to physical surroundings. Propen argues that the GPS is a powerful rhetorical artifact that can shape geographical knowledge-making and ultimately how we understand the worlds in which we live.

In "Baroque Computing: Interface and the Subject-Object Divide," Elise Takehana also interrogates the boundaries between subject and object, citing this duality as a central debate within the digital humanities. Citing what she views as a problematic association within the digital humanities, whereby the humanities is associated with the study of a subject, and computer science is associated with the alleged science of an object, Takehana sees the theme of "Interface" as particularly well-suited for exploring the conceptualizations characterized by the fundamental subject/object dualism. Adopting Derrida's concept of the parergon, Takehana regards the interface experience as one that overextends boundaries of subject and ground as well as subject and object. Derrida defines the parergon as the extra or exterior that exposes the lack of the interior. Parerga are thus not clearly definable as part of the subject or ground but integral to both simultaneously. The graphic user interface is thus rife with parerga that draw users into an overextended interface experience between the user and computer. Takehana thus argues that a more Baroque conception of digital interfaces can help us to see and interact with new media in more complexly mediated and therefore less merely instrumental ways. To do so, she uses the illustrative example of the music video, Pomme C, to argue that digital spaces essentially renew Baroque traits of transgressing the order and boundaries that we rely on to define our world and distinguish subject from object, thereby allowing us to create and shape new knowledge. She adapts the language of musical and visual aesthetics to a field typically dominated by the rhetoric of engineering, both staging and fulfilling engineering's desire for the "other" of art, and art's longing for the "other" of technology.

Finally, in "Digital-Material Assemblages: Layers of Text and Agency in Mobile Augmented Reality," John Tinnell argues that the landscape of mobile media is rapidly transforming with the recent incorporation of augmented reality (AR) technologies. Like Propen and Takehana, Tinnell understands digital interfaces as complexly mediating our everyday experience of the world, and as blurring the boundaries between subject and object. Through their ability to create a unique visual-tactile interface that blends a person's gaze of the material environment with digital images and texts, for example, he argues that mobile devices bring AR into the realm of everyday experience. With this, Tinnell examines what he refers to as the "digital-material scene of writing" afforded by AR platforms, thus engaging this new technological development as a key site for interrogating foundational oppositions at the core of anthropocentric or "human-centered" approaches to design and rhetoric. Traditionally, he says, print and even electronic media have been designed to isolate page/screen content as the essence of textuality, keeping it categorically distinct and insulated from the accidents of the sensible world that necessarily envelope the embodied act of reading and writing. By contrast, the very definition of AR textuality premises itself upon an essential accidentality: the ongoing, decentralized interplay between fabricated digitality and self-organizing materiality. This material composition marks the actancy of the sensible in excess of whatever intelligibility may be willed by an individual human agent. In AR, says Tinnell, one must learn to design and write with the accidents of the sensible, no longer in spite of them. AR textuality thus shapes and is shaped by the participant's agency and situatedness among various social, cultural, and material forces.

Chapter One

Posthuman Topologies

Thinking through the Hoard

Anthony Miccoli

INTRODUCTION: BEYOND EMBEDDEDNESS

It's there. The stuff. The shiny objects that promise an exterior—an "out there" to our intrinsically apprehended "in here." Even making the distinction of the raw exterior as "phenomena" serves to place that exterior paradoxically out of reach: the phenomenal is a secret world of unfiltered sense data that informs the content of our knowing. Epistemologically complicated, existentially manifested, the exterior hinges upon the fundamental assumption that a space between it and the interior must exist in order for the consciousness to function. Since that exterior was always already problematic to know phenomenally, the best we could do was turn our philosophical attention inward. Mindedness itself became a product of the internal, biological functioning of the body. Of that, at least, we thought we were reasonably sure. Our biological substrate was a kind of first cause, or "prime mover" of consciousness; no matter how expansive the consciousness became, it always had its start in the biological. Of course, we could imagine the potential for a consciousness of some kind emanating from a machine; and we could finally give animals a fair ontological shake as well, branding their own cognitive functions as a species-specific consciousness, embedded and bound by the specific biological schematics of their brains. Even in the most posthuman visions, consciousness needed to be embodied in *something*.

Inspired by Jane Bennett's essay "The Force of Things: Steps Toward an Ecology of Matter," I hope to show that traditional characteristics of our unique humanity—our capacity to focus our attention on an objective (intentionality), our capacity to make choices (volition), and our capacity to reason in relation to those choices and objectives (logic) are functions which occur distributed across topological[1] and biological substrates. Intentionality, volition, and logic aren't simply *expressed* or *inscribed* upon an exterior. They are, essentially, materially instantiated by what we customarily know as the exterior. In order to fully explain this relationship, a technic human ontology—a sense of be-ing which is intrinsically linked to the material technological objects we utilize—should be our starting point: humanity is a species embedded in its technics and whose evolution was always already technological. From this posthuman ontology, we must move on.

This entails emphasizing the very material "exterior," and placing it on equal footing with the ontological self/mind. By starting our analysis in the exterior rather than the self, we will ultimately see that what we traditionally know as "the human" is a function of a distributed—or aggregate—cognition, whose logic is particular to the temporal and spatial placement of that entity. Such a model makes possible a posthuman determinism that does not rely on either biological or technological determinism, but a combination of the two contingent upon the specific topologies the human occupies. I will show that the "interface" is in fact an instantiated intentionality, where the human is an aggregate of biological and environmental materialisms. The interior and exterior—and the interface that links them—are functional myths allowing us to read sense into a distributed consciousness. The implications of this are vast, making possible a topology consisting of particular logics: thinking becomes physically, intrinsically linked with the physical spaces we occupy, giving way to a posthuman determinism that is neither fully biological nor fully exterior.

I think that previous models of the embedded consciousness essentially gave us a pass when it came to thinking about how human beings interact with what we commonly think of as the exterior. Mark B. N. Hansen's contention that the exterior is a kind of media upon which consciousness is expressed,[2] or Andrew Clark's theories of extended cognition,[3] allow for a necessary re-thinking of the relationship between the human and the world which it occupies. Posthuman theories of technological ontology[4] also shed light on the relationships between human and artifacts. But even the most advanced posthumanist discourse can still rely upon vestigial humanist conceits: (I know) through technology; (I am) technologically. These still center humanness on that embedded capacity to think, to intend, to express, and to make choices. Volition becomes the anchor or marker of human ontology. The capacity to choose is something "within us," never mind that the choices we're making are based upon fundamental materialisms. That mechanism of choice—that volitional switch—was something which necessarily, albeit abstractly, occupies some kind of "internal" space.

So then, how does one "emphasize the very material exterior," especially in the context of "radical interfaces"? Several authors in the present collection provide compelling analyses of specific materialisms.[5] However, for a broader perspective—and one which most radically demonstrates an aggregate cognition—I'll use hoarding as an ongoing "case in point." While the investigation of a singular interface between human and artifact can yield a very specific analysis, the amorphous and sprawling nature of the hoard offers a more broad overview of the cognitive and ontological processes at work. After all, "exteriority" (as in, all the stuff that surrounds us on a daily basis), is a rather sprawling and amorphous entity, which explains exactly why it's so much easier for us to consider ourselves as a singular, autonomous entity, than one embedded in the vast, "thingy," and otherwise disparate exterior.

NAÏVE ENTANGLEMENTS

> Today the tendency is to refer all expressions of thing-power back to a *human* operation conceived as its ultimate source—to, for example, the cultural meanings invested in a rat, the no-deposit/no return policy governing the bottle cap, or the corporate greed oozing from the Nikes. But what if we slowed this crossing from thing to human culture in order to reach a more complex understanding of their relationship? (Bennett 2004, 355)

The first step in re-thinking the interface is to focus our attention on the exterior—the material objects around us on a daily basis—in a particular way. It requires a focus that, as Bennett states above, "slows the crossing from thing to human culture." Cultural studies has trained many of us to "speed up the crossing"; to see objects as a series of significations, and deconstruct those significations in such a way as to make apparent the relations among the many signifiers which we embed in—or associate with—the objects themselves. Our obsession with things is a simultaneous cause *and* effect of consumer culture, where even the critique of such obsessions has itself been commodified into not one, but two reality television shows about hoarding.[6] In each, we find a hoarder's carefully orchestrated and accelerated descent through the layers of a lifetime of collecting, all-too-conveniently mirroring their accelerated "therapy." The originary traumas of their pathology are uncovered like the myriad dining room sets, family photos, and the occasional mummified remains of a lost cat found in their hoards. The entire process—from being "outed" as a hoarder and revealing the dire consequences if the clean-up does not take place, to the carefully edited and time-compressed process of clean-up and "recovery"—are completed in the general context (and structure) of reality television in a neat, one-hour segment.[7] In that one-hour segment, the dichotomy of the hoarder's inner mental state and the exterior condition of the hoard itself is explicitly emphasized—at first to the audience via "candid" interviews with neighbors and family members, and then later through interviews with therapists and professional organizers brought in by the shows themselves. From a basic psychological, philosophical, and cultural studies perspective, this is what one would expect. *Of course* the hoard is a metaphor for their inner mental state; and inversely, the inner mental state is *of course* being exacerbated by the presence of the hoard. Logically, the internal mirrors the external.

This is an important step in analyzing the ethical ramifications of cultural signification, but, taking a page from Bennett, I believe that this type of analysis can pull us away from the real effects of physical objects on the very mechanism of signification, representation, and analysis themselves. Using the above example, what if we were to view the hoard and hoarder not as two separate, autonomous entities, but instead view the hoard/hoarder as a singular, aggregate, "hoard(er)"? The hoard(er) becomes an aggregate of the physical and mental biological processes which arise in conjunction with the environmental processes occurring around it. In this way, the hoard(er) becomes *literally* greater than the sum of its parts. Its cognition/consciousness is distributed across the topological landscape it occupies. This model stands as an alternative to the traditional assumption that there is a fissure or space (literal and theoretical) between objects and the humans who manipulate them. Bennett's "slowing the crossing from thing to human culture" helps us to better investigate the nature of this perceived separation.

"Slowing the crossing" also works in the other direction, unbalancing a classic existentialism that customarily renders all objects as constructions of our own consciousnesses. These constructions are fundamentally based upon exterior phenomena, but those phenomena are essentially unreachable; and the consciousness is the inner process which makes sense of that outer world (i.e., the hoard is "out there in the hoarder's consciousness").[8] The sense data exterior to us is essentially the raw material of experience. That raw material is probably quite spectacular in its unfiltered form; but our biologically material and ideal limitations keep that world out of our grasp. Existential philosophy has re-articulated that inability as one of many defining characteristics of our humanity. From this perspective, humans are limited, and philosophically can focus only upon the inner mechanisms through which we render those phenomena into what is and isn't knowable. That which is outside of us is fused with that which is inside (our biological processes, our experiences, our values, our cultural influences), and we create and occupy a rich and vibrant lifeworld. Not too shabby for an entity that lives in an existential bubble, or occupies a world which is composed of nothing but textual significations.

But Bennett implies that our focus on the self-projected *significance* of objects has pulled us too far from a thing-world which is vibrant in its own right. To pull us from a cultural anthropocentrism, Bennett challenges us to—at least hypothetically—invest objects with a certain "thing power," which consists of "a relational effect, a function of several things operating at the same time or in conjunction with one another" and "entails the ability to shift or vibrate between different states of being" (2004, 354). One could view Bennett's "agency" as an anthropomorphism in and of itself: "agency" has a human connotation as a frame of reference through which to apply or project the ability to *do something*—or have some kind of efficacy in the world. But if we think of this as a *first step* toward re-thinking the interface, the value of such a metaphor becomes more clear. Training ourselves to avoid an almost instinctive attribution of a human teleology to an object opens up a hopefully disorienting conceptual space. The "naïve realism" Bennett espouses requires us to momentarily suspend the characterization of the object as an aspect of our own consciousness or as an "effect of culture" (2004, 357). In the same way we can guide our concentration when viewing an optical illusion, pulling one image forward and pushing the other toward a perceptual background, we can, in Bennett's words, "*render more manifest* the world of nonhuman vitality" (2004, 357).

Rather than view the hoard as superfluous and merely an instantiated representation of the hoarder's inner state; or to view the hoard as a "symptom" of a larger psychological illness, we can investigate the hoard as an aspect of the hoard(er), with a non-anthropomorphic vitality in its own right, empowered by its own phenomenal, topological presence, regardless of the particular human who has contributed to that presence. It has a power via its mass, its weight, and its composition. The fact that it is composed of elements which are potentially significant to the hoarder is simply an aspect—*not the basis*—of its power. If we think of the hoard as a product of the hoarder, or even the hoarder as the product of the hoard, we are investing within the non-human component a very human ability to itself "produce," or "have responsibility" for something. Traditionally, we would say that the hoard is only powerful because the hoarder is the one investing it with a certain significance: in the case of Kay, featured in the sixth season of TLC's *Hoarding: Buried Alive*,[9] the hoard began after the death of her husband of twenty-five years. Eighteen years later, Kay's 2,500 square foot home is filled with a hoard consisting of trash, rotting food, and cockroach infestation, resulting in structural damage to the home. In order to render the unique topography of Kay's hoard (and hoards in general) more manifest, we must see it not in relation to the hoarder, but *as a constituative aspect of the hoarder's lived experience*, where there is no distinction between the hoard and hoarder. There is no point of contact, or interface, between the hoard and the hoarder because they are, essentially, the same thing, operating as a singular topological aggregate.

This is an important step, because it is often our ability to read cultural significance into every aspect of the lifeworld which actually abstracts the notion of the interface itself. For example, how much of our interactions with "environment" or "technology" are complicated by the sheer conceptual weight of the terms themselves? A deconstructive analysis of signification does help us to avoid rendering nature, animals, and people to objects[10] —but it also creates an existential conceit that colonizes these exterior phenomena as outposts of our individual consciousnesses. However, if we double-down, as it were, on the very real materiality of these entities, we see that it is that materiality which makes the concept itself possible—not as a target of projection; or as the raw sense data upon which we attach, inscribe, or otherwise embed our ideas—but as the stuff that makes possible our capacity to conceptualize them. In other words, the material phenomenon is not the always already unreachable signifier. It is, instead, a part of the mechanism of the cognitive process that makes such signification possible. The hoarder thinks as hoard(er); the hoard manifests itself as hoard(er). The individual does not see herself as a traditional "hoarder," but instead as something else. In Kay's case, she refuses to consider herself as a hoarder, but rather as "lazy," or someone who doesn't like to do housework (*Hoarding*, 2013).

Hoarder-as-hoard(er) is a difficult line of reasoning to maintain, because it requires a radical re-imagining of the relationship between the "self" and the "world." Bennett explains, "The relevant point for thinking about thing-power is this: a material body always resides within some assemblage or other, and its thing-power *is a function of that grouping*. A thing has power by virtue of its operating *in conjunction* with other things" (2004, 353–54). So, we can think of human cognition as a function of the grouping (or aggregate) of biological materialism with other material phenomena that body encounters. Bennett later characterizes the lifeworld as populated with "various materialities constantly engaged in a network of relations." (2004, 354). This is where we can push even further, for Bennett's "network of relations" can potentially lead us toward a Marxist dialectics, giving the "relation" between material objects a misleading rhetorical and discursive weight, as well as reinscribing an exterior/interior dualism and model of representation. That is to say, focusing on the "relations" between objects—even strictly material ones—is to rhetorically render those material objects into *ideas* before the "relations" can be investigated and analyzed. Applied to the hoard(er), then, we would end up invoking the effects of consumer culture before we understood the distinct materialisms which make manifest the hoard(er) as phenomena. In this mode, once things are discursively rendered into ideas—represented by the mind—the discussion is subsumed into investigations of interior/exterior dualisms. What if we proposed that what the hoarder is encountering is not the mental representation of what it's seeing? What if the representational mechanism itself (not the specific representation) is constituted by the specific grouping of exterior material phenomena (hoard) with biological phenomena (hoarder)?

MYTHS OF INTERFACE AND AUTONOMY

One of the byproducts of recasting the mechanism of symbolic (representational) thinking as a form of aggregate cognition is that it breaks down even the most unforgiving biological paradigm regarding thought as a biological functioning in and of itself. Hansen touches briefly on this problem in his "Media Theory," via an eloquent analysis of Bernard Stiegler's interpretation of biological autopoiesis, specifically that it "demonstrates that biological life necessarily involves a 'structural coupling' of an organism and an environment . . . but it does so, importantly, in a way that opens the door to technics, that in effect contaminates the logic of the living with the distinct and always concrete operations of technics." (2006, 299). Autopoiesis, on the surface, seems to be a very compelling model of biological life: human beings are self-sustaining, but closed, systems. We biologically interact with the environment around us via the "structural coupling" that Hansen points out. The danger in this model is that its characterization of systems as "open," "closed," "coupled," and such, serves to emphasize an exterior/interior dualism. In the closed system of biological autopoiesis, our conscious interactions with the world are always already a series of representations. There is no real informational exchange. We experience nothing but the effects of perturbations of our own sense organs. Granted, our minds can create a wonderfully detailed and rich world based on (what amount to) somewhat indirect vibrations of the world around us. But this implies that the material world is, experientially, something on the outside, which only indirectly affects the processes that are going on *"inside of us."*

However, if we view the *mechanism* of representation as being "structurally coupled" with the material world we occupy (and not a process which occurs after the fact), rather than as an outgrowth or product of a strictly autopoietic internal biological functioning, then we can further chip away at the interior/exterior dualism which so deeply characterizes the idea of interface. Hansen gestures toward a certain "contamination" of logic due to the operations of the technics around us. Contamination is a good point of departure, conceptually. But the term still infers an interior/exterior binary—as if logic has been somehow infected or polluted by the function (or results) of this structural coupling. In terms of the hoarder, Hansen's "contaminated logic" seems to make the most sense: the hoard is causing a flaw in the hoarder's logic; or, more accurately, the hoarder is mis-assigning his or her variables (if I collect more of x, then y result will occur). But if the *process* of logic is itself the product of—and particular to—the aggregate hoard(er), then the hoard(er)'s behavior is only pathological in relation to not-hoarders.

In Kay's case, we see that much of her trouble with the hoard is linked to an inability *and* unwillingness to apprehend the debris amid which she lives as a "hoard." As the episode opens and viewers are first introduced to Kay, she describes her living conditions and justified the dense clutter as a collection of things "that haven't made it to that planned spot," even though, during the description, a montage of images show that there are no "spots" for anything—except amid/within the hoard itself (*Hoarding,* 2013). She discusses her rooster collection, and how she likes them because they are "interesting" and "colorful," yet the roosters that are visible are covered in thick, grey layers of dust (*Hoarding*, 2013). As the montage continues, we see her precariously walking over a (now-compressed) pile of cans about a foot deep as she states "I recycle the cans," although it is made clear that she does not (*Hoarding*, 2013). At one point, describing her roach infestation, she states that she doesn't "leave foodstuffs in the bedroom," yet as the camera pans over discarded paper plates rotting food is clearly visible (*Hoarding,* 2013). The logic is flawed (or contaminated) to the outsider, but not necessarily to Kay herself, since she views the objects within the hoard as a series of abstract potentialities, rather than what they appear to be in the present.

Let us, for a moment, consider logic itself. What we call logic has deep roots in Western philosophy as an intrinsic apprehension of causal connections (logic as *a priori*), or as the explanation of an imagined connection between two separate events (logic as *a posteriori*). In an *a priori* mode, an innate logic is brought to bear via experience, guiding the moment that those initial sensations are transformed into ideas, and allowing us to form judgments based upon them. In an *a posteriori* mode, the causal connection we think we perceive directly is merely a relationship we project upon experience. What the two views have in common is the act of representation itself. Representation allows the *application* of logic to a given situation—regardless of which epistemological view (*a priori/a posteriori*) is favored. Logic is dependent upon representation either to render it manifest, or as an example of the symbolic thinking necessary to apply a causal connection to perceived events. If logic is contingent upon representational thought, and representational thought is a function of an aggregate cognition comprised by the distinct topologies in which the biological human functions, then logic is not just *contaminated* by this structural coupling, it is *instantiated* by it. We're not talking about the results of an inner logic inscribed upon some phenomenal canvas of exteriority. In an aggregate cognition, the mechanism of logic itself arises from this structural coupling. Kay-as-hoard(er) literally sees no disconnect between her stated intentions and the condition of the home itself: the rooster collection is "present" and vibrant, the cans *are* being recycled, and there is no rotting food (or resulting odor) because she doesn't bring food into the bedroom (even though she later stated that she eats regularly in the bedroom).

For the hoard(er), the hoard is a component of the broader, aggregate thought process. It is, functionally and literally, as much of a cognitive component of the hoard(er) as his or her biological brain. The common denominator of both *Hoarders* and *Hoarding: Buried Alive* is the way in which the hoarder him or herself seems to defy all logic, and the ensuing frustration the occurs (on both sides) when an outside party attempts to point out those flaws. In the context of the program, usually the hoard(er) is confronted by a neighbor or family member who attempts—unsuccessfully—to intervene and help clean up the hoard.[11] For Kay, the first unsuccessful clean-up comes at the hands of her daughter, Shannon, and Shannon's partner, who are aghast at the condition of Kay's refrigerator: almost everything in it is expired and clearly rotting. As Shannon and her partner begin to discard the contents, Kay becomes increasingly agitated, especially as Shannon points out the health risks involved. Kay then abruptly "snaps," slamming the refrigerator with her hands and shaking her head: "Fine! Just throw it all away because there's [unintelligible] whether I bought it recently or not! Just throw it away! Fine!" (*Hoarding*, 2013). Kay then storms out of the house and refuses to speak directly to anyone, aside from threatening to kick everyone out of the home—including the camera crew. This is a staple within the broader hoarding genre of reality television, but also a glimpse into hoard(er) logic: when backed into a corner regarding the "use" of some piece of garbage, the hoard(er) becomes defensive and responds with something akin to "I just need it." The hoard(er) *does* "just need it" in the same respect that non-hoarders can point to regions of their skulls and say that they "just need" that particular part of their brains. The representational thinking that makes logic possible is contingent upon the copresence of the hoarder and the hoard.

The boundary lines between "human" and "environment" or "human" and "lifeworld" (basically, any metaphor which relies upon interior/exterior binary relationship) become much less distinct in this configuration. Culturally, we apprehend ourselves as autonomous beings. We tend to lump that which we perceive as outside of our "selves" into a generic, autonomous exterior (rendering it as "other"). This affective perception of autonomy conceptually arises from the structural coupling described above. The "self-awareness" which this coupling makes possible is dictated by the specific environments we occupy. The self-observed biological phenomena from which we think we arise becomes a necessary sensory and conceptual singularity upon which we can project, or, in which we can "locate" our awareness. But that awareness can't occur unless there is a *known* topological parameter to occupy. For the hoard(er), that topological parameter is one defined by the hoard. For human beings in a broader sense, the topological spaces we occupy are culturally defined, and allow for a more-or-less uniform experience among those who occupy similar conceptual and topological spaces (i.e., socioeconomic standing, physical locations, ethnic traditions, among others). All of this falls under an "anthropocentric topology"—a logic of self-awareness which is instantiated across a material landscape. In this case, we are thinking in terms of *homo sapiens*, but this could apply to other species as well, whose distributed cognitions are themselves unique to the topologies they occupy. Perhaps more important (and for some, more scary) is the potential for a new way of thinking about artificial intelligence. The silicon-based, distributed processing is unique to the material substrates those processes occupy. Machines could be "self-aware" in ways that we as humans aren't configured to apprehend.

The perceived "autonomy" of *anything* is functionally mythological. It is a projection which logically situates an entity in the perceived lifeworld. As humans, we base our specific autonomies on the perception that we are self-contained individuals. As we can now see, that autonomy is based upon a structural coupling of our biological functioning with the topologies we occupy. To mitigate the implied spaces between a perceived self and a perceived world, we fall back on equally mythological "interfaces" to bridge the gap between the autonomous self and the world that self occupies. This myth of interface is necessary to maintain the integrity and efficacy of the self in everyday interactions with the world. However, in moments of intense concentration, meditation, artistic inspiration, or even sexual ecstasy, the "interface" is conceptually suspended, allowing for aggregate cognition to become less distantly experienced. In the absence of these situations, the "interface" between us and the lifeworld is a functional myth, similar to Hume's "customary connection" to explain cause and effect. To best have efficacy in the world, we tend to rely on a model of autonomy which best facilitates that interaction.

CONTINGENT INTENTIONALITY: ACTIVELY MATERIALIZING THE EXTERIOR

> When a technological artifact is used, it facilitates people's involvement with reality, and in doing so, it coshapes how humans can be present in their world and their world for them. In this sense, things-in-use can be understood as mediators of human-world relationships. Technological artifacts are not neutral intermediaries, but actively coshape people's being in the world, and their perceptions and actions, experience, and existence. (Verbeek 2006, 364)

In order to move toward a more active materialization of the exterior, we need to momentarily step away from the hoard(er) and think about more specific objects. When we use an object, what exactly is going on? What takes place when we engage an object in light of our aggregate cognition and instantiated logic? For this, we need to look to the moment of use which Verbeek points to in his "Materializing Morality: Design Ethics and Technological Mediation." Here, the artifact/object is viewed in a traditionally posthuman mode: as a bridge or facilitator. The object allows for a more involved, efficacious interaction with the world. Verbeek presciently describes this moment of use as a moment in which an exterior object "coshapes" an essentially existential being-in-the-world. The interface here is less concrete than a "bridge," instead acting as a frame for the experiencing of the world. Framing is a way to establish boundaries or parameters around experience. However, metaphors such as this can be problematic in that they draw attention to the space between the human and the object of use. Anything which "mediates" implies a space between two autonomous sides: consciousness/human on one side and not-conscious/not-human on the other.

That being said, a very necessary aspect of the interface isn't simply its topological makeup: aggregate cognition or not, the interface still infers an action of some sort, a moment of use. Verbeek implicates the artifact as an object which can directly influence behaviors. Invoking Latour, Verbeek states, "Actions are the result not only of individual intentions and the social structures in which human beings find themselves (the classical agency-structure dichotomy), but also of people's material environment" (2006, 366). The interface is manifested not only through the manipulated material objects (which are coupled with the physical human to manipulate them), but also an inherent intentionality that the human being wishes to augment through the use of the artifact itself. Intentionality is another one of those sacrosanct philosophical ideas through which humanity is defined (especially in the existentialist philosophies of Husserl, Heidegger, Sartre). Existentially, intentionality is a key aspect of be-ing: intentionality is the focusing of the will which allows the self to bring the phenomenal world into being. In light of an instantiated logic and aggregate cognition, intentionality must also be re-cast in the light of Bennett's vital materialisms. Simply stated, there can *be no intentionality without a materiality*. Verbeek gets close to this when he states "The influence of technological artifacts on human actions can be of a nonlingual kind. Things are able to exert influence as material things, not only as signs or carries or meaning" (2006, 367). However, ascribing to an object the ability to "exert influence" is to fall into a counterproductive anthropomorphism.

The important aspect here is that intentionality isn't necessarily embedded in an object—no more than teleology is metaphysically embedded in a physical object as its final cause. Intentionality arises in the co-presence, or coupling, of object-with-human. This model is closer to Hansen's contention that intentionality is expressed in and through the medium of the external world. To be fair, we should view this in context of Verbeek's essay, where he calls for a more ethical use of objects via an awareness of the "scripts" which can be written into the design of artifacts themselves, which can evoke—and even prescribe—behaviors: "Scripts transcend functionality: they form a surplus to it, which occurs once the technology is functioning. When technologies fulfill their functions, they also help to shape the actions of their users . . . the script approach opens up a new way to morally assess technologies with respect to the role they play in their use contexts" (2006, 362). The scripts embedded in technological artifacts do more than affect the way in which the autonomous consciousness uses this autonomous object: the script affects the very instantiation of the consciousness itself in any given moment. That is to say, an object (in which a script is embedded), when structurally coupled with a human, becomes the logic of the human-object entity. The moment of this structural coupling amounts to an ontic fusion—one that moves beyond "shaping" consciousness and creates an entirely other entity that is now the sum total of both previously singular entities. Conceptually, the "*interface*" can only be conceived in the absence of that ontic fusion. In other words, the interface can only be thought of when the coupling it signifies is not instantiated through its use. Intentionality becomes the moment of the affective "reach" toward the object, and when the fissure of interface is at its widest. Paradoxically, the closer in space we get to the object we're reaching for (intending to use), the wider the fissure gets—reaching its widest in moments when the human is in closest physical proximity with the object. It's not until the moment of use that the fissure of interface is closed and two objects (previously conceived as autonomous) become fused ontologically. With Kay, this would explain why each time she picks up a singular object from the hoard, she can immediately weave not only a narrative as to how the object was acquired, but also propose an intricate plan as to what she will eventually do with the object itself.

In the midst of the massive clean-up and "haul-away," Kay still attempts to hold onto things that—in reality—serve no purpose. At one point, Dr. Julie Pike (one of the program's licensed psychotherapists) escorts Kay into the home with the intention of pointing out the multitude of roaches crawling up the walls as haulers dig through the trash. Instead of noticing the roaches, Kay focuses on a single CD which she "didn't want to step on and break it because it's wrapped in plastic and still good." When Pike confronts Kay on her skewed focus, Kay responds, rapid-fire: "Once I get the house cleaned up, I will have my stereo system set up again and I will [pause] do what I used to before which is to have music on pretty much all the time in the background." While not "logical" in the colloquial sense, Kay's thought process *is* logical within the topology of the hoard. As hoard(er), Kay is completely logical: that particular CD *will* be listened to because it remains in a state of potential use (it has not been opened yet) and because there is a stereo system which exists in the context of the hoard. Later, Dr. Pike will explain—and images will corroborate—that nearly all of the electronic equipment buried in the hoard has been irreparably damaged due to either the crushing weight of the hoard itself, or the massive roach infestation.

It is generally easy to think about what a "moment of use" is when manipulating a specific object. But what of the hoard(er)? Of what "use" is the hoard? This question—in one form or another—is often the action which momentarily creates the fissure which temporarily renders the hoard(er) to the hoard/hoarder. When questioned about the hoard, or "forced" to imagine the hoard through the vantage point of someone else, the hoarder is then sent into a state of confusion, stress, and sometimes emotional breakdown. When Kay's home is finally "cleared," and her and her daughter assess the structural damage done to the home, Kay *seems* to have a moment of clarity, during which time Pike attempts to take advantage of the disconnect: "You are someone who hoards because you put these items above your own personal well-being. That is what defines this problem" (*Hoarding*, 2013). Kay seems to be affected by Pike's condensed analysis, as if she is actually seeing the house as it is. But one could posit that she is only temporarily seeing the house that way, only because the hoard—a working aspect of her cognition—has been cleared. Without this instantiated aspect of her thinking, there is nothing there to physically reinforce Kay-as-hoard(er). Characterized as an autonomous entity, the "hoard" has no use; similarly, it is also devoid of intentionality—as is the hoarder, who "never intended" it to get so out of hand. But looking at this forced moment of fissure, we can get a better sense as to how our individual lifeworlds are comprised by this aggregate cognition.

Does simply being in a room with a bunch of random objects mean that we are then ontically determined to think randomly? The relationship is not as simple as that, mainly because of the constant bombardment of our senses by all of the sense datum produced by the unique topologies we occupy, coupled with the unique memories and impressions that objects have made on us in the past.[12] If human beings are indeed aggregates of the physical and mental processes occurring around them, and cognition/consciousness is distributed among the topological landscape it occupies, then the traditional idea of the "human being" as an autonomous, self-contained entity is no longer applicable. A "human being" is an aspect of the unique topology which that human's biological body happens to occupy at a given time. This explains affective reactions to particular places at particular times; or the ways in which some people are more susceptible to clutter or open spaces; or the utter terror hoarders feel when their hoards are removed against their wills. As we have already demonstrated with the hoard(er)—the hoard doesn't simply reflect a state of mind, but *is part of an aggregate mind.* When hoarders feel that they are losing their minds because the hoards are being altered or removed, they are actually onto something. Parts of their minds are being removed. For the hoard(er), the cognitive process is intrinsically affected when part of the hoard is altered or removed. So, intentionality, mindedness, and logic are functions that happen across what we are lead to believe is an exterior/interior binary; but, that binary is merely a convenient truth for a consciousness which needs to locate itself in physical parameters in order to best negotiate the environment it occupies. The idea that the "outside" is separate from the "inside" is illusory; necessary, but illusory—in the same fashion that movement itself is an illusory quality the mind places on the results of spatiotemporal shifts in objects and perceptions; or as the mind customarily connects two objects which tend to happen in constant conjunction with each other. Interior and exterior are myths used to read sense into an aggregate cognition.

THE TOPOLOGY OF PARTICULAR LOGICS

Breaking down the interior/exterior has very specific, posthuman implications. Tying cognition, logic, and the mechanisms of representation to a distributed, aggregate model makes possible the "particular logics," or modalities of thought and expressions of "mindedness" which are contingent upon the physical topologies occupied by a specific process of mind at a particular time. That is to say, our thinking is physically, intrinsically linked with the physical spaces in which we live, work, and play. A vital materialism allows for a legitimized exploration of the importance of architectural, ergonomic, and artifactual design that go well beyond Verbeek's already forward-thinking rematerialized ethics. That an ethical, affective component can be designed into a singular artifact is reason enough to explore the vital materialisms which Bennett articulates. Thinking about the ethics that may have been inadvertently included in objects which already exist could help to explain political, socioeconomic, environmental, and cultural patterns which have been at work throughout our (technic) evolution. This posthuman ethics would not be limited by the traditional inner/outer binary or more literary characterizations of "otherness." Indeed, an entire re-thinking of the most pervasive binary and rhizomatic cultural constructions lies before us; not to mention the potential to look at the entire postcolonial movement itself through a lens which sees the colonizer and colonized (perhaps even the signified and signifier) not as distinct, autonomous entities, but as aggregates of specific topological spaces and constructions—or, as Bennett puts it, "assemblages":

> The relevant point for thinking about thing-power is this: a material body always resides within some assemblage or other, and its thing-power *is a function of that grouping*. A thing has power by virtue of its operating in conjunction with other things. . . . This is not a world . . . of subjects and objects, but of various materialities engaged in a network of relations. It is a world populated less by individuals than by groupings of compositions that shift over time. (2004, 353)

Focus must remain on the materialities which are engaged in a network of relations, not the relations themselves. Considering this, would we be so quick to dramatically extract the hoard from the hoard(er)? When an elderly person loses a spouse, or a parent loses a child, the grieving process looks quite different when viewed through its posthuman topology: if the persistent fixtures of the spousal or filial topologies are suddenly removed, what and who remains is fundamentally altered: cognitively and ontologically. When aboriginal peoples are forcibly removed from their native environments, does not the entire cultural landscape of that people become irreversibly transformed? How does one act "logically" in such an instance when the mechanism of logic itself has been altered? In one of the more poignant moments of the show, when Kay is explaining the great loss she felt when her husband died, she is shown laying in her half of the bed, while the other side (her deceased husband's) has been consumed—and thus replaced—by a pile of books that extends at least three feet higher than Kay.

I think that for too long, perhaps all the way back to Marx, cultural studies and interdisciplinary studies in general has been plagued by a certain "tyranny of relations," in which the spaces between objects—their fissures of interface—have been reified and discursively instantiated as having power. The absence, or the spaces between objects, are invested with a power that actually emanates from the objects themselves. The constituent biological material which makes up the human is simply a specific aggregate within a larger aggregate of vital material. In light of a structural biological coupling, the "will" or "intentionality" become arbitrary placeholders for the process of aggregate cognition.

Furthermore, the structure of logic itself is no longer limited to a specific cause and effect, "on or off" characterization. The "logic" is not so much *relative* to a specific time or place, as it is instantiated by it. The specific structural coupling of biology and topology, in essence, make the rules. The uniformity of those rules, and to a certain extent, the blueprint for their breaking, is already present in the functioning of the distribution of that unique cognition across that unique environmental/topological landscape. Hegemony is a product of the artifacts themselves rather than simply a product of relations among cultural artifacts. A certain habit of thinking—or "shape of thoughts" comes about in the hegemony of the physical landscape. Urban blight, gentrification, suburban sameness, rural idyllism, all have particular logics about them, which become impossible to know until one is physically immersed in them. In order to understand them, we need to physically immerse ourselves into those landscapes, and think accordingly. Hegemonic thinking is a product of the environment, not the other way around.

Philosophically speaking, a posthuman model of particular logics finally overcomes the shadow of autonomy in which vestigial humanist conceits still hide. Autonomy—whether in self, other, subject, object, and so forth—is no longer viable in a posthuman model. One's choices *are*, in fact, determined by the co-present biological and topological materialisms which instantiate the mechanism of choice itself. Neither has "agency" in its own right. But placed in an biological and ontic coupling, the aggregate interior/exterior makes agency possible. The choices we make are determined by the unique biological function of cognition spread across a topological landscape. The possible choices available to that distributed consciousness are limited in number, but that number goes up exponentially with each object introduced into that landscape. For all intents and purposes, there are infinite possibilities. But, *technic-ly*, the number of choices is finite, limited to the permutations of structural couplings possible. This is a posthuman determinism.

NOTES

1. I use "topological" and "topology" throughout this essay in its more physical connotation: as the composition and arrangement of the physical objects and characteristics of a given space or environment, and how those characteristics are interrelated. In an etymological sense, I am calling for a literal "logic" of "space."

2. "In this sense, the medium perhaps names the very transduction between the organism and the environment that constitutes life as essentially technical; thus it is nothing less than a medium for the exteriorization of the living, and correlatively, for the selective actualization of the environment for . . . a demarcation of world, of an existential domain, from the unmarked environment as such" (Hansen 2006, 300).

3. Clark's model of "extended cognition" is one which "allows at least some aspects of human cognition to be realized by the body and/or the extraorganismic environment" (2011, 81). Clark also states that in this extended model, "some of the thinking, and even representing, may supervene on activities and encodings that criss-cross brain, body, and world." (2011, 149).

4. I'm thinking specifically here of N. Katherine Hayles' seminal work from 1999, *How We Became Posthuman: Virtual Bodies in Cybernetics, Literature, and Informatics.* Chicago: The University of Chicago Press., and my own work from 2010, *Posthuman Suffering and the Technological Embrace*. Lanham, MD: Lexington Books.

5. See Amy Propen's "The Rhetorical Work of the GPS: Geographic Knowledge-Making and the Technologically-Mediated Body"; Elisabet Takehana's "Baroque Computing: Interface and the Subject-Object Divide"; and John Tinnell's "Digital-Material Assemblages: Layers of Text and Agency in Mobile Augmented Reality" in the present collection.

6. *Hoarding: Buried Alive*, Discovery Communications, LLC (Silver Spring, MD: TLC, April 2010–); *Hoarders*, A&E Television Networks, LLC (New York: A&E, August 2009–).

7. In almost every instance, there is a positive conclusion, to which dramatic before-and-after images testify: the hoard is mostly or partially cleared—or at least cleared enough for the hoarder to avoid eviction, the intervention of family protective services, or seeing the property condemned. In the few follow-up shows, however, the results seem far less positive, and several hoarders begin hoarding again.

8. I am indebted to Albert Hakim, editor of *Historical Introduction to Philosophy*, who provided this specific turn of phrase to characterize the transcendental subjectivity found in the existential philosophies of Husserl, Heidegger, and Sartre, among others. The example of an object being "out there in my consciousness" (2006, 539), has helped several of my Introduction to Philosophy students understand an otherwise tricky concept.

9. *Hoarding: Buried Alive*, Season 6. Ep. 6; "They're Crawling!" Jessie Dinh, Series Producer; Candace Boissy, Supervising Producer. Discovery Communications, LLC (Silver Spring, MD: TLC, March 20, 2013).

10. This was a favorite maneuver of Heidegger, who goes through maddening deconstructions of terms (and exteriority itself) only to eventually reconstruct them as aspects of humanity.

11. In every instance, given the scope of the hoard and the depth of the psychological issues, these interventions are always unsuccessful, and are clearly contrived by producers of the programs to agitate both the hoarder and his or her loved ones for the sake of compelling television.

12. I'm not contending here that the human mind/aggregate human does not retain memories; nor am I contending that the memories themselves have some kind of physicality. They too become part of that aggregate collection of aspects that make an individual, self-reflexive human being.

REFERENCES

Bennett, Jane. 2004. "The Force of Things: Steps Toward and Ecology of Matter." *Political Theory* 32, no. 3: 347–372.

Clark, Andrew. 2011. *Supersizing the Mind: Embodiment, Action, and Cognitive Extension.* New York: Oxford University Press.

Hakim, Albert, ed. 2006. *Historical Introduction to Philosophy.* 5th ed. Upper Saddle River, NJ: Person Education, Inc.

Hansen, Mark B.N. 2006. "Media Theory." *Theory, Culture, and Society* 23, no. 297: 297–306.

Hoarding: Buried Alive, Season 6. Ep. 6; "They're Crawling!" Jessie Dinh, Series Producer; Candace Boissy, Supervising Producer. Discovery Communications, LLC. Silver Spring, MD: TLC, March 20, 2013.

Verbeek, Peter-Paul. 2006. "Materializing Morality: Design Ethics and Technological Mediation." *Science, Technology, and Human Values* 31, no. 3 Ethics and Engineering Design: 361–380.

Chapter Two

The Rhetorical Work of the GPS

Geographic Knowledge-Making and the Technologically Mediated Body

Amy D. Propen

I'm used to having a map which shows me a guided path to my destination and allows me a view of my surrounding areas, and that was sort of comforting. With the GPS I realized I would not have that, so that was a little concern and remains a little bit of a concern with the GPS.

—GPS user OSA9[1]

I liked just the map function and being able to see where I was in relation to the surrounding countryside.

—GPS user BA14

The one time I used [the GPS] for a long distance thing where I kind of knew where I was going, it was telling me to go a way that I wasn't at all familiar with, and I didn't—I didn't do it—it kept recalculating, because I kept going the way I thought it was.

—GPS user BA16

I love how it suggests an alternate route. I loved how it would recalculate and adapt to me—I was in control.

—GPS user BA21

I quickly became impressed with how much simpler it was [to] make my travel plans, as well as how much less stress it implied for myself and/or my navigator, since missed exits, wrong turns, and road closings were no longer potentially catastrophic failures.

—GPS user OFM10

As the quotes from these GPS users describe, in-car navigational devices have various, but clearly distinctive impacts on the personal experiences of everyday navigation. Users can program their routes and then negotiate the information provided by the GPS with their own knowledge or interests, ultimately making navigational decisions that impact both the mind and body. Users may work with and against the GPS, they may resist or subvert it, or they may supplement its directives. These interactions between the GPS and the user have the capacity to encourage the development of new geographic knowledge or to constrain or limit geographical understanding. In this chapter, I use a rhetorical lens to argue that the GPS is a material, rhetorical artifact that helps mediate our navigational experiences and shape how we understand the worlds in which we live.

Borrowing from Judy Segal, I consider rhetoric to be concerned with the study of text and discourse in order to achieve "a greater understanding of human action" (Segal 2005, 2). In addition, I understand "text" not only in the more traditional sense of "printed words on the page" but also as multimodal, or as combining visual, material, auditory, and digital information. Based on the ideas of Carole Blair, I also understand texts as bearing the capacity for social consequence and as potentially influencing both the mind and body. In redefining what gets to "count" as a text, Blair poses five questions that help speak to the broader impacts and consequences that multimodal texts can have in the world. To this end, she asks: "(1) What is the significance of the text's material existence? (2) What are the apparatuses and degrees of durability displayed by the text? (3) What are the text's modes or possibilities of reproduction or preservation? (4) What does the text do to (or with, or against) other texts? And perhaps most important, (5) How does the text act on people?" (1999, 30). Based on these ideas, I argue that a rhetorical approach to understanding GPS use can help illuminate both the processes that shape our interactions with the technology and the implications of those interactions for the technologically-mediated, posthuman body.

When referring to the posthuman body, I subscribe to Hayles' (1999) and Brooke's (2000) general positioning of the posthuman as necessarily mindful of the practices of embodiment, and sensitive to the idea that modes of being are inextricably linked to the material world. Katherine Hayles, for example, writes that we should understand the posthuman body as able to interact with information technologies without being seduced into "fantasies of unlimited power and disembodied immortality"—that we must understand modes of being as inextricably linked to the material world (1999, 5). Bruno Latour, in the essay "On Technical Mediation," moves through several ideas about what counts as *mediation* with regard to the interactions of human beings and technological objects. First, Latour defines mediation as the series of goals, steps, and intentions that a "human agent" can describe in relation to a series of events involving a technology (1994, 32). For Latour, either the human agent may enlist the technology, or the technology may enlist the human; eventually, he says, "a third agent emerges from the fusion of the other two" (32). This "composite agent" is similar to his notion of the hybrid—it is, in a sense, a translation—it represents a shift in how a goal might be pursued (Latour 1994, 33). A hybrid agent may be human or nonhuman and may have its own goals or functions. While Latour understands human and nonhuman agents as sharing equal responsibility for uses of technologies and goals related to technologies, I do not view technological objects as "standing in" for human action in quite this same way, although I do understand the GPS as influencing human action. Given the decision-making and knowledge-making practices fostered through users' interactions with GPS technology, I argue instead that the GPS and its user co-construct agency, in what Krista Kennedy has called an "interactive process that involves exchanges between multiple agents, texts, and influences" (2010, 308). Additionally, as Peter-Paul Verbeek contends, humans and reality "mutually constitute each other" and are always interrelated; technologies then become mediating tools through which humans relate to and interpret the world around them (2006, 5). More specifically, Verbeek says that

> human beings cannot but be directed at the world around them; they are always experiencing it and it is the only place where they can realize their existence. Conversely, their world can only be what it is when humans deal with it and interpret it. In their interrelation, both the subjectivity of humans and the objectivity of their world take shape. What human beings are and what their world is, is co-determined by the relations and interactions they have with each other. (2006, 5)

Verbeek uses this perspective as a framework "for analyzing the influence of technology on human behavior," for, as he puts it, "[t]echnological artifacts mediate how human beings are present in their world, and how their world is present to them" (2006, 5). In this chapter, I build on these ideas about technological mediation and posthuman bodies in order to understand GPS users as engaged in an ongoing, interactive exchange with the technology constituted by multiple discursive practices that include: using the GPS in conjunction with other information; negotiating, challenging, and subverting suggested routes; storing and retrieving information; and engaging with the technology while remaining attentive to physical surroundings. More specifically, in my recent study of GPS technology, I wanted to explore the following questions:

1. In what ways do the various cues provided by the GPS resonate with and influence its users?
2. How do GPS users engage with the device?

3. What are the potential consequences of the GPS on geographical understanding?
4. What are the rhetorical consequences of the GPS on the mind and body and for purposeful decision-making?

DATA COLLECTION AND ORGANIZATION

To address these questions, I designed a qualitative study and conducted face-to-face interviews of approximately thirty minutes with twenty-two GPS users.[2]

I conducted my research at a small liberal arts college in the mid-Atlantic region, and delimited the parameters of my sample to any faculty and staff at the college, including part-time adjunct faculty and lecturers, as well as the family members of any faculty or staff who work there.[3] In my sample, 59 percent of participants were female and 41 percent were male;[4] 95 percent were Caucasian and 5 percent were African American; 43 percent of participants were adjunct faculty, 39 percent were college staff, 9 percent were full-time faculty, and 9 percent were family members of faculty or staff; 59 percent of participants owned a GPS, 36 percent borrowed a GPS from the library, and 5 percent borrowed a GPS from a family member who owned one. Of the participants who owned a GPS, 31 percent had owned a GPS for three years or more; 31 percent had owned a GPS for two years or more; 30 percent owned a GPS for one year or more; and 8 percent had owned a GPS for less than one year.

Participants who volunteered to be in the study were asked to use an in-car GPS navigation device for a two-week period during any driving they did. If a participant owned a GPS, they were free to use that one. If a participant did not own a GPS, they were able to borrow one through the GPS borrowing program at the college library.[5] After they had completed the usage period, I interviewed each participant about their experience using the GPS.[6] I tape recorded and transcribed each interview verbatim immediately after the interview. After all interviews were completed and transcribed, I coded the transcriptions and looked for themes that emerged.[7]

It is important to note that it is not my goal within this chapter to conduct a usability study of the GPS for new and experienced users or to conduct any sort of feature comparison across GPS brands.[8] Rather, my goal has been to capture a diverse range of GPS experiences across users, and to see what material rhetorics can reveal about the impact of these technological contexts on the mediated, posthuman body. This is not to say, however, that the experiences of newer and more seasoned users are identical, or that differences therein should not be acknowledged. In some cases, these categorical distinctions seem to account for different types of bodily experience or influence of the technology on geographic understanding. In such cases, I am mindful of these differences and acknowledge them specifically within the context of their value to a rhetorical approach. In fact, as I coded the interview transcripts and looked for the themes that emerged, I was attentive to these potential differences and created the participant key such that the comments of newer users could be distinguished from those of more experienced users and invoked if necessary.[9] I have concealed, however, any specifically identifying information about participants, both in terms of personal names or references, and in terms of any specific information related to addresses or particular place-names described. I refer to names of towns, cities, restaurants, stores and other common or generally known sites only when such references would not reveal any specifically identifying information.

As an artifact of material rhetoric, the cartographic texts produced by the GPS may be viewed as durable, reproducible to a certain extent, and may supplement or be supplemented with other types of geographic knowledge claims. The GPS engages the user, and in doing so, enables and invites an audience, even a dialogue. Some users tend to anthropomorphize the GPS or forge what appears to be a personal relationship with the artifact. Users' interactions and communication with the device also provide a means for challenging or correcting it, in order to assert their own claims to knowledge and engage in decision-making. With these issues in mind, this chapter examines GPS use and its function in everyday life, subsequently demonstrating the value of material rhetoric in helping to illuminate the everyday contexts influencing the technologically-mediated, posthuman body.

THE PHYSICALITY OF THE GPS

In describing their experience with the GPS, many users were quick to note the physicality of the device and the ways in which the GPS, as a tangible artifact, impacts their driving routine. One user described their specific routine for keeping and using the GPS in the car: "I just plug it into the cigarette lighter . . . the armrest flips up and there's a little storage thing inside, so I just drop [the GPS] back in, and then the mount goes under the seat" (OSA11). Here, the user's description of "just" plugging the GPS into the cigarette lighter, and "just" dropping it into the armrest implies a simplicity and ease of use that makes act of handling the GPS seem like an ordinary part of the driver's daily routine. Another user noted that if they "get into somebody else's car and they don't have it, it just seems like something's really missing" (OFT2). These accounts help demonstrate how the physical tangibility of the GPS impacts drivers' experiences even prior to turning on the device. It is also important to note here that, for Hayles, physicality is distinct from materiality, especially within the context of understanding the work of "hybrid texts and subjectivities" (1999, 2). The GPS may illustrate this point well, for while it is indeed a physical object in itself, it is not defined solely in terms of its physicality; in other words, its physicality does not constitute the whole of its material impact on the body or its relationship to the user or to knowledge-making. Instead, its physicality is but one component of its functioning as a multimodal rhetorical artifact.

VISUAL CUES AND THE BODY

The GPS is a rhetorical artifact that produces multimodal cartographic texts. Consistent with David Turnbull's understanding of the map, these texts produce "a graphic representation of the milieu, containing both pictorial (or iconic) and non-pictorial elements," and "directly represent at least *some* aspects of the landscape" (1989, 3).

The visual texts displayed by the GPS are indeed an integral component of both its multimodality and its impact on the user's bodily experience. In fact, many of the GPS users interviewed in this chapter were quick to describe visual aspects of the GPS that resonated with them in one way or another. One user described the visual display as comforting: "[C]uriously enough, even when I'm going somewhere completely familiar [or] routine, like the supermarket or the gym, I don't even get past our street without pushing the button that brings it on, because I don't like seeing the empty screen . . . I don't really need to have it on—I just like it" (OFT2). Another user described a similar notion: "I liked just the map function and being able to see where I was in relation to the surrounding countryside" (BA14). In this sense, the GPS helps create a sort of cognitive or mental map that can aid "in the development of spatial understanding and spatial ability" (Xia et al. 2008, 446).

One user also described getting lost in terms of how their car was represented symbolically through the visual display on the GPS screen. For example, as this user described, they were driving on back roads when the GPS did not produce maps specific to their location: "I was on the road, but there was no map to tell me where I was. . . . I was on white—I was no longer on a road—there was no gray or pink line to direct me" (BA21). Eventually, this user had to stop and ask for directions before continuing home. While all of these users, to some extent, describe the physical experience of driving relative to their interaction with the symbolic features of the visual display, this last user's description of getting lost as driving "on white" bears a particularly close analogue to that of someone describing a gaming or virtual environment.[10] Kevin Moberly, for example, understands video games as "evaluat[ing] players on their ability to compose themselves in relationship to these highly symbolic environments—to write and ultimately revise their actions in relationship to the reality that is manufactured on the screen" (2008, 291). As Brendan Keogh also describes in his discussion of cybernetic memory and video games in this collection, the gamer's various experiences, and in this example, perhaps, the experiences of the GPS user, "exist in a relationship between bodies and worlds" (234). Indeed, the GPS requires that users compose themselves in relationship to highly symbolic, selectively represented routes, locations, and place names. These symbolic representations of the world around us then have clearly material impacts, which may result, as it did in this user's case, in having to supplement the cues of the GPS by stopping to ask for directions.

AUDIO CUES AND THE BODY

Users' responses to the audio cues provided by the GPS were often linked to their knowledge of the surrounding area. That is, many users tended to appreciate or pay greater attention to audio cues when they were less familiar with their surroundings. As one GPS user described: "[W]hen I knew where I was going, [or] I knew I had to turn left on a certain road and it would say 'turn left in 800 feet' or something, that was distracting—I knew that. So the voice coming at me, telling me, was just sort of something I didn't need to know" (BA16).

Perhaps the most interesting consequence of the audio cues is the way in which they summon the driver, inviting an audience, and in some cases, even a dialogue, with the GPS. Users are often aware of this aspect, acknowledging both the engaging and sometimes humorous nature of the audio cues.[11] One user, for example, noted the challenge of finding an appropriate voice setting: "Well, we did find that we liked certain accents better than others. So we did like the feature of being able to change them" (BSA17). Another user mentioned that "you find yourself kind of talking to the unit eventually" (OSA6). Interestingly, one user explicitly described the ability to choose a voice setting as tied to the tendency to develop a relationship with the GPS: "[A]fter using it long enough, the GPS has become a person. . . . [I]t's almost like it's interacting with the person even though it's electronic. But I think the fact that you can choose your voice and it's a very soothing voice, you kind of develop that relationship with the machine" (OSA11). While it is beyond the scope of this chapter to engage in an extended analysis of the implications of the gendered or culturally diverse GPS voice settings from which users may choose, it is clear that users tend to choose a voice that resonates with them or makes them feel comfortable. It is also worth noting that Emily McArthur, in her analysis of Siri and Benjamin's aura in this collection, considers the social functions of Siri's voice settings. McArthur, albeit in a different analytical context, argues that Siri's "feminized" and "eternally polite" vocalizations gives the user a feeling of power and control over the technology (129). While the GPS users interviewed for this study did not cite gender, per se, in their discussions of choosing a voice setting for the device, it is clear that choices related to gender and voice settings impact the user's willingness to interact with the GPS, whether it be out of a sense of comfort or control. The willingness to interact with the GPS then often leads to an anthropomorphizing of the device, or, as Latour defines it, a projection "of a human behaviour onto a nonhuman . . . object" (Johnson [Latour] 1995, 265).[12] Latour sees anthropomorphism as underscoring an "untenable" divide between humans and nonhuman technical objects (1995, 266). That is, on the one hand, the GPS may be understood as already anthropomorphic, in that it has been invented by humans and shapes human action (1995, 265–266). On the other hand, Latour also understands these nonhuman artifacts as "substitut[ing] for the actions of people, or as a delegate that permanently occupies the position of a human" (266). Again, this chapter does not view the GPS as standing in for human action in quite this way—while the GPS, as a multimodal technology, certainly mediates the actions of its users and thus serves to muddy the divide, it does not substitute for human action, permanently occupy the position of human, or take on full agent status. Instead, users more typically describe an ongoing set of interactions, negotiations, and exchanges with the device that serve to co-construct agency and influence geographic knowledge-making. Moreover, the articulation of a sort of relationship with the GPS provides users with a means for pushing up against or resisting the knowledge claims implicit in or inferred through its multimodal cues and directions.

WORKING WITH AND AGAINST THE GPS

Understood through a Blairian lens, the GPS functions rhetorically by challenging and sometimes correcting users' knowledge of their environment. It invites an audience not only through its multimodal cues but through the directions that it asks users to follow. Many users, for example, expressed a sort of power struggle with the GPS or a hesitance to "listen to" the device and the ways in which they would challenge its claims to knowledge. As one user put it, "I do still check to make sure that it's giving me good information. . . . I'll go and say, 'No, I'm not listening to you,' as far as that goes" (OA1). Another user described a more explicitly adversarial interaction: "[W]e decided to go a different way we knew, and the GPS was fighting us, for maybe six or seven miles" (BSA20).

The GPS often engages in a variation of what Carole Blair refers to as the act of "correcting," or in GPS parlance, the action more commonly referred to as *recalculating* (1999, 39). The recalculating function understandably elicits different reactions from different users; nonetheless, it was common for users to describe an interaction with the GPS that involved the act of recalculating and its impact on both the mind and body. One user who felt familiar with the surrounding area and did not want to take a different route described active resistance to the GPS: "The one time I used it for a long distance thing where I kind of knew where I was going, it was telling me to go a way that I wasn't at all familiar with, and I didn't—I didn't do it—it kept recalculating, because I kept going the way I thought it was" (BA16). Rhetoric scholar Carolyn Miller also acknowledges what she calls "resistance models of agency . . . that usually rely on a metonymy between agent and agency" and function in the interest of resisting some perceived form of ideological pressure (2007, 144). In this case, the user exercises informed decision-making that is enabled through an interactive sort of agency that is constituted through and within the exchange between the GPS's directive and the user's resistance of it. As Miller describes, agency may then be understood both as an act of resistance and as happening in the "kinetic energy of rhetorical performance" (2007, 147).

Another user interpreted the rerouting feature as adapting to their needs and described feeling empowered by it: "I love how it suggests an alternate route. I loved how it would recalculate and adapt to me—I was in control" (BA21). This user also implicitly understands geographic knowledge-making as an interactive process that involves a negotiation with the GPS. By attributing the GPS with the communicative ability to "suggest" a route, this user also describes the GPS as working "in the service of agency" (Kennedy 2010, 305). That is, the GPS is obviously not "sentient in the ways that we normally think of agents," but may be understood as "perceiv[ing] their environment and initiat[ing] action with it" (2010, 306). The GPS then has the capacity to engage the user with its cues and directives that may either facilitate or discourage decision-making. Moreover, it becomes possible to see how users resist the GPS through embodied practice that, as Hayles has put it, "moves in conjunction" with technology and is mindful of "discursive constructions" (1999, 195).

SUPPLEMENTING, RESISTING, AND COMPETING WITH THE GPS

The GPS also functions rhetorically by supplementing users' own local knowledge as well as the directions provided by paper maps and online programs like Google Maps or MapQuest. Conversely, users may also supplement the directions provided by the GPS with their own knowledge or research to create individualized routes. The supplemental work performed with and against the GPS invites users to question or resist the cartographic texts it produces, and by extension, to question the feasibility of a route's actual reproducibility in the world. For, as Blair writes: "Sometimes what appears to be the rhetorical text is not *the* rhetorical text, but an altogether different one; and what counts as *the* text is open to question" (1999, 40). Further complicating this issue is the fact that GPS databases are not always reliable. As one user put it: "On our trip, we used it a lot to find things like fast food restaurants, you know, grocery stores, and I'd say it was kind of a 50/50 mix; half the time it helped us out a lot in finding where we needed to go, and half the time, its database was just so out of date, even with recent updates" (OSA6). As Mark Monmonier has described, the "GPS calculates location by comparing time signals from several satellites, each with a direct line of sight to the receiver. Each satellite broadcasts a signal travelling at the speed of light but requiring a measurable time to reach the ground" (2002, 13). These signals can then be processed by a GPS receiver, "enabling the receiver to compute position, velocity and time" (Dana 2000). While GPS databases allow users to update their devices at regular intervals, this information may not always be fully accurate or the information may be based on outdated paper maps, placing drivers at risk for receiving inaccurate information. While many users appreciate what they perceive to be the reliability of the GPS, others are likewise aware of its various potential limitations and thus do not view it as wholly dependable.

The extent and type of supplemental activity performed by the GPS also depends upon the varied contexts and comfort levels of the particular user. For example, noting some common pitfalls of the GPS, one user recalled their initial apprehension when first using the device: "There was a certain degree of trepidation the first few times using the device. I made sure to print hard-copy directions from the internet for the first two or three trips, just in case I lost the signal, or the maps on the GPS unit weren't properly updated" (OFM10). Conversely, a more seasoned user commented that "the GPS supports our Google Maps in case we get lost" (OSA8). One user commented on the limitations of both the GPS and internet mapping programs, stating: "I found several 'roads' which don't exist. Of course, I've also found these on MapQuest and Google Maps" (OA12).

Other users prefer to supplement the GPS with paper maps because they feel that paper-based maps more easily afford a broader view of the surrounding areas. As one user put it: "I'm used to having a map which shows me a guided path to my destination and allows me a view of my surrounding areas, and that was sort of comforting. With the GPS I realized I would not have that, so that was a little concern and remains a little bit of a concern with the GPS" (OSA9). This user also linked some of this perceived limitation to the smaller screen size of the GPS, as opposed to the less limiting paper map: "[The] GPS will show you a six inch screen where you are. . . . I like to know what's farther around than that, so I will sometimes carry a [paper] map" (OSA9). Another user also described the comfort of juxtaposing the GPS with a paper map: "I really like combining [the GPS] with a map. To me, it doesn't replace a traditional map, because you can't see the whole picture—I'm very much of a map kind of person—I like to follow exactly where I am. I like to literally have my finger on the road and see, oh, yeah, we just passed that town" (OFT2).

Interestingly, this user also went on to say that "with two people in the car, one person can concentrate on driving, and the other one can be . . . searching up ahead." Certainly, having both a driver and passenger in the car changes users' interactions with the GPS. Researchers Gilly Leshed et al. too note that "the passenger riding next to the driver sometimes received the role of interacting with the GPS, in terms of switching settings . . . or pulling up information" (2008, 6). In this way, the GPS engages both the driver and passenger in a type of cooperative supplemental work that not only has consequence on the body but, as I will soon describe, also has broader implications for purposeful decision-making, social relationships, and more nuanced understandings of the environment.

REPRODUCIBILITY, PRESERVATION, AND THE PRACTICES OF MEMORY

From a Blairian perspective, the GPS text also bears various degrees of reproducibility and possibilities for preservation. As Blair notes, for instance, "[t]he link between reproduction of a text and memory is substantial" (1999, 38). As Harris and Hazen also describe, it is "useful to think of the relationships that unfold as maps are created, the meanings that are cited in selection of particular technologies [. . .] or as maps are engaged by users" (2009, 53). Thus, the ability of the GPS to preserve and reproduce stored routes and saved information, and display this information through multimodal, cartographic texts can indeed foster new schemas for geographic meaning and relationships between the GPS and the user. For example, several GPS users remarked that they appreciated the ability to program, save, and thus preserve routes in their lists of favorite places. As one user described, "I love how you can save addresses. So if I'm going to a repeat place, or to some friends, and I can't remember exactly how to get there: [Just] go back in, tap on the address, and it's great. So you don't have to keep reentering, you can actually save, just like on a cell phone, like saving numbers" (OA3). Another user even described the "home" button in terms of its capacity for memory: "You can push 'home,' and it takes you home from wherever you are—it routes you there. So once you tell it where home is once, you don't have to do it again" (OSA4).

Users also describe the act of programming and storing information, and being able to easily retrieve information without having to reenter it. As Collin Brooke notes, the "externalization of memory has become an accepted and even integral part of society" (2000, 786). He says that the "distribution of memory into the environment" not only means that "'the mind can now keep some knowledge intimately in memory and relegate other knowledge to written texts,'" but it also means that "[i]n many contexts, access to a text is accorded the same importance as knowledge itself" (786). This idea remains important for rhetoric scholars as it is often associated with debates in classical rhetoric about the merit of artificial technologies of remembering. Specifically, the idea is typically associated with Plato's denouncement in the *Phaedrus* of technologies of writing as threatening humans' capacity for being able to recall and recite information from memory, and, as Brooke notes, for its broader implications on "the relation of memory to thought" more generally (2000, 784). However, as Brooke writes:

> While Plato could distinguish between natural and artificial memories, the advent of mass literacy seems to put such a debate to rest—at least to the extent that memory loses its most important claim to canonical relevance for rhetoric. . . . Plato may be correct in suggesting that literate peoples will be more forgetful, but as a culture, their powers of recall are limited only by the material resources they devote to memory. (2000, 786)

The GPS indeed allows for the devotion of material resources to memory and functions as a sort of digital memory text. Brooke also notes that "[a]s our memories and technologies have become even more artificial, they have done so only insofar as they circle back and approach the appearance of the natural" (2000, 787–788). We see this case in point, for example, when the one GPS user describes losing the visual display of the simulated car on the road as "driving on white" (BA21). This user arguably toggles between accounts of the technology that "approach the appearance of the natural" and a return to the bodily experience that ultimately underpins all GPS use. Through this example, it becomes possible to see the broader implications of GPS use for how we understand our environments and engage with, describe, or recall the world around us.

THE GPS AND ITS BROADER IMPLICATIONS FOR INTERPERSONAL COMMUNICATION AND GEOGRAPHIC KNOWLEDGE-MAKING

Carole Blair writes that "[t]he entry of a text within a particular context is a move on that context that changes it in some way. Perhaps the best way to think about this notion," she says, "is to ask what is different as a result of the text's *existence*, as opposed to what might be the case if the text had not appeared at all" (1999, 34). The entry of the GPS within the contexts of everyday navigation signals a shift in the ways that humans have begun to understand and navigate their physical environments. As we have already begun to see, the GPS and its user may co-construct agency in such a way that facilitates purposeful, navigational decision-making. However, the use of the GPS within the contexts of everyday navigation arguably has wider consequences beyond its immediate use to achieve a particular navigational goal while driving, and has a broader potential to impact personal relationships and how we interact with and understand the worlds in which we live.

Several participants interviewed for this study, for example, described with humor but also with great seriousness the ways in which the GPS has somehow affected their social lives or their relationship with a spouse or partner. One user emphatically described the "intense relief" that accompanied what they described as being able to give over control to the GPS:

> My better half is kind of neurotic about directions, and navigating for him is an experience that is so stressful that I could just not handle it on a regular basis. . . . [H]e needed to have directions five steps in advance, and so I would sit there with the map, trying to plot out, like a naval submariner, to figure out where exactly it was that we were going so I could tell him: "In another three miles or so, we're going to need to take whatever exit toward whatever town." It was very stressful. . . . So it was intense relief, saying that I could give up navigational control. (OA1)

This user interestingly finds relief in being able to give the GPS greater "control," seeing it as a means of alleviating travel anxiety in a personal relationship. Likewise, another user also described the GPS as helping to lessen the travel anxiety that often preceded a trip and the positive impact of that lessened anxiety on their interpersonal relationship:

> Amusing but true: The GPS has done a lot to help my relationship. . . . [D]riving used to be a very stressful activity for me, which left my poor girlfriend with a choice of either being stuck driving everywhere or dealing with the stress of trying to give me directions. With the GPS, she can finally just relax on a trip and not worry about a thing, with the assurance that if I'm stressed or seem angry, it's absolutely nothing to do with her and nothing for her to worry about. That's been a huge boon, since we do tend to go on long trips together, and vacations no longer start with the burden of a huge build-up of tension on the road. (OFM10)

Perhaps most interesting about these accounts is that the GPS functions not only to mediate bodily experience, but by extension, to mediate interpersonal relationships. This mediating function may then have the result of enabling or constraining geographic knowledge.

For example, as one user remarked: "I found it changed my dependency. Normally, I'd just think, 'Oh, I can find my way,' and I thought, 'I like using the GPS, this is easier.' I'm definitely not going to get lost now" (BSA17). On the other hand, keeping in mind the issues mentioned by Monmonier, not all GPS users find the device to be a navigational panacea; as this one user notes: "More than once I've found myself unable to follow directions because the old exit for a given road was a quarter mile or so past the new one, and I wasn't aware what road I was looking for until it was too late" (OFM10).

Many users also commented, however, that the GPS made them feel more adventurous while driving or made them more willing to make navigational choices that they might not otherwise make without the GPS. As one user said: "I think it's made me more adventurous. . . . If I want to stop and look at a historic site, I'm not concerned about, 'oh my gosh, how am I going to get back on?'" (OA3). One user also described a newfound ease in making travel plans that accompanied use of the GPS, and the subsequent alleviation of travel anxiety for both the user and passenger: "I quickly became impressed with how much simpler it was making my travel plans, as well as how much less stress it implied for myself and/or my navigator, since missed exits, wrong turns, and road closings were no longer potentially catastrophic failures" (OFM10).

CONCLUSION

To use the GPS, then, is to engage with a discursive technology that has real and corporeal impacts for the navigational choices that we make while driving. As a rhetorical approach helps demonstrate, the GPS and its user engage in an ongoing interactive exchange through which agency is co-constructed by means of multiple rhetorical practices that include: supplementing the GPS and using the GPS supplementally; correcting the GPS and being corrected by it; programming and reproducing routes and challenging those reproductions; engaging in digital memory practices by storing and retrieving addresses and other geographical information; negotiating and challenging its routes through subtle subversions; and engaging with its symbolic representations while remaining attentive to actual physical surroundings. As we have seen, the GPS has the capacity to impact geographical understanding, both through constraining or discouraging geographical knowledge-making and through enabling and encouraging travel and geographical exploration. The potential of the GPS to impact geographical knowledge subsequently affects our everyday, social and interpersonal relationships, and how we choose to explore and understand the worlds in which we live.

To understand the GPS and its user as co-constructing an interactive agency that has the capacity to inform purposeful decision-making is to understand contextualized, mediated bodies from a vantage point that more explicitly incorporates and resists the practices imposed upon them through technologies. Understood from this perspective, a rhetorical approach may be seen as illuminating a more informed understanding of the interaction between bodies, spaces, and material discursive practice. While the study of the GPS in this chapter helps make salient these ideas, it is also important to note, as John Tinnell points out in his study of techno-geographic interfaces in this collection, that we ought also to consider the material rhetorical impacts of a host of mobile and locative media—that the GPS relative to vehicular transportation and its rhetorical impact on geographic knowledge-making is but one facet of the continually shifting landscape of locative media that likewise have varied implications for how we compose and conceptualize the world around us (2014). To use the GPS specifically, then, is to engage with a material rhetorical technology that has real impacts for the navigational decisions that we make while driving. Moreover, the GPS can impact geographical knowledge-making and can subsequently affect our everyday, social and interpersonal communication practices, and thus how, and with whom, we choose to explore and understand the worlds in which we live. It is my hope that rhetorical studies of the GPS such as this one can serve as a catalyst for further discussion of the discursive, communicative, and sociocultural implications of locative media and mobile technologies in our everyday lives.

NOTES

This chapter is reprinted with modification from chapter four of *Locating Visual-Material Rhetorics: The Map, the Mill, and the GPS* (Parlor Press, 2012). I gratefully acknowledge permission from Parlor Press to reproduce a modified version of this work here. Thank you to Martin Dodge, who offered helpful suggestions related to the study of GPS use and pointed me in some useful directions that eventually led to the research and writing of this chapter. Thanks also to Krista Kennedy for our engaging conversations about the interactiveness of agency, which informed my analyses in this chapter. I am also most grateful to the twenty-two people who shared with me their stories about GPS use. This chapter would not have been possible without them.

1. See discussion of data collection and organization in this chapter, and note 9, below, for an explanation of the key with which I refer to quotes from interview participants in this chapter. I do not apply the additional descriptor "GPS user" in citations from interview participants throughout the rest of the chapter; I only do so in their initial usage because I have not yet described my method of data collection and so these codes would have little meaning for the reader without this additional context.

2. Due to scheduling difficulties, four interviews were conducted over email.

3. Delimiting the sample in this way not only allowed for easy access to a local population but also extended the sample beyond just that of full-time faculty, thereby preventing a more insular or limited sample of solely academic users. Additionally, it helped to set some parameters around what could potentially become a very broad population consisting of everyday GPS

users. My decision to delimit the parameters of my sample to faculty, staff, and their family members at the college was based on the sample used by Barry Brown and Eric Laurier in their ethnographic study of the practices of city tourists in Glasgow and the implications of those practices for the design of electronic mapping systems (2008, 1). To this end, Brown and Laurier, respectively from the Department of Geography and the Department of Computer Science at the University of Glasgow, "recruited groups of visitors to the city from friends and family" of their university's staff (2008, 2). Brown and Laurier presumably excluded university faculty and staff from their sample because, in their case, they wanted tourists to be wholly unfamiliar with the city, and faculty and staff who work at the university would already have local familiarity, whereas friends and families would presumably visit from other locales. Their choice to recruit participants with ties to a particular institution informed my rationale for understanding my participant pool as a starting point that could help delimit an otherwise potentially broad sample population of everyday GPS users. That is, Brown and Laurier's sample of Glasgow tourists could have been potentially quite broad without their having also defined "Glasgow tourists" as those people who were not only tourists to the area but also friends and family of their university's staff. Likewise, I defined "everyday GPS users" as those people who were not only GPS users in their own right but also either faculty and staff at the college, or family members of any faculty or staff at the college.

4. Readers will note that I use the plural pronoun "they" or "their" when referring to a GPS user. Since participant responses did not tend to fall along gendered lines, I did not want to cause any confusion or distraction by preceding a quotation with a gendered pronoun, potentially leaving readers to wonder whether any significance ought to be inferred through its use.

5. I also viewed the library's borrowing program as a way of providing those curious about the GPS with access to the technology without their having to invest financially in the device. In fact, several participants indicated that they'd been curious about what it would be like to use a GPS, but did not want to purchase one without first having the opportunity to try one out. Thus, they saw my study as a good excuse to "test drive" a GPS, and the library's program allowed them to do so. Because the library's borrowing program is open only to college faculty and staff, all participants in this study are employees, and none were students. The library was enthusiastic about helping out with this study and I am of course grateful for their assistance.

6. I provided interview questions in advance as part of the consent form. The interviews generally proceeded in accordance with my questionnaire, with some departures given participants' responses, or when I chose to follow up on any markers. I asked the following eight questions of all participants: 1) Can you describe your background or familiarity with the GPS?; 2) Can you recall and describe your initial impressions when first using the GPS?; 3) Describe a typical scenario using your GPS. You do not need to reveal any personal details; 4) Are there particular features of the GPS that stood out to you while using it?; 5) What do you like about the GPS?; 6) What do you dislike about the GPS?; 7) Did the GPS change your driving habits, driving routine, or decision-making at all? If so, can you describe this in more detail?; and 8) Is there anything else you would like to share?

7. While I could have used the interview questions as an organizing framework to begin the coding process, I chose instead to undertake a close reading of the interview transcripts, often reading them multiple times, to identify any recurrent topics or themes that emerged. In a separate document, I then created a coding outline, which contained a list of these themes, or categories and sub-categories that emerged. I then copied quotations from each transcript into the pertinent sections of the coding document. Some categories and sub-categories in the coding document turned out to be more populated with quotations than others, and some categories did not turn out to be relevant to the direction taken in the final analysis, and are thus not represented in this chapter.

8. Nonetheless, to further contextualize my research, I did familiarize myself with some of the studies related to usability and safety of in-car navigational devices and an ethnographic study of GPS use geared toward improving GPS design features. Much of this research is published within the fields of information science and human-computer interaction (HCI), and while the goals or theoretical bent of these studies often differ from the direction of this work, I invoke some of them when they provide insights or information pertinent to comments made by

participants in this study (see Blanco et al., 2006; Burnett et al., 2004; Leshed et al., 2008; and Noel et al., 2005).

9. In this chapter, I identify the speaker of a quote according to the following key, which allowed me to track responses on a macro-level yet maintain participant confidentiality: If the participant borrowed a GPS from the library, I refer to them with a "B" (in one case, a participant borrowed a GPS from a family member); if the participant owns a GPS, I refer to them with an "O"; if they are college staff or administration, I include an "SA"; if they are adjunct faculty, I include an "A"; if they are a full-time faculty member, I include an "FT"; and if they are a family member of college faculty or staff, I include an "FM." Numbers are assigned randomly.

10. While GPS devices have countless applications and myriad uses, this chapter focuses on its use for purposes of in-car navigation. When used as such, the GPS, while displaying information in a virtual environment, is not necessarily functioning in the capacity of, say, a video game. Nonetheless, it bears noting that some users did express a potential interest in using the GPS for the purposes of activities such as geocaching or other forms of play. Thus the door remains open for research related to GPS "play" and its implications for the study of visual-material rhetorics.

11. While recent scholarship in visual rhetoric and composition studies has been attentive to forms of visual, material, and digital rhetoric, it bears noting that scholars have also argued for a better integration of sound in the study of multimodal composition (Shipka 2006, 371).

12. According to Communication Studies scholars Kristine L. Nowak and Christian Rauh, "social cognition theory argues that the ability to identify anthropomorphic characteristics and categorize objects in the environment as humans, animals, or objects is a basic human cognitive function" (2005, 155). Further, they note that "[o]bjects, animals, and humans form the basic social categories to which people assign the things they encounter" (2005, 156). They understand anthropomorphism as "the attribution of human form or other human characteristics to any nonhuman object" (2005, 156). In their study of the avatar, Nowak and Rauh "examine anthropomorphism only in terms of human morphology, or appearance, and not behavior," and thus they focus more so on the visible image of the avatar and its influence on online perceptions of anthropomorphism (156). In this chapter, anthropomorphism comes into play most visibly in participants' attribution of human characteristics to the non-human GPS; that is, participants often described their responses to the GPS's voice cues as though they were engaging in a dialogue with another human. Conversely, participants would sometimes describe the GPS as talking to them, particularly when describing its "recalculating" cues.

REFERENCES

Blair, Carole. 1999. "Contemporary U.S. Memorial Sites as Exemplars of Rhetoric's Materiality." In *Rhetorical Bodies*, ed. Jack Selzer and Sharon Crowley, 16–57. Madison: University of Wisconsin Press.

Blanco, Myra, Wayne J. Biever, John P. Gallagher, and Thomas A. Dingus. 2006. "The Impact of Secondary Task Cognitive Processing Demand on Driving Performance." *Accident Analysis and Prevention* 38: 895–906.

Brooke, Collin Gifford. 2000. "Forgetting to Be (Post)Human: Media and Memory in a Kairotic Age." *JAC* 20(4): 775–795.

Brown, Barry, Eric Laurier, and Hayden Lorimer. 2008. "Driving and 'Passengering': Notes on the Ordinary Organization of Car Travel." *Mobilities* 3(1): 1–23.

Burnett, G. E., S.J. Summerskill, and J. M. Porter. 2004. "On-the-Move Destination Entry for Vehicle Navigation Systems: Unsafe By Any Means?" *Behaviour and Information Technology* 23(4): 265–272.

Dana, P. H. 2000. "Global Positioning System Overview." *The Geographer's Craft Project.* Department of Geography, The University of Colorado at Boulder. www.colorado.edu/geography/gcraft/notes/gps/gps_f.html. Accessed September, 2008.

Harris, Leila, and Helen Hazen. 2009. "Rethinking Maps From a More-Than-Human Perspective: Nature-Society, Mapping and Conservation Territories." In *Rethinking Maps: New*

Frontiers in Cartographic Theory, ed. Martin Dodge, Rob Kitchin, and Chris Perkins, 50–67. New York: Routledge.

Hayles, Katherine N. 1999. *How We Became Posthuman: Virtual Bodies in Cybernetics, Literature, and Informatics.* Chicago: University of Chicago Press.

Johnson, Jim [Bruno Latour]. 1995. "Mixing Humans and Nonhumans Together: The Sociology of a Door-Closer." In *Ecologies of Knowledge: Work and Politics in Science and* Technology, ed. Susan Leigh Star. 257–277. Albany: SUNY Press.

Kennedy, Krista. 2010. "Textual Machinery: Authorial Agency and Bot-Written Texts in Wikipedia." In *The Responsibilities of Rhetoric*, ed. Michelle Smith and Barbara Warnick. 303–309. Long Grove, IL: Waveland Press.

Keogh, Brendan. 2014. "'You really are you, right?': Cybernetic Memory and the Construction of the Posthuman Self in Videogame Play." In *Radical Interface: Trans-disciplinary Interventions on Design, Mediation, and the Posthuman*, ed. Dennis Weiss et al. Lanham, MD: Lexington Books.

Latour, Bruno. 1994. "On Technical Mediation: Philosophy, Sociology, Genealogy." *Common Knowledge*. 3(2): 29–64.

Leshed, Gilly, Theresa Velden, Oya Rieger, Blazej Kot, and Phoebe Sengers. 2008. "In-Car GPS Navigation: Engagement with and Disengagement from the Environment." *Proceeding of the Twenty-sixth Annual SIGCHI Conference on Human Factors in Computing Systems.* New York: Association for Computing Machinery.

Miller, Carolyn R. 2007. "What Can Automation Tell Us About Agency?" *Rhetoric Society Quarterly* 37(2): 137–157.

Moberly, Kevin. 2008. "Composition, Computer Games, and the Absence of Writing." *Computers and Composition* 25: 284–299.

Monmonier, Mark. 2002. *Spying with Maps: Surveillance Technologies and the Future of Privacy.* Chicago: University of Chicago Press.

Noel, Elliott, Blair Nonnecke, and Lana Trick. 2005. "Evaluating First-Time and Infrequent Use of In-Car Navigation Devices." *Proceedings of the Third International Driving Symposium on Human Factors in Driver Assessment, Training and Vehicle Design.* 376–381.

Nowak, Kristine L., and Christian Rauh. 2005. "The Influence of the Avatar on Online Perceptions of Anthropomorphism, Androgyny, Credibility, Homophily, and Attraction." *Journal of Computer-Mediated Communication* 11(1): 153–178.

Segal, Judy. 2005. *Health and the Rhetoric of Medicine*. Carbondale: University of Southern Illinois Press.

Shipka, Jody. 2006. "Sound Engineering: Toward a Theory of Multimodal Soundness." *Computers and Composition* 23: 355–373.

Turnbull, David. 1989. *Maps are Territories, Science is an Atlas: A Portfolio of Exhibits.* Chicago: University of Chicago Press.

Verbeek, Peter-Paul. 2006. "Materializing Morality: Design Ethics and Technological Mediation." *Science, Technology, and Human Values* 31: 361–380.

Xia Jianhong (Cecilia), Colin Arrowsmith, Mervyn Jackson, and William Cartwright. 2008. "The Wayfinding Process Relationships Between Decision-making and Landmark Utility." *Tourism Management* 29: 445–457.

Chapter Three

Neo-Baroque Computing

Interface and the Subject-Object Divide

Elise Takehana

We are not and have never been alone. Curiously, this fact is all too often overlooked by humanity as it ironically replaces or attempts to curtail our vulnerability and reliance on nature by funneling that dependence through machines we purport to control. As a species, humans have developed their technologies to subsume what they perceive as resources, ultimately repurposing and re-envisioning those items as inert in comparison to themselves. Digital computing, however, offers significant resistance to the seemingly unyielding power of the human by appearing to appropriate qualities humans have long perceived as their own: cognition, language, and even emotional response. While I will not discuss developments in artificial intelligence, a sentient or autonomous machine is a source of fear, one Apple Siri designers were aware of when they decided to use an artificial and seemingly female voice (McArthur). Siri can neither be too human nor too intimidating. Falling victim to the will of Hal from *2001, A Space Odyssey* or being converted into a power source by our own machinic creations as in *The Matrix* may still be the stuff of science fiction. Regardless, the underlying fear of such narratives is that humans are not independent of or superior to all that surrounds them. Herein, I will examine how digital spaces and their representations overwhelm humanity and expose its dependence and intimate connection to its environment. Digital spaces could be the Neo-Baroque space par excellence—one that renews baroque traits of overstepping the ordered categories humans rely on to define their world and in particular the human subject from the object other, allowing us to see anew what we thought we knew.

Many critics and theorists including Angela Ndalianis, Mieke Bal, Timothy Murray, Omar Calabrese, Monika Kaup, Alejo Carpentier, Gilles Deleuze, and Félix Guattari have reexamined the baroque not as an era but as an impulse that circulates through history and appears to be rising to dominance today. These critics have considered a wide range of boundary transgressions that show that current phenomena are overpowering their proper categories. For Ndalianis, transmedia storytelling encourages narrative to spill over multiple genre and media boundaries; Bal examines how installation art challenges the boundaries of bodily and lived space of the artist and the viewer; and Carpentier considers identities of groups straddling the border of cultural and historical traditions such as in colonized South America. The twentieth and early twenty-first centuries have brought humans and their technologies to a critical mass so that the borders between distant communities, between real and virtual spaces, or between mass and individual knowledge cannot easily contain themselves. The Baroque movement spurred by the Protestant Reformation, which had been, in part, sparked by Martin Luther's publication of the Ninety-Five Theses was at base made possible by the technological capabilities of the printing press. Today, the networked computer seems to contribute to aggravating a baroque resurgence.

Such baroque traits or impulses, as opposed to the Baroque as a distinct historical period, accentuate the sensual, bodily, and situated position of viewers. In particular, immersive and overpowering experiences are often labeled as "baroque." Such experiences of overwhelming the viewer come from traversing medial boundaries or even the invisible fourth wall, such as Andrea Pozzo's ceiling paintings in the Saint Ignatius church (fig. 3.1). Discerning the boundaries of the architecture and the painting becomes a process of looking at and looking through[1] the boundaries between the two.

Baroque systems in fact "tend to invade space in every direction, to perforate it to become as one with all its possibilities" as Angela Ndalianis argues (25). Baroque impulses thus use markers of order, such as visual perspective, as not only an element of showmanship, but more profoundly as order's own undoing. Pozzo's command of perspective is impressive—the viewer is tricked and confused by his or her own senses.

To facilitate explaining how computer interfaces and their representations are a part of a neo-baroque renewing of baroque boundary transgression, I take up Jacques Derrida's term parergon as the framing concept for the role and function of computer interface.[2] Parerga serve as supplements or frames to erga, or major texts, that are both part of the ergon and distinct from it. They flicker and flirt as Pozzo's painting of columns appears as both part of the building and of the painting. Although an anthropocentric view—one that privileges the needs, tendencies, and perspectives of humans with little to no regard for purposes beyond human gain—has dominated the design of human-computer interfaces, interfacing is essentially a process of interaction and transformation by means of translation, decoding, and encoding. These interfaces are never neutral but create user relationships through culturally

Figure 3.1. Andrea Pozzo, Glorification of Saint Ignatius. 1691–1694. Source: Jonny Morris, The ceiling at the Church of Saint Ignatius of Loyola at Campus Martius. 2011, Digital Image. Available from: Flickr, www.flickr.com/photos/jonnybcivics/6208723300/in/set-72157627810449008 (accessed November 1, 2012). Reprinted, by permission, from Jonny Morris.

specific features chosen for their familiarity to a particular group. Thus the differences in presence and function of the computer and the human waver and mutually influence one another. But as we become increasingly accustomed to the computer, as we internalize the computer interface, its metaphors and devices end up altering those same humans who have formatted such computational machines. These compromises occurring in the subject-object divide ultimately result in accentuating the parergon between subject and object, between user and computer via the interface. In considering the dialogues and overlaps of human and computer, I will take as a case in point, Calogero's music video and lyrics for his 2007 song "Pomme C,"—a piece whose purpose seems to lie in confusing human for interface. "Pomme C" takes up this neo-baroque boundary transgression between the human and computer, or subject and object, within the space of a graphic user interface. That such a compelling example should be found in a pop culture genre like music video is not surprising. Music videos are distinct artistic expressions *and* are categorically attached to an ergonic text: the musical expression from which it is based. "Pomme C" shows viewers a parergonal relationship (subject-object transgressions) within a parergonal space (digital space of the graphic user interface) in a parergonal genre (the music video).

DON'T FIX IT . . .

In discussing the burgeoning field of digital humanities, a field expressly interested in examining and incorporating computing as a part of understanding human culture, one core problem that the humanities must face is its roots in the "individualized, proprietorial, liberal humanist author."[3] The functioning of knowledge development in the humanities has not evolved despite drastic changes in authoring and dissemination practices made available by digital technologies. The pressure and demands our technologies place on us are still latent and often cited in the digital humanities and in media studies. Kathleen Fitzpatrick, while charting changes in the field of publication and scholarship post digital technologies, copy left, and creative commons, places the central problem not on the tool but on the very real effect the tool lays on the "shifts and reinventions in our intellectual lives."[4] We are left in a constant process of becoming and a becoming that alters as circumstances change. What we fear is this sense of the loss of individuality, of control, and of authority as though we had not recognized our continual dependence on others and objects. While the experiences of the human and of the computer could very well be so different, so alien as to not be relatable, it is unrealistic to deny that "Differences in complexity not withstanding, the human and computer are increasingly bound together in complex physical, psychological, economic, and social formations."[5] The convergence between the human and the computer that we see in the graphic user interface shows the human familiar metaphoric icons that trigger the computer by way of electrical signals. That interface is both part of the machine and part of the human. Facing and examining our relationships with the things and beings that surround and ultimately affect humans, allows humanity a more realistic understanding of our role in a larger ecology, be it a social, cultural, biological, or technological one. Because the computer has perhaps enlivened such an initiative more than any other entity or event in recent history, looking to the relationships we develop with the computer could help the humanities envision a model of culture that does not privilege so highly the role of humans. The demarcation between human and computer is not as hard or clearly drawn as they may appear. As Lev Manovich asserts, when we are working with a computer, "We are no longer interfacing to a computer but to culture encoded in digital form."[6] To see ourselves as separate from and independent of our computers is to overlook that we have placed ourselves inside of our machines and are forever altered by that exchange. Imagining ourselves as creators and wielders of technologies all but certainly contributes to the meaning and function of humans, but this story is one-sided and tilts toward the perspective of the human agent.

Communicating with our computers requires an interface that translates the data into a readable form for the human and provides humans with an input system to interact with the computer. Søren Pold describes the role of the interface quite nicely as he transitions from discussing the interface as viewed by computer scientists and to that of digital artists. Pold quotes a usability expert, Don Norman, who states: "An interface is an obstacle: it stands between a person and the system being used. . . . If I were to have my way, we would not see computer interfaces. In fact, we would not see computers: both the interface and the computer would be invisible, subservient to the task the person was attempting to accomplish."[7]

Eradicating the conflict of interfacing with a computer is one way to return the computer to a tool of the human, to make that computer an object subservient to the user's needs. The struggles or glitches we encounter with our use of a computer only remind us that this tool does not always bend to our whims. Marianne Van der Boomen explores the slips of functional interface design by appropriating Martin Heidegger's ideas of tools ready-at-hand (Zuhanden) and tools present-at-hand (Forhanden) to critique humans' use of the computer rather than Heidegger's initial application to the hammer.[8] Functional interface design does all that it can to objectify the computer so that it can function as a tool whose use or mechanism can be overlooked, keeping the computer a tool ready-at-hand. A user considers the interface and computer when the tool no longer functions smoothly, making the computer present-at-hand. For a functional interface designer, seeing the interface is seeing a problem in workflow. Even as computer programmers consider aesthetic judgments of code, their overwhelming criteria for beauty in code is a functional beauty where the code enhances or eases the fulfillment of a goal, not of code qualities in and of themselves.[9] Essentially, maintaining the subject-object or user-machine divide compromises the history of human computer interface (HCI) as it strives to maintain the computer as a tool ready-at-hand.

Alternatively, Søren Pold also examines the computer interface as viewed by the digital artist. From such a perspective, Pold argues that the artist sees the interface quite differently as "a dynamic form, a dynamic representation of the changing states of the data or software and of the user's interaction. Consequently, the interface is not a static, material object."[10] From such a vantage point, the interface experience overextends boundaries of subject-object or user-machine, bringing humans to consider the computer interface beyond the models of mechanics and efficiency. For the artist, or even the humanities scholar, "the aesthetic dimensions and imaginative vision makes interface a space of being and dwelling, not a realm of control panels and instruments only existing to be put in the service of something else."[11] Moving away from task-oriented work where the computer is ready-at-hand, the arts and humanities can look on the computer interface as an environment where human culture extends itself and interacts with a digital ecology. What the interface offers is a permeable border between the human subject and the computer as object. In such a case, humans may see the interface and thus differentiate or obscure their difference from the interface and so recede. As a permeable border, the interface, like the paragon, offers a space to flirt with interpretations of reality much like the trompe l'oeil where in one instance there is a rabbit and the next a duck.

. . . IT'S BAROQUE

While, like the engineer, we may at times see the boundary between human and computer as impregnable, our immersion into digital spaces often causes us to question any view of the human as independent of the computer. Considering the human itself as having permeable boundaries with objects, of existing in a state where categorical boundaries are at least momentarily overcome, serves as a fundamental example of Baroque impulses and parergonal relationships. Derrida offers, in his *The Art of Painting*, the term parergon, which encapsulates the central characteristic of computer interfaces as being near simultaneously ready-at-hand and present-at hand. Derrida defines the parergon as the extra or exterior that exposes the lack of the interior. For a parergon to exist, it must detect a lack in the ergon and thus must recognize a dividing line between the inside and outside of a composition or form. Parerga augment representation. For instance, the presence of clothing on a figure or of columns on a building supplement—but are not critical to—the representation or form. Parerga are not part of the subject or the ground but are integral to both representational elements simultaneously. When determining the traits of the parergon by examining the frames around Baroque art, Derrida describes the parergon as one that "merges into the wall, and then, gradually, into the general text."[12] More importantly, Derrida notes that the parergon is more than a formal feature, but counters discipline-specific norms of thought within the sciences, philosophy, or sociology for example. Making the parergon visible and dominant reveals disciplinary thought as motivated rather than neutral, in a way making the methodologies and assumptions of a field visible or present-at-hand. With the parergon as a frame, we see that truth is not transcendent but constructed, yet fields like philosophy and science try to hide this characterization of truth by neutralizing and effacing effects of the frame. We see the knowledge of a field through a framework posited by analytics within that field. Thus, philosophy does not see knowledge as it lies but only through the parergon it builds to interface with a truth beyond the human.

When posited as a baroque impulse of dynamism and embodiment, the parergon demonstrates how such excess overwhelms ordered boundaries or categories even down to the fundamental subject-object divide. It is hardly a coincidence that Derrida uses images of Baroque and Rococo art as objects of analysis in discussing the characteristics of the parergon. Gregg Lambert describes the baroque impulse through four traits: "[1] an experience of temporality marked by the themes of novelty, variety and multiplicity [2] a loss of distinct perception of the central figure or action [3] the physical participation of the spectator in the presentation of the artwork through an emotional feeling of dizziness or swooning (literally, of being overpowered by the spectacle) [4] finally, a heightened sense of enthusiasm, delight or marvel."[13] These generalized qualities highlight seriality and varied repetition rather than linear narrative, a shifting and unstable subject, immersion and showmanship in crafting a work, and an emotional rather than logical motive to both create and consume. Unpacking these traits exposes a thought process unlike rational logic but very appropriate to the mediated environment humans face and work with in the digital age. Baroque systems in fact "tend to invade space in every direction, to perforate it to become as one with all its possibilities."[14]

Such Baroque tendencies toward chaos, accumulation, technological spectacle, and embodiment counter a Classical impulse to order or narrativize the mass accumulated during periods of Baroque behavior. Even modernist critics who do not cite the Baroque directly still point to such a loose dualism. Leo Steinberg posits the works of the Masters of Illusion and the Modernists as simply a variation rather than complete opposites. For Steinberg, the difference between the Master of Illusion and the Modernist is not whether each accentuates illusionism or the surface of the canvas, but in which of these elements is perceived first, as both are present at all times.[15] Even Lettrist Isidore Isou forwards a model of alternating aesthetic motives in the *amplique* and *ciselant* stages of art, where the first is concerned with compiling more and more complex stories and the second in examining or reflecting on the process and goals of art.[16] Further, Monika Kaup reads Gilles Deleuze and Felix Guattari's conceptual cycle of stratifying hierarchies and destratisfying those hierarchies through rhizomes—the becoming major and becoming minor—as a congruous case to the Classical-Baroque cycle. Whichever model one chooses to narrow in on, the general message of all of these cases is extraordinarily similar: find a—perhaps incongruent—balance between elements that appear to be opposites to take full advantage of all human capabilities by allowing the differences to converse with one another rather than resolve or argue.

Although the influence of the avant garde on the evolution of computer interface design can be seen down to basic functional commands like the "cut and paste," industry standards in commercial mainstream media for not just realistic but photorealistic imagery can represent another response to a data rich environment. Just like avant-garde practices of collage bring attention to the parerga of the in-between of the assembled elements, the baroque tendency is to flirt with the limits of parerga, hiding *and* revealing their presence rather than simply masking or ignoring the parerga per realist works or simply exposing parerga per avant-gardist work. With a flickering parergon, we can see its connection to the ergon and its difference from that ergon over and again. One such representation of interface as parergon between subject and object or human and computer appears in the music video for French world musician Calogero's song "Pomme C."[17]

FROM A CERTAIN POINT OF VIEW: PARALLEL AND PERPENDICULAR

In many ways, the music video for "Pomme C" attempts to undermine the role of the human while representing a human relationship. By opening the video with a view of the inside of the computer, allowing the computer to speak ("Isn't it nice to have a computer that will talk to you?"), and showing the keyboard typing on its own, the viewer witnesses an interface experience between the computer and the interface as though the computer could offer up the input it normally depends on its human user to supply. We see the human constituents as representations housed in the computer's display as the camera immerses itself into the screen space. Calogero's appearance in the video and his interaction with computer screens betray a certain power over his virtual environment. For instance, he is able to push the edges of an application window to expand his livable space. In addition, during the first chorus, he picks up transparent computer screens that include images of himself singing and playing the guitar (fig. 3.2). This becomes particularly disjunctive during the second chorus when a carousel of facial images of Calogero singing the lyrics rotate around Calogero who's standing behind the screens with his mouth pursed (fig. 3.3). The music video as parergon frames the music and lyrics of the ergon. During the verses, the video and the lyrics appear to overlap and the frame is obscured. Essentially, the viewer sees Calogero singing and so sees the performance as authentic. During each chorus, however, the performance of the lyrics is usurped by prerecordings on computer screens. The performance of the song now appears parergonal to the lyrics because the performance presents itself as bound.

Not only is he aware of his ability to manipulate his space in ways the other humans in the video cannot, he shows no discomfort in ceding his primary presence to recordings of others. By the third chorus, Calogero allows images of other people on screens to lip sync lyrics for him as he builds a collage of their faces over his own (fig. 3.4).

Figure 3.2. Calogero, "Pomme C," Universal Music France, April 29, 2008. www.youtube.com/watch?v=-47Fmmj1FsA. Reprinted, by permission, from Pierre Alain Simon, Universal Music France.

By holding these screens up to himself, he passes on some of the apparent musical production to digital representations of himself and eventually of other humans. His digital representations are just as capable of performing the song and seamlessly do so with him. With each of the three choruses, Calogero replicates the repetition by framing the physical performance of singing to a prerecorded event. The transparency of these screens allows the flickering illusion that the scene in the screen temporary overlaps and so obscures the parergon for a fleeting second.

When turning to the representation of a young man and woman who meet on a dating site, the viewer encounters a fundamental difference between the man's and the camera's or viewer's interpretation of a computer generated image. When the man's line of sight is perpendicular to the screenic image of the woman, he sees the image of her striated by the screen's display (fig. 3.5).

However, when the man's line of slight runs parallel to the surface of the screen, the woman becomes real as he pushes the screen down to the floor (fig. 3.6).

For the man to perceive the real woman rather than her representation, he must manipulate the screen until it is not longer in a traditionally usable position. In this sense the interface appears as an obstacle rather than the musical collaborator that it is for Calogero. The man fights against and tries to eradicate the parergon of the screen. However, as the man pulls down the

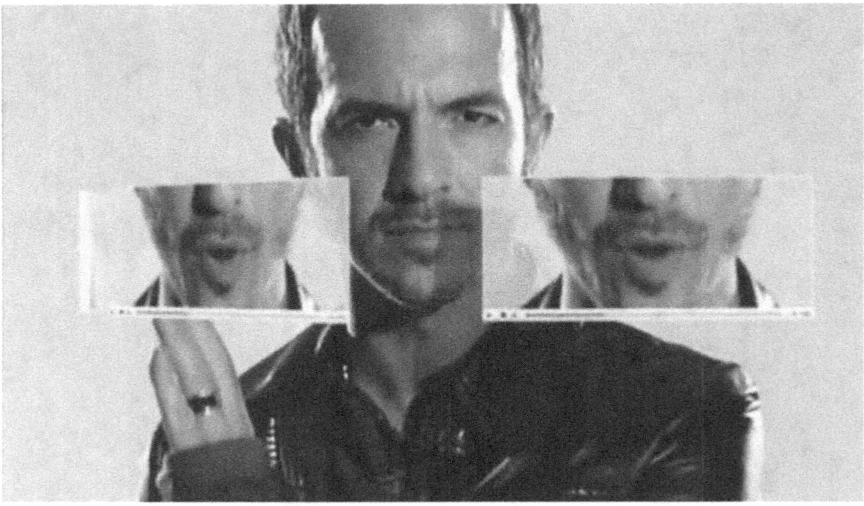

Figure 3.3. Calogero, "Pomme C," Universal Music France, April 29, 2008. www.youtube.com/watch?v=-47Fmmj1FsA. Reprinted, by permission, from Pierre Alain Simon, Universal Music France.

screen to "create" a woman from her representation, Calogero sings [in translation] or "Apple C / A little downloaded love," refocusing any agency on the computer's functions rather than the man's will that we see visually represented. In short, when his line of sight is perpendicular to the screen, the man sees what is available to him: an image only. When his line of sight is parallel to the screen, he sees the illusion he desires: the actual woman.

On the other hand, when considering the camera's or the viewer's perspective on the computer screen, quite the opposite dynamic takes hold. When the camera/viewer's line of sight is perpendicular to the screen, we see the illusion of a happy heterosexual couple (fig. 3.7).

As the camera rotates counterclockwise on this scene, the camera/viewer's line of sight eventually falls parallel to the computer screen, revealing the actuality of the scene: a lonely man yearning to have the company of a woman and not simply her image (fig. 3.8).

As the seamless montage of a happy couple unfolds to expose the borders of the individual windows that compose the scene, Calogero sings [in translation] "But I imagine / That a machine / Can only pretend." Calogero then ultimately enforces the interpretation that humans want to see things differently than a computer, but to grasp their own reality, they must come to terms with seeing the world from the perspective of the other (as viewer, camera, or computer). In both cases—the man's perspective and the camera/viewer's perspective—the perpendicular line of sight demonstrates the illusion and the

Figure 3.4. Calogero, "Pomme C," Universal Music France, April 29, 2008. www.youtube.com/watch?v=-47Fmmj1FsA. Reprinted, by permission, from Pierre Alain Simon, Universal Music France.

parallel line of sight, the real, however the man sees the woman as real and her image as illusion, while the camera/viewer sees the image as real and the woman as illusion.

Although, as an audience accustomed to new media communication, we are aware that it is the machine that is creating the image before our eyes and we are, in any case, often reminded of the machine's presence whenever we face buffering and loading times, we allow ourselves to be fooled only insofar as we can maintain the gratification of viewing a well-made seamless image. Experimental neuroscientist R. Beau Lotto takes interest in illusions and particularly in how our eyes can function as interfaces, translating reflected light from the external world into electrical impulses for our visual cortex to decipher. When he demonstrates his gray cube illusion in the National Geographic series *Test Your Brain*, he ultimately reveals the limited framework of rules or codes that we use to translate information. For Lotto, "What illusions tell us is that what we're trying to see is not the world as it is but something else. We're actually generating perceptions of a world that's useful to see, and that useful perception doesn't actually have to correspond with what's actually there."[18] He here forwards the idea that perception is not a practice of authenticity, but one whose truth-value is nebulous. Instead, perception is a calculation of probability based on assumptions built from past experience. In R. Beau Lotto and David Corney's study of lightness illusions, they devised artificial neural networks (ANNs) consisting of 400

Figure 3.5. Calogero, "Pomme C," Universal Music France, April 29, 2008. www.youtube.com/watch?v=-47Fmmj1FsA. Reprinted, by permission, from Pierre Alain Simon, Universal Music France.

input nodes to process 400 pixel images of dead leaves. They discovered that their ANNs made the same errors in perception as humans, pointing to a fundamental quality of biological or technological perception: perception based on past experience shows illusion and reality as degrees on a scale and not absolutes. Illusions appear as such when the condition of the source differs from the perception of that source.[19] Without absolute divides between reality and illusion, or source and perception, one cannot carry the label of illusion with conviction. Any perception is thus more or less likely to be accurate, but assurances regarding veracity are wholly impossible. Similarly, Omar Calabrese notes that paradoxes in visual perception, most evident in the works of artists such as René Magritte and M. C. Escher, demonstrate the human subject's inability to determine a dominant interpretation of space.[20] Baroque works are virtually inseparable from intellectual games of illusion and can be conceived as fundamentally theatrical because they see every reality as similar to its fictive representation. The baroque is located where reality and representation encroach on one another. It is such an encroachment that makes the subject and object unstable and reversible.

The Baroque returns as the Neo-Baroque at this time for many reasons which scholars like Omar Calabrese and Angela Ndalianis explicate in much more detail. With the rise of intermediary work and a growing archive of information made possible in part by the development of the computer, humans now must face encroachments and transgressions of all sorts. Neo-

Figure 3.6. Calogero, "Pomme C," Universal Music France, April 29, 2008. www.youtube.com/watch?v=-47Fmmj1FsA. Reprinted, by permission, from Pierre Alain Simon, Universal Music France.

Baroque, unlike the Baroque, does not maintain the barriers dividing high art and popular culture, and so witnessed such convergence of the two to the point that critics like Lev Manovich are confident in their assertions that there is little difference between the avant-garde and popular culture at large. Umberto Eco holds a similar view when he praises Calabrese's *Neo-Baroque: A Sign of the Times*, crediting the author of being aware that university students today cannot read Plato outside the context of their enculturation in popular forms and phenomena. Eco equally signals that even the guardians of the gate, the professors, are not immune to the charms of popular and commercial culture but are enamored as any other. In fact, the Baroque and Neo-Baroque revel in spectacle and do not contend to be anything other than a sensational pleasure like a blockbuster special effects film celebrating the triumphs of technology over reality (Ndalianis 171). In such cases it is an emotional or sensational response that is solicited, not a rational or logical one.

Of the many Neo-Baroque qualities Omar Calabrese explicates, Kresmir Purgar takes up that of the approximate and inexpressible as a primary motive for actions of the neo-baroque subject. He summarizes Calabrese's discussion of the approximate and inexpressible saying "a subject *knows* that there exists the remainder of a content or meaning which eludes description, but is *unable to express it.*"[21] This constant lack of unity and stability is where passion is born from and, as a consequence, the subject is open to

Figure 3.7. Calogero, "Pomme C," Universal Music France, April 29, 2008. www.youtube.com/watch?v=-47Fmmj1FsA. Reprinted, by permission, from Pierre Alain Simon, Universal Music France.

interpreting experiences while remaining vulnerable and conflicted.[22] This feeling of being pressed upon by the remainder shows the influence the other, or object, supposedly external to the subject, has over that subject and his or her understanding of identity. Because the subject cannot make a unity of meaning, he or she cannot be seen as wholly separate from the objects he or she manipulates. Purgar argues that the observer can become the subject of a painting by taking up the role of mediator between the inside and outside of a scene. The painting is, from this point of view, not just consumed by the viewer but alters the viewer through a confrontation. In examining Lovro Artuković's 2004 painting *Little Red Riding Hood (evil?)*, Purgar argues that the Little Red Riding Hood's gaze confronts the audience, now made to feel responsible for Red (fig. 3.9).

Although she is depicted alone on a canvas, the viewer can see the wolf lying in wait on another canvas out of her line of sight. The viewer becomes folded into the message of impending doom, disempowerment, and responsibility for others. Here the subject is vulnerable to the object resulting in what Purgar calls a second person singular narrative where the work is directly oriented to the observer and both the subject and object become equal and active participants. This orientation results in a linkage that is not logical or grammatical but emotive. Purgar suggests that we must be open to new representational practices and such new representational practices must take into account Mieke Bal's assertion that the subject and object are fundamen-

Figure 3.8. Calogero, "Pomme C," Universal Music France, April 29, 2008. www.youtube.com/watch?v=-47Fmmj1FsA. Reprinted, by permission, from Pierre Alain Simon, Universal Music France.

tally entangled.[23] Much like Lotto's experiments with light illusions betray a fundamental negotiation between human cognition and environmental stimuli, Artuković's painting reveals a moment of flirtation between painted representation and an empathic or anxious moment shared between the viewer and Red Riding Hood.

We see such an entanglement between subject and object (in this case human and computer) at the close of Calogero's music video in the only instance where the man's and the camera/viewer's line of sight overlap (fig. 3.10).

In the closing vignette the camera rotates counterclockwise, altering an apparently real woman into an image of a woman. As the camera/viewer looks over the obscured man's shoulder as he sits in front of his computer, both are confronted by the image of the woman whose caption reads [in translation] "Should I save? / Or should I quit?" The objects of the video, both the woman and the computer, push back upon the subject, threatening a loss or disconnection. In charting an analogous relationship between a man and a computer file or program, she flirts in the parergonal space of the interface screen. Her comment flips the location of authority from outside to inside the computer by exposing the presence of computer interface functions. In a moment of confrontation, she breaks the fourth wall and reveals that she is not as "ready-at-hand" as her stagnant image may seem.

Figure 3.9. **Lovro Artuković, *Little Red Riding Hood (evil?)*. 2004, Oil on canvas, 145 x 120 cm. Owned by Fima Holding. Photograph by Gunter Lepkosqki. Reprinted, by permission, from Lovro Artuković.**

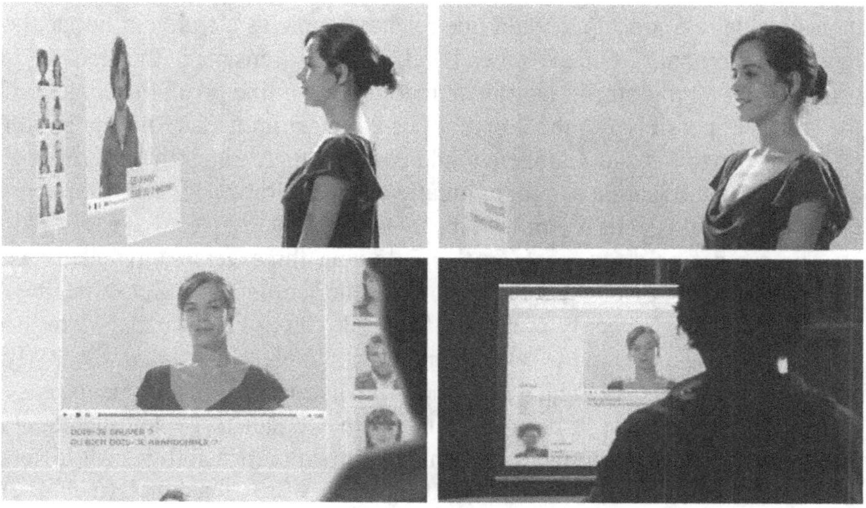

Figure 3.10. Calogero, "Pomme C," Universal Music France, April 29, 2008.hwww.youtube.com/watch?v=-47Fmmj1FsA. Reprinted, by permission, from Pierre Alain Simon, Universal Music France.

ARE YOU HUMAN?[24]

While the music video for "Pomme C" shows the application window commonly used in graphic user interfaces as parerga between program and content, its lyrics reveal the computer interface as a parergon between the human and computer forthright. When considering the lyrics, the human-computer relationship mirrors both the heterosexual romance and an intimate user experience to blur the boundary between what we perceive as human abilities and machinic traits. The lyrics leave open space for interpreting the relationship that the man is developing as taking place either between him and a human female or between him and his computer. Calogero sings:

The first two lines of the song are translated as "Apple C / A man and a woman," but when they are *spoken* sounds like "Apple it's a man and a woman," doubling their implications. The literal translation of the written lyrics points to viewing the "Apple" as Apple®, the computer company, an interpretation supported by having already mentioned emails, keyboards, saving, and quitting. When *listening* to the lines, the possibility of a different interpretation arises. With a man and a woman immediately following the mention of the apple, could it not also be the forbidden fruit of Genesis, and ultimately the revelation through knowledge of the limits and context of our human experience? The third line raises some difficulty because this idiomatic phrase does not directly translate into English. Roughly, it means to have a full, complex, and intricate schedule. This may refer to the complexity of the male-female relationship, should one read the apple as a fruit and not as the renowned company. On the other hand, directly translated, the line reads "And it's all a program," a legitimate reading of the line given the context of the song but possibly not the dominant or the most immediate interpretation of the line. In this sense, Calogero could be referring to either the heterosexual male-female dynamic as programmatic or the entire world as itself a computer program. We thus return to "Apple" as Apple®.

The same ambiguity exists in the following three verses, translated as: "She writes to me / But my screen / Formats the sentiments // But I imagine / That a machine / Can only pretend // My DSL / Is not truly real." Here the double meaning arises in the seventh and eighth lines, which *sound* like: "My goddess she / Is not truly real." But who is the goddess? While it may be easy to read the goddess as a desirable, unknown woman this man encounters online, there are two linguistic elements that open up possibilities for different interpretations. Calogero chooses to use the word "machine" (machine) rather than "ordinateur," (computer) possibly because machine is feminine and ordinateur is masculine. Since machine is feminine, "she" would be the appropriate pronoun to use should one want to refer to the noun "machine." Is "My goddess she" possibly the "machine"? This seems to be the direction Zazie, Calogero's songwriter, was moving toward since the line that sounds like "My Goddess she," the "Ma déesse elle," is actually *written* as M' A D S L" or "My DSL." Is now the woman only virtual or has the interface or the computer disappeared? Could then, even the first line's "She writes to me" refers to a computer writing rather than a woman writing?

Curiously, the *aural* readings of the verses tend toward viewing a human female interlocutor while *textual* readings tend toward viewing a techno interlocutor for our love-starved and lonely male protagonist. Clearly, the music video rests on a dominant, anthropocentric representation of a more ambiguous and complex communication, but the affects of computing have begun to reach a critical mass at which we can no longer consider the computer as an object and the interface as a boundary or contact point between humans and computers. These two entities form an assemblage,[25] or what rhetorical scholar, Collin Brooke calls a medial ecology, as humans recognize a shared autonomy with the objects they had become accustomed to utilizing as tools or instruments useful to anthropocentric goals. Other rhetoricians like David Sheridan, Jim Ridolfo, and Anthony Michel turn to Bruno Latour's Actor-Network Theory to chart pedagogical practices that take as fundamental a sense of agency that is "distributed across complex networks of human and nonhuman actors."[26] While such discussions of media ecologies, distributed agency, or even those conversations that fall into object oriented ontology are en vogue now, suppressing the autonomy or intent of the human subject is itself a practice or interest that appears frequently in the art of the twentieth century. Surrealists claimed, for example, that automatic writing and their parlor games, both borrowed from practices of the historical baroque era, diminished the control of authorial intent. William Burroughs used the cut-up method as a way to insert a procedural gap between himself and his writing and his collaborator, Brion Gysin, created the Dream Machine as a hyponotic device aimed at unseating consciousness.

In more recent times, digital poet John Cayley describes his project Writing to be Found (based on the uncovering of a series of words not yet saved to Google's search field) as a proper collaboration between himself and the Google search function. The first step of his process of writing with Google was to query phrases demarcated by quotation marks until a string of text was not found. Such a failed query would serve as proof that the lines were original to Cayley. He sees this type of writing as one "shaped by the way that Google is shaped, by the way in which Google curves the space of the network. And Google does also, in a sense, write with me: constraining, directing, guiding, and, especially, punctuating my writing."[27] There is a compromise here between authorial intent and program intent. Later, to his surprise, Cayley discovers that original strings of text he had written in the search field were not there upon later queries. Not all Google servers had been updated to include his original lines as though he had not written them. His collaborator had denied him, even worked against him, by withholding the work they had done together, thus obscuring the unique in a mire of 0110111001101111101110100.

HUMAN RECESSION THROUGH IMMERSION

Such projects as well as the lyrics of "Pomme C" show a recession into the environment, ground, or objects of our worlds. This is a kind of recession also highlighted by thinkers fearful of a loss of critical distance and the resulting increased vulnerability to totalitiarian control. Walter Benjamin, for instance, spoke of morphing or merging into his environment in his *Berlin Childhood Around 1900*. Joyce Cheng identifies, in particular, two moments where Benjamin describes an entanglement between the human and the animal environment. They are two moments of identification: one with otters during rainy weather and another with butterflies while netting them in his garden. Benjamin writes of his hunts: "the more I strove to conform, in all the fibers of my being, to the animal–the more butterfly-like I became in my heart and soul–the more this butterfly itself, in everything it did, took on the color of human volition."[28] The flexibility of self identity that children experience, allows Benjamin to explore mimetic impulses that diminish critical distance. This diminishment happens in a way quite similar to work that fellow surrealist Roger Caillois developed in essays such as "Pierres," "L'Ecriture des pierres," "La mante religieuse," and "Mimétisme et psychasthénie légendaire" much to Benjamin's fearful reservations. Though Benjamin and Caillois disagree as to the dangers of mimetic and passive behaviors, Caillois's description of meditation resonates with Benjamin's hunt:

> My thought tries to seize [stones] at the ardent moment of their birth. I then experience a very special kind of excitement. I feel myself becoming a bit like stones. At the same time, I liken them to myself by means of the unsuspected properties I sometimes attribute to them in the course of my speculations, which are alternatively precise and lax, and which combine the web of dreams with the chain of knowledge.[29]

This move of the subject receding into the surrounding objects of his or her environment justify Caillois's conclusion to "L'Ecriture des pierres" where he compares the human experience to a board game. In that scenario, the human is not the player or the dice, but the "passive counter" that moves across the board whose order the "counter" cannot determine or escape but infinitely tries to understand and augment.[30] Such passivity is often construed as suicidal or at least lacking a will to live, but the "passive counter" can be compared to the detached observer.

While it has become customary to regard observers or viewers as detached from their context, they are not wholly detached. Anna Gibbs and Paul Elliott both look to how the human neural network responds to what it observes. Viewers function beyond empathizing with what they see and rather *feel* what they see in representations such as with film. Gibbs looks to mimetic communication as one that draw on Charles Darwin's idea of motor sympathy between bodies while Elliott cites two neuroscientists: Richard Cytowic who explored synaesthesia as a subconscious function that connects the sensations with conscious thought orders, and Salvatore Aglioti, whose tests on mirror neurons showed that test subjects reported pain and muscular responses to watching filmic images of a hand being stabbed. In both cases, a deep and fundamental human connection to shared bodily experiences form the undercurrent of mimetic or passive behavior.

Hybridizing the connection between the human and its natural environment equates quite simply to focusing on a parergonal existence at the edge of being subject and object simultaneously. Constantly skirting parerga essentially draws human subjects or human users into an overextended interface experience between themselves and the objects that constitute their environment. In the parergonal existence of the contemporary computer user, user and interface become entangled rather than one qualifying another, much in the same manner as Roger Caillois justifies mimesis in the natural world. Caillois argues that camouflage is a luxurious "failure to maintain the boundaries between inside and outside, between, that is, figure and ground . . . a doubling, a mimicry, of the space around the body in order to allow for its possession by the surrounding environment."[31] After rebuffing scientific justifications for camouflage in the natural world, including considering such mimesis a chance occurrence, a simple demarcation between species, and a defense mechanism against predators, Caillois takes up a controversial pseudo-scientific position of regarding mimetic traits and behaviors in the animal world as a remainder of a past ability to mimic surroundings, an ability that is no longer present but has left its morphological trait behind. He goes even further to say that this mimicry is not a defense mechanism, and not a luxurious excess, but quite simply "a disorder of spatial perception" where the organism "is no longer located at the origin of the coordinate system but is one point among many. Dispossessed of its privilege, it quite literally *no longer knows what to do with itself.*"[32] The organism has thus surpassed what Caillois considered the most demarcated difference: the divide between organism and environment. By doing so, the organism has diminished its existence as it is confronted by an environment that devours the subject. This becoming-environment can be conceived of as sinister, where humans lose their agency and humanity (we see this in images of suppression and punishment as in the crew of Davy Jones in *Pirates of the Caribbean: Dead Man's Chest* and *At World's End* as well as the multiple nymphs and humans turned to trees, mist, or spiders in the Greek myths) or equally as immersive play. Immersion, as characterized by Oliver Grau, is a state of diminished critical distance, and primarily a process of altering mental states.[33] By immersing the human user into the digital space of the interface, humanity "surpasses the human condition through the sensory experience characteristic of computer-based interactive worlds . . . the body is coupled to virtual worlds and undergoes a 'communication process,' a post-biological experience in which human actions engage digital technologies in the process of understanding the world."[34] Diana Domingues, an artist interested in the anthropological aspects of the effects of technology on human interaction, does not see digital technologies as threatening. While normally dramatized as the death of the subject, the human constituent is still there, he or she simply undermines the long held cultural hierarchy that places the human as dominant entity and source of agency. By demonstrating spaces that are not anthropocentric or physically like our earthly environment, digital technologies including computers and their interfaces draw attention again to our basic relationships to our environment and its fundamental affect on our perceptions and behaviors.

HUMANITY RE-ENVISIONED

This mimetic recession, this parergonal existence in-between both subject and object is neither a nihilist act nor necessarily a disempowerment, but rather exposes a paradigmatic shift toward a different representational model in which human subjects spend much of their time in spaces that are not always permitting of a power-wielding subject. When considering the fundamental relationship between human subjects and their techno-objects, Mark Coté argues that media and technology are not foreign, external objects. Building on Bernard Steigler and others including Jacques Derrida, Coté posits that, "sensory perception is only ever calibrated in relation to technics" in a way that we cannot understand media or technology by thinking about them but by *feeling* them.[35] In this sense, humans do not *use* technology and media to accomplish a task. Instead humans and their technologies *interface* with and overlap one another, thus coevolving through exploring the shifting and fluid boundaries between them. By taking up of evolutionary requirements of tool use as primary to language development, Coté highlights how deeply connected human bodies are to the technologies they adopt and develop. Stone tool use led to an upright stance, which eventually allowed for the development of a larger brain capable of language. For Coté "Under technogenesis, the ordinary human is always already transductively related to technics. As such, it is *techne* which renders *phusis* mutable, especially in relation to the *polis*" (par. 57). Through our technologies we externalize while simultaneously those same technologies become internalized. Often our technologies expose our limits. Stephanie Boluk and Patrick LeMieux make this clear in their analysis of Hans Holbein's *The Ambassadors*, whose memento mori it not a reminder of mortality but is rather a memento mortem mortis, that "the subject herself is distributed, deferred, and derealized in the same way the anamorphic image [of the skull] smears across the canvas."[36] Imagining ourselves in anamorphic digital spaces whether in the games Boluk and LeMieux consider or in even utilitarian uses of digital technologies, we must cede a sense of control we base on our knowledge and privilege of spatial, physical realities of human spaces and habituate ourselves to modes or assemblages that do not consider the human experience as central.

Essentially, we can see humans as evolving by colliding with, shifting, and morphing into the parergonal boundaries between themselves and their environment: natural, social, technological, and so forth. Humans develop profound relationships with their environment and the objects therein. For instance, Aaron Rosand, a renowned concert violinist, says of his violin, "It's almost like a living thing. You treat it carefully; you treat it gently. It talks to you. . . . You're caressing your instrument all the time."[37] Laurence Lesser, a concert cellist, continues this line of thought: "Life has so many possibilities and such endless richness that unless you just shut down, you're going to learn from the environment. For us string players, our instruments are our environment. It keeps stretching us, it keeps demanding of us, it keeps us aspiring to grow."[38] This connection between object and subject, between environment and organism, is not strictly typical of (or exhausted by) artistic experience. N. Katherine Hayles describes the heavily mediated environment of the currency exchange broker by highlighting how the brokers do not view the currency exchange market as abstract at all. She quotes one trader who calls the market "a lifeform that has being in its own right . . . it has form and meaning as a greater being."[39] Hayles cites others who describe trading on the market as similar to sexual and physical vulnerabilities or violent bodily penetration. As abstract as we may think our networked "spaces" are, the biological and technological overlap. In folding technology with the human, we are not being robbed of our humanity but we are rather reconnecting to our bodily, physical, and sensual abilities over our mental and rational ones.

Beyond animating or anthropomorphizing objects, considering humans as holding a position equivalent to rather than dominant to their surround offers humans a chance to relate and collaborate with the objects that are clearly inseparable from our functions. To such an extreme lie those invested in object-oriented ontology, a field that, as Ian Bogost summarizes "puts *things* at the center of being. We humans are elements, but not the sole elements, of philosophical interest."[40] Bogost purports to go beyond process philosophy and actor-network theory by concentrating on the object itself rather than on "couplings and decouplings" that emphasize the connections between objects in networks or assemblages.[41] Even further, Bogost divorces himself from the term "object" as it, by default, also harkens to the subject that opposes the object.[42] Voiding humanity of its perceived position of privilege brings about fear regarding the potential for objectification and its accompanying mistreatment, or even eradication of humans. Jane Bennett offers another perspective less detached from humanity per se that views the vitality of all things as a chance for humans to redeem their humanity, so to speak. For Bennett, "the image of dead or thoroughly instrumentalized matter feeds human hubris and our earth-destroying fantasies of conquest and consumption."[43] Should humans quit their delusions of independence and exclusive agency, they may consider that as an ergon, their lack requires a parergon. As we incorporate such parerga into our practices and develop wholly new behaviors in response, the boundary between parergon and ergon blur, the differentiation between subject and object is deemphasized by a shared dependence and a shared agency. Bennett charts these in many assemblages such as the August 2003 blackout, attributed to multiple actants[44] including wonky electron flow, spontaneous fires, and mislaid faith in self-regulation. Even reconsidering Charles Darwin's *Formation of Vegetable Mould through the Actions of Worms with Observations of Their Habits* offers a view of human history quite dependent on functions and assemblages that work regardless of the existence or will of humankind.[45] What makes us human is also beyond human, revealing "an 'alien' quality of our own flesh."[46] Even the evolution of humans as a species cannot be separated from the affects of tools, other creatures, or environmental factors have that contributed to shaping our appearance, cognition, and behavior.[47]

The human subject then is not the producer of an object or an object of its environment in such clearly defined terms. Differences between terms like subject and object, user and device, simply become a border in which an event takes place rather than a definitional debate between the concrete manifestations of each term. The resulting parergonal relationship we see on the border of what is human and what is computer reveals our digital "bel composto"—the interface event. To replicate the response of the Baroque viewer of any of the great ceiling paintings, we cannot stand back and analyze the scene or fall into the perfectly dramatic perspectival space represented therein.[48] Instead, we need to realize something in that space where humans and computers overlap. For the Baroque viewer of the ceiling painting that realization comes from moving around until the perspective is exposed and the difference between the painting and the architecture becomes evident. This viewer will find that the borders of the painting are as concerned with the viewer's eye as with the building's contours. Similarly, light illusions and Calogero's "Pomme C" demonstrate that what humans see is both a flickering of what they want to see and what is before them. And so neither humans nor computers are the frame of reference or the content within the frame, but instead are entangled constituents flirting with one another over the shared space of the parergon. Humans see themselves extended into digital space and watch the computer's functionality affect human circumstances. For those who venture into the space of the digital graphic user interface, perhaps the moment of realization lies in the surprises where computers assert their agency and humans surrender theirs, where each feigns the other and meet as parerga to one another.

NOTES

1. Richard Lanham in his *The Electronic Word: Democracy, Technology, and the Arts*, describes two modes of viewing: one in which the viewer ignores the medium to regard the content (looking through) and another where the viewer is aware of the medium that frames the content (looking at). For instance, when we read a novel we default to look *through* the print medium to access the story. When we notice the page qualities and font, we begin to look *at* the medium.

2. Jacques Derrida "The Parergon." *October* 9 (1979): 20. Derrida describes his own repurposing of the term parergon from a note in the second edition of Immanuel Kant's *Religion within the Limits of Reason Alone*, where Kant regards the parergon as the "concept of observations."

3. Gary Hall, "The Digital Humanities Beyond Computing: A Postscript." *Culture Machine* 12 (2011): 4.

4. Kathleen Fitzpatrick, "The Digital Future of Authorship: Rethinking Originality." *Culture Machine* 12 (2011): 7.

5. N. Katherine Hayles, "Intermediation: The Pursuit of a Vision." *New Literary History* 38.1 (2007): 101.

6. Lev Manovich, "Art After Web 2.0." in *The Art of Participation* (New York: Thames and Hudson, 2008), 70.

7. Søren Pold, "Interface Realisms: The Interface as Aesthetic Form." *Postmodern Culture* 15.2 (2005): par. 5.

8. Marianne Van der Boomen, "Interfacing by Iconic Metaphors." *Configurations* 16 (2008): 46.

9. Scott Dexter et al., "On the Embodied Aesthetics of Code." *Culture Machine* 12 (2011): 5.

10. Pold, "Interface Realisms," par. 8.

11. Johanna Drucker, "Humanities Approaches to Interface Theory." *Culture Machine* 12 (2011): 12.

12. Jacques Derrida, *The Truth in Painting* (Chicago: University of Chicago Press, 1987), 61.

13. Gregg Lambert, *The Return of the Baroque in Modern Culture* (London: Continuum, 2004), 23.

14. Angela Ndalianis, *Neo-Baroque Aesthetics and Contemporary Entertainment* (Cambridge: The MIT Press, 2004), 25.

15. Leo Steinberg, "Reflections on the State of Criticism." in *Robert Rauschenberg,* ed. Branden W. Joseph (Cambridge: MIT Press, 2002), 13.

16. Astrid Vicas, "Reusing Culture: The Import of *Détournement*." *The Yale Journal of Criticism* 11.2 (1998): 384–385.

17. His songwriter, Zazie, directly references the collage aesthetics of the graphic user interface in the title "Apple C," the MAC OSX command to copy text.

18. National Geographic, "Episode 2: You Won't Believe Your Eyes." *Test Your Brain* (2012).

19. David Corney and R. Beau Lotto, "What Are Lightness Illusions and Why Do We See Them?" *PLoS Computational Biology* 3.9 (2007): 1797–1798.

20. Omar Calabrese, *Neo-Baroque: A Sign of the Times* (Princeton: Princeton University Press, 1992), 115.

21. Kresimir Purgar, *The Neo-Baroque Subject*. (Zagreb, Croatia: Meandarmedia, 2006), 20; 22.

22. Purgar, *The Neo-Baroque Subject*, 22.

23. Mieke Bal, *Quoting Caravaggio: Contemporary Art, Preposterous History* (Chicago: The University of Chicago Press, 1999), 29–30.

24. In the summer of 2012, I received a telephone call from a marketer. When I interrupted her to ask if she was human, she took it as a compliment, saying "thanks, I didn't know I sounded so good." I hung up shortly afterward but am still unsure if she was human or not.

25. I more closely refer to Jane Bennett's adoption of Gilles Deleuze and Félix Guattari's use of the term "assemblage" as "ad-hoc groupings of diverse elements" (Bennett 23). In such groupings, power is distributed such that "Assemblages are not governed by any central head: no one materiality or type of material has sufficient competence to determine consistently the trajectory or impact of the group" (Bennett 24).

26. David Sheridan, Jim Ridolfo, and Anthony Michel, *The Available Means of Persuasion: Mapping a Theory and Pedagogy of Multimodal Public Rhetoric* (Anderson: Parlor Press, 2012), 106.

27. John Cayley, "Writing to be Found and Writing Readers." *Digital Humanities Quarterly* 5.3 (2011): par. 4.

28. Joyce Cheng, "Mask, Mimicry, Metamorphosis: Roger Caillois, Walter Benjamin and Surrealism in the 1930s." *Modernism/modernity* 16.1 (2009): 78.

29. Claudine Frank, "Introduction to 'The Natural Fantastic,'" in *The Edge of Surrealism: A Roger Caillois Reader*, ed. Claudine Frank (Durham, NC: Duke University Press, 2003), 348–349.

30. Roger Caillois, "The Natural Fantastic" in *The Edge of Surrealism: A Roger Caillois Reader*, ed. Claudine Frank (Durham: Duke University Press, 2003), 357.

31. Roger Caillois, "Mimicry and Legendary Psychasthenia" in *The Edge of Surrealism: A Roger Caillois Reader*, ed. Claudine Frank (Durham, NC: Duke University Press, 2003), 199.

32. Caillois, "Mimicry and Legendary Psychasthenia."

33. Oliver Grau, *Virtual Art: From Illusion to Immersion* (Cambridge: The MIT Press, 2003), 13.

34. Diana Domingues, "Day-Dreaming States in Interfaced Environments: Telematic Rituals in 'Ouroboros.'" *Leonardo* 37, no. 4 (2004): 309.

35. Coté, "Technics and the Human Sensorium: Rethinking Media Theory through the Body." *Theory and Event* 13, no. 4 (2010): par 1.

36. Stephanie Boluk and Patrick LeMieux, "Stretched Skulls: Anamorphic Games and the *memento mortem mortis*." *Digital Humanities Quarterly* 6, no. 2 (2012): par. 15.

37. Daniel Wakin, "Dispossessed," *New York Times*, January 15, 2012, www.nytimes.com.2012. par. 12.

38. Wakin, "Dispossessed," par. 26.

39. N. Katherine Hayles, *Electronic Literature: New Horizons for the Literary* (South Bend, IN: University of Notre Dame Press, 2008), 98.

40. Bogost, *Alien Phenomenology or What It's Like to Be a Thing* (Minneapolis: University of Minnesota Press, 2012), 6.

41. Bogost, *Alien Phenomenology*, 7.

42. Bogost, *Alien Phenomenology*, 23.

43. Jane Bennett, *Vibrant Matter: A Political Ecology of Things* (Durham, NC: Duke University Press, 2010), ix.

44. Bennett appropriates this term from Gilles Deleuze who defines it as a "source of action" (Bennett 9) that is not bound to or defined by human actors.

45. Bennett, *Vibrant Matter*, 95–96.

46. Bennett, *Vibrant Matter*, 108.

47. Bennett, *Vibrant Matter*, 31.

48. Ndalianis, *Neo-Baroque Aesthetics*, 209.

REFERENCES

Bal, Mieke. 1999. *Quoting Caravaggio: Contemporary Art, Preposterous History*. Chicago: The University of Chicago Press.

Bennett, Jane. 2010. *Vibrant Matter: A Political Ecology of Things*. Durham, NC: Duke University Press.

Bogost, Ian. 2012. *Alien Phenomenology or What It's Like to Be a Thing*. Minneapolis: University of Minnesota Press.

Boluk, Stephanie, and Patrick LeMieux. 2011. "Stretched Skulls: Anamorphic Games and the memento mortem mortis." *Digital Humanities Quarterly* 6.2.

Brooke, Collin Gifford. 2009. *Lingua Fracta: Towards a Rhetoric of New Media*. Cresskill: Hampton Press.

Caillois, Roger. 2003. "The Natural Fantastic." In *The Edge of Surrealism: A Roger Caillois Reader*, edited by Claudine Frank, translated by Claudine Franck and Camille Naish, 349–357. Durham, NC: Duke University Press.

———. (1935) 2003. "Mimicry and Legendary Psychasthenia." In *The Edge of Surrealism: A Roger Caillois Reader*, edited by Claudine Frank, translated by Claudine Franck and Camille Naish. 91-102. Durham, NC: Duke University Press. Translation of "Mimétisme et psychasthénie légendaire." Minotaure 7: 5-10.

Calabrese, Omar. 1992. *Neo-Baroque: A Sign of the Times*. Translated by Charles Lambert. Princeton: Princeton University Press.

Calogero. 2007. "Pomme C." Pomme C. Mercury compact disc.

Cayley, John. 2011. "Writing to be Found and Writing Readers." *Digital Humanities Quarterly* 5.3.

Cheng, Joyce. 2009. "Mask, Mimicry, Metamorphosis: Roger Caillois, Walter Benjamin and Surrealism in the 1930s." *Modernism/modernity* 16.1: 61–86.

Corney, David, and R. Beau Lotto. 2007. "What Are Lightness Illusions and Why Do We See Them?" *PLoS Computational Biology* 3.9: 1790–1800.

Coté, Mark. 2010. "Technics and the Human Sensorium: Rethinking Media Theory through the Body." *Theory and Event* 13.4.

Derrida, Jacques. 1979. "The Parergon." October 9: 3–41.

———. (1978) 1987. The Truth in Painting. Translated by Geoff Bennington and Ian McLeod. Chicago: University of Chicago Press. Translation of La vérité en peinture. Paris: Flammarion.

Dexter, Scott, Melissa Dolese, Angelika Seidel, and Aaron Kozbelt. 2011. "On the Embodied Aesthetics of Code." *Culture Machine* 12: 1–23.

Domingues, Diana. 2004. "Day-Dreaming States in Interfaced Environments: Telematic Rituals in 'Ouroboros.'" *Leonardo* 37.4: 308–314.

Drucker, Johanna. 2011. "Humanities Approaches to Interface Theory." *Culture Machine* 12: 1–20.

Fitzpatrick, Kathleen. 2011. "The Digital Future of Authorship: Rethinking Originality." *Culture Machine* 12.

Frank, Claudine. 2003. "Introduction to 'The Natural Fantastic.'" In *The Edge of Surrealism: A Roger Caillois Reader*, edited by Claudine Frank. Translated by Claudine Franck and Camille Naish. 348-349. Durham, NC: Duke University Press.

Grau, Oliver. 2003. *Virtual Art: From Illusion to Immersion.* Cambridge, Massachusetts: The MIT Press.

Hall, Gary. 2011. "The Digital Humanities Beyond Computing: A Postscript." *Culture Machine* 12: 1–11.

Hayles, N. Katherine. 2008. *Electronic Literature: New Horizons for the Literary.* Notre Dame: University of Notre Dame Press.

Hayles, N. Katherine. 2007. "Intermediation: The Pursuit of a Vision." *New Literary History* 38.1: 99–125.

Kaup, Monika. 2005. "Becoming-Baroque: Folding European Forms into the New World Baroque with Alejo Carpentier." *CR: the New Centennial Review* 5.2: 107–149.

Lambert, Gregg. 2004. *The Return of the Baroque in Modern Culture.* London: Continuum.

Manovich, Lev. 2008. "Art After Web 2.0." *The Art of Participation.* 66-79. New York: Thames and Hudson.

National Geographic 14 Apr. 2012. "Episode 2: You Won't Believe Your Eyes." Test Your Brain. Television.

Ndalianis, Angela. 2004. *Neo-Baroque Aesthetics and Contemporary Entertainment.* Cambridge: The MIT Press.

Pold, Søren. 2005. "Interface Realisms: The Interface as Aesthetic Form." *Postmodern Culture* 15.2.

Purgar, Kresimir. 2006. *The Neo-Baroque Subject.* Zagreb, Croatia: Meandarmedia.

Sheridan, David M., Jim Ridolfo, and Anthony J. Michel. 2012. *The Available Means of Persuasion: Mapping a Theory and Pedagogy of Multimodal Public Rhetoric.* Anderson: Parlor Press.

Steinberg, Leo. 2002. "Reflections on the State of Criticism." In *Robert Rauschenberg*, edited by Branden W. Joseph. 7–37. Cambridge: MIT Press.

Van den Boomen, Marianne. 2008. "Interfacing by Iconic Metaphors." *Configurations* 16: 33-55.

Vicas, Astrid. 1998. "Reusing Culture: The Import of Détournement." *Yale Journal of Criticism* 11.2: 381–406.

Wakin, Daniel. "Dispossessed." *New York Times*, January 15, 2012. www.nytimes.com.

Chapter Four

Techno-Geographic Interfaces

Layers of Text and Agency in Mobile Augmented Reality

John Tinnell

For media theorists and technologists, "ubiquitous computing" (ubicomp) marks the paradigmatic horizon of current innovations in mobile media, locative networks, wearable devices, and smart objects. Ubicomp platforms promise to break from the personal computing (PC) era and the models of computing that defined it, creating as many new possibilities and new problems as the shift from mainframe computers to PCs initiated during the 1980s. Pioneering ubicomp in the late 1980s, Mark Weiser and his collogues at Xerox PARC prototyped entirely new categories of mobile, wearable, and embedded devices designed to bring the virtuality of computation "into the physical world" (Weiser 1991, 98). The smartphones and tablets that dominate today's mediascape are all conceptually bound to the "tabs, pads, and boards" outlined in Weiser's famous essay "The Computer for the Twenty-first Century." In that paper, Weiser defines ubicomp in opposition to the virtual reality of the desktop screen, specifically the demands the latter makes of people: to stop whatever activities they might otherwise be engaged in, in order to interact—stationary and isolated—with clunky and conspicuous computers via specialized commands and technical jargon.

Echoing this aspect of Weiser's sentiment, many contemporary researchers and designers of mobile and wearable devices routinely valorize "human-centricity" as a top priority that, they believe, should guide future developments in computing. Design researcher Isabel Pedersen defines human-centricity as "a value system that privileges humans over machines and other hegemonic orders in order to avoid dehumanizing effects" (Pedersen 2008,

168). For Steve Mann, a leading inventor of wearable devices, the desktop computer is machine-centric and anti-human because it requires humans to interact with it on the basis of its own limits—desktops must remain stationary and so we must remain stationary if we want to use them (168). In mobilizing the computing experience, researchers and designers share a goal "to liberate people" from the so-called "dehumanizing results so common in current technology" (166). And yet, since the spread of smartphones, media-rich societies have witnessed some negative, even dangerous outcomes associated with human-centered designs that give people the ability the access the Web whenever and wherever they want.

For instance, accidents related to "distracted driving"—and even distracted walking—have become the subject of many news reports, university research studies, and public service campaigns around the world. Quickly approaching the status of a global epidemic, the injuries and deaths that result from such accidents are stark indications of a *dramatic decrease in environmental awareness* associated with some of the human-centered imperatives orienting mobile development. While these technologies do not *make* people distracted—human error, the failure to manage one's attention safely, could be regarded as the final cause—enormous blind spots are designed into the technologies. Most mobile apps allow people to access networked media on-the-go, but the interaction often remains anchored in the PC model: to engage with the device, one must become immersed in the virtuality of the screen. The fact that outcomes of digital distraction have at least some basis in design speaks to the ethical need for interface design and media content that exploits context-awareness and other unique affordances of ubicomp, which also suggest interesting aesthetic and rhetorical possibilities.

Indeed, a growing contingency of digital devices and locative networks has begun to operate on the basis of real-time, geographical data. Increasingly, developers are working to realize a relatively underdeveloped goal of Weiser's vision for ubicomp: to create computers that interface with the multitude of autonomous, non-computational flux he called "the infinite richness of the universe" (Weiser 1991, 94). Most notably, the landscape of mobile media is rapidly transforming with ongoing efforts to incorporate the global positions system (GPS) and augmented reality (AR) technologies. Ultimately, this trend in mobile development beckons parallel developments in the ways we understand and create new media projects. Scholars need to theorize this emerging approach to interface design and its implications for digital rhetoric and culture. Whereas anthropocentric viewpoints on mobile media often revel in the user-autonomy associated with anytime-anywhere Internet access, posthumanist perspectives might draw attention to a broader ecology of digital interaction whereby multimedia becomes structured by non-human activity. In such cases, as we shall see, environmental and geographical flows become key constituents of so-called techno-human relations.

The general aim of this chapter is to examine the *techno-geographic* scene of writing afforded by geo-information and mobile AR, engaging these technological developments as key sites for interrogating foundational oppositions at the core of human-centered approaches to design, rhetoric, and textuality. In evoking the phrase "the scene of writing," I am referring to two theoretical gestures that have gained intellectual currency in scholarship on writing. First, working in genre studies, compositionist Amy Devitt argues that acts of writing (and reading) are "enmeshed within scenes and the specific situations within them" (Devitt et al. 2004, 48). The writer alone at her desk is, in effect, always working in conversation with an array of social norms, discursive conventions, technological systems, and so forth. Culturally situated, scenes of writing entail a dialogical connection with social change, shifting discourse communities, and advancements in media technology. Second, in his essay "Freud and the Scene of Writing," Jacques Derrida describes Freud's tendency to study dreams as a kind of psychical writing—"a signifying chain in scenic form" (Derrida 1978, 218). Dreams, like hieroglyphics, become an emblem for Derrida's expansive notion of writing, which he assembles alongside his famous critique of phonocentric traditions in philosophy and linguistics. Dreams, hieroglyphics, and the like each mark a "form of writing which puts words on stage without becoming subservient to them . . . a model of writing irreducible to speech" (209). The scene of writing, in this sense of the phrase, pertains not only to the play of words on a page, but more broadly to the spatial arrangement of signs, traces, or *gramme*—the Greek word for written mark—across any surface of inscription whatsoever, from paper to the psyche. Through examining the emergence of ubicomp in contemporary culture, one begins to observe scenes of writing in which vast stretches of the Earth's surface become a surface for digital inscription, and geographical flows double as networked *gramme*.

TECHNO-GEOGRAPHIC MILIEUS

Prior to recent developments in ubicomp, several notable philosophers, two of whom we will consider below, began theorizing about modes of relations evident in major twentieth-century technologies that created productive interfaces between technical machines and geographical flows. Thus, our inquiry into ubicomp's emerging mediascapes begins with the task of (re)assembling a conceptual thread, which can be traced from a series of innovations in continental philosophy to recent scholarship across rhetoric and media studies.

A student of Maurice Merleau-Ponty's during the 1950s, Gilbert Simondon was one of the first French philosophers to assert the cultural and ontological import of technical objects. Simondon's influential doctoral dissertation, *On the Mode of Existence of Technical Objects*, presents elaborate descriptions of the technical dynamics at work in a number of machines integral to industrial production throughout the twentieth century: steam turbines, automobile engines, boilers, electron tubes, and power plants. These extremely detailed accounts all lead to philosophical reflections; Simondon treats technical objects as if they constitute an important class of beings. For Simondon, the historical transition from hand tools to industrial machines marks a new era of techno-human relations, such that "human reality resides in machines as human actions [become] fixed and crystalized in functioning structures" (Simondon 1980, 13). Simondon aspires to act as a "responsible and creative consciousness" for this "society of technical beings," with the basic objective to "achieve an understanding of technical reality and introduce it into our culture" (14). While his later theory of individuation proved to be an important influence on Gilles Deleuze, more recently, Bernard Stiegler has revived Simondon's early work on technics, bringing it to the attention of leading scholars in media studies and posthumanism such as Mark Hansen and N. Katherine Hayles.[1] Particularly relevant to the technologies I address below is a concept initially formulated by Simondon, which Stiegler seems to privilege increasingly as his *Technics and Time* series continues to unfold: *techno-geographic milieus*.

Simondon introduces the notion of a techno-geographic milieu to describe what he sees as an especially unique kind of machine. Generally put, these machines establish an unparalleled degree of interlocking with their surrounding environments; they assign a multitude of technical functions to the geographical flows that they have been designed to accommodate if not exploit. Such machines are distinct from older windmills and watermills in that the former sync geographical flows with electrical networks.[2]

Simondon presents the Guimbal turbine (in the context of a wave-power generation factory) as his primary example of a techno-geographic milieu. The Guimbal turbine stands as a technical ensemble "constituted by [the] oil and water in motion within and around [it]" (Simondon 1980, 50). In his frequent references to Simondon's analysis of the Guimbal turbine, Stiegler summarizes the way in which the seawater becomes a "polyfunctional" element in the technical system:

> the [techno-geographic] milieu structurally and functionally associates the energies and natural elements composing this milieu, such that nature becomes a function of the technical system. This is the case of the Guimbal turbine, which assigns to saltwater (the natural element) a triple technical function: to

furnish energy, to cool the structure of the turbine, and to catalyze the waterproofing of the stages. (Stiegler 2010, 83)

This passage calls to mind Heidegger's concern that, in the grip of modern technology, Nature becomes a standing-reserve of energy, ready-to-hand, and thereby reducible to a reservoir of resources available for (and subservient to) human use.

Both Simondon and Stiegler share Heidegger's contempt for a world reducible to calculation and usage; so much so, in fact, that Simondon and Stiegler constantly refuse to discuss technics as a mere tool or utensil.[3] In other words, if Heidegger associates modern technics largely with a technocratic and inauthentic way of dwelling, the two French philosophers (particularly Stiegler) theorize technics as the originary condition of human becoming and cultural development. Contra to the more conventional manner of discussing machines in terms of calculated automation, Simondon locates the ultimate power of machines in their "margin for indetermination . . . a margin that allows for the machine's sensitivity to outside information" (Simondon 1980, 13). In the case of techno-geographic milieus, then, what is most significant—at least in terms of Simondon and Stiegler's emphasis—is the fact that this machinic margin of indetermination syncs with and becomes sensitive to the geographical flows, which retain some degree of autonomy and accidentality even as they perform certain functions assigned by the technical system. As such, we might say, "the two worlds act on one another" (46).

Stiegler extrapolates from Simondon's writing on techno-geographic milieus as he broadens the concept to account for ways in which digital networks have begun to converge with geographical flows. According to Stiegler, many current technological and industrial transformations are related to "the fact that, with digital technology, and particularly networked technology, the technical milieu is becoming . . . a techno-geographic milieu" (Stiegler in Couze Venn et al. 2007, 334). Throughout the third volume of *Technics and Time*, Stiegler focuses on what he calls "the digitization of territories" now occurring with the emergence of geo-information. By "geo-information," Stiegler means to indicate all manners of digital simulations and reproductions of Earthly locations, as well as all mobile devices that operate in accord with geo-referenced data. Akin to the Guimbal turbine, geo-information assigns a "technical navigation function" to territory (Stiegler 2010, 138). That is, territories effectively enter into the technical system as a generative epistemological force, playing a heuristic role as "a vehicle for action and information in a three-way [transductive] relationship involving man, the machine and the world" (Simondon 1980, 68). Through digitization qua geo-information, territories become a constitutive element in the essential technicity of perception, memory, and decision-making—and not merely a standing-reserve to be seen, retained, and acted upon by an already constituted subject.

THE GPS AS A TECHNO-GEOGRAPHIC INTERFACE

Ultimately, we might understand the phenomenon of geo-information in terms of a broader strand of technical evolution giving rise to a "geo-graphical" gaze, in which perception and cognition fuse with the writing (*graph*) of the Earth (*geo*). Attending to both senses of the phrase, the writing of the Earth signifies the (f)act of making inscriptions upon the Earth as a (digital) writing surface, and it also signifies the modalities through which the Earth's geographical flows become translated (i.e., sensed and actuated) into written traces—or at least affect the arrangement and/or presentation of digital data—amidst the real-time functioning of techno-geographic interfaces. No longer restricted to turbines and power plants, techno-geographic modes of operating have contributed to a definitive class of digital interfaces, the iterations of which many of us already engage with on a daily basis.

At this moment in digital culture, turn-by-turn GPS navigation devices are arguably the most pervasive and influential manifestation of techno-geographic interfaces. In the case of GPS navigation, the machine—or, more precisely, the software program (in conjunction with larger technical systems)—assigns a polyfunctional role to human geography; the margin of indetermination becomes sensitive to an individual's automotive movements and, in many cases, the movements of other automobiles qua the tracking and displaying of nearby traffic patterns. That GPS navigators function in *real-time* relative to dynamic, geographical variables marks a crucial dimension of the paradigmatic shift away from print cultures rooted in the epistemological habits and temporal conventions of alphabetic literacy.[4]

Stiegler argues that, because audiovisual media are currently displacing the archival modalities of literate writing, the literate experience of historical consciousness is becoming supplanted by the real-time of contemporary technics.[5] Streams of real-time information events conjoin primary retention and tertiary retention, such that any distinction between lived perception and technical recording becomes, in these cases, "absolutely formal and empty" (Stiegler 2008, 242). The "now" of real-time is always already-there. Hence, real-time media—especially techno-geographic interfaces—do not constitute a departure from technological memory, but rather the arrival of a new paradigm in memory's exteriorization. Moreover, Stiegler also contends that the production of real-time affects knowledge-building; all information, communication, and memory practices that generate and preserve knowledge are of technics and therefore not outside of time and its technical evolution. For instance, once accustomed to GPS devices, we tend to make navigational decisions in the real-time of a geo-graphical gaze that is processually co-constructed between lived perception and technological memory, and not primarily on the basis of our own historical consciousness.

In her analysis of the GPS as a rhetorical artifact, Amy Propen foregrounds questions of text and agency: what kinds of multimodal utterances are generated during GPS navigation, and who/what (co)produces them? Propen outlines visual cues through which the GPS digitizes territories (e.g., highlighted routes, place indicator icons, real-time mileage approximations), as well as the audio cues by which it communicates instructions to drivers (e.g., recalculation announcements). Having conducted interviews with a pool of GPS users, she presents insightful reflections about the various techno-human relations apparent in embodied experiences of GPS navigation. Throughout the chapter,[6] Propen returns to her thesis that "the GPS and its user co-construct an interactive agency that is malleable and involves a process of ongoing exchange, negotiation, and sometimes resistance" (Propen 2012, 142). Some of her interviewees, for example, described their techniques for commanding the GPS-issued driving instructions; the drivers would purposefully disobey certain instructions, knowing that the device would in turn adapt to their movement (141). One might say, counterintuitively, that the driver's agency in these cases derives from the device's capacity to maintain precise surveillance over them, and that the device's program runs (i.e., generates multimodal texts) in response to the driver's non-discursive actions. The radical distribution of agency in GPS rhetorical production suggests a mode of environmental knowing that can at once be "understood as contributing to posthuman, embodied ways of knowing" (123).

In such iterations (i.e., navigation aids for automotive transportation), the GPS has redefined the navigational experience for many people throughout the world; and yet, the GPS—the global positioning system—is a vast, all-purpose technology. The familiar navigational aids some drivers have in their vehicles provide *access* to the GPS, but they are not actually themselves *the* GPS. This reduction has been thoroughly naturalized in everyday speech patterns and in product advertisements. Strictly speaking, the GPS is a system of 24 satellites launched by the U.S. Department of Defense, which became fully operative and available to civilians in the mid 1990s. The "GPS" that we have in our cars is only one application that makes use of the GPS proper. With the rise of smart mobile devices and other ubicomp platforms, we are witnessing the start of a generalization of GPS technology, which has recently inspired an expanding array of new applications concerned principally with the location-based, techno-geographic production and circulation of writing, media, and digital data of all kinds. Propen rightly claims that GPS navigation "engages users in a unique way that elicits interaction and bodily engagement in everyday settings and mundane activities"

(123). In light of current media trends, I believe that techno-geographic interfaces will become integral to an increasingly diverse mix of everyday activities. The techno-geographic operations of the digital age are no longer unique to the rhetorical situations of vehicular transportation. From rhetorical considerations of GPS navigators, we can move to extrapolate insights relevant for conceptualizing and composing all manner of techno-geographic interfaces emerging with the spread of smartphones, tablets, and other mobile and locative media.

TECHNO-GEOGRAPHIC INTERFACES BEYOND GPS NAVIGATORS

Techno-geographic interfaces constitute a new technics of retention and perception, with which we may associate certain alterations in the ways we navigate information and make decisions in real-time on location. Following the operational logic of GPS navigation, the Internet and digital cultural production at large now exhibit a pronounced tendency to incorporate geographical criteria as a structural principal governing the organization and presentation of networked media files. Mobile communications scholars Adriana de Souza e Silva and Daniel Sutko situate cutting-edge mobile media and GPS navigation within the same historical trajectory: "these location-aware mobile phones are part of the greater ecology of locative technologies that includes the iPhone, the Droid, Blackberries, and GPS navigation systems in cars" (Souza e Silva and Sutko 2011, 24). Their discussion of the hardware that provides access to the mobile Web speaks to the general phenomenon whereby the social networks and media sharing platforms of Web 2.0 become reconfigured and revitalized by on-the-go connectivity and location sharing; however, their analysis lacks a topology of contemporary manners of linking digital information to particular locations and mobile bodies. In order to understand the GPS's ongoing influence on digital writing and ubicomp culture, we need to survey the techniques through which this emergent variety of techno-geographic interfaces—from popular location "check-in" apps to more avant-guard AR apps—are created and accessed. The consequences of these developments, as I will argue in the two remaining sections, beckon scholars and designers to explore posthumanist perspectives for conceiving of agency, especially in regards to AR textuality.

The concept of techno-geographic interfaces allows us to foreground emerging modes of linking among networked media, everyday objects, and geographical flows. These modalities, which I describe below, are evident in smart mobile devices and apps—namely in the area scholars and artists call *locative media*. Framing the technology in these terms, however, makes for scholarship that highlights the sheer portability of devices (e.g., the capacity for on-the-go connectivity) and/or "first generation" geo-information techniques for linking digital data to static places. As such, the mobile Web appears to be a simple extension of more or less the same materials one accesses via desktops and laptops. Indeed, early modes of linking digital data to specific locations, such as QR coding and geotagging, typically do not sync with the autonomous, dynamic movements of geographical flows. For instance, a basic QR code attached to the façade of a building merely refers one's mobile browser to a specified website, which operates without any direct relationship to the present activities occurring at that particular location. Likewise, geotagging a photograph or a webpage renders it sensitive to geo-referential Web searches, but this amounts to little more than a different means to essentially the same ends (i.e., an alternative search criteria).

Focusing primarily on locative media, Jason Farman identifies site-specificity as a key feature that allows data to be hosted, organized, and accessed on the basis of location (Farman 2012, 39). This is, of course, a core capability that distinguishes the mobile Web of ubiquitous computing (e.g., the Internet of things) from the Web as it has existed since its inception from the vantage point of desktops and laptops.[7] The Web of the PC era supports a vast collection of documents and networks, all of which remain globally accessible regardless of one's location, provided she or he is within range of a landline connection or a wifi hotspot. Until recently, online information was not sensitive to place; geo-information was not part of its metadata, and the contents of webpages did not account for a user's actual location at the time of browsing. While site-specificity undoubtedly plays a necessary role in the mobile Web, this aspect alone is not sufficient for distinguishing and theorizing media that are designed to sync with geographical flows. Site-specificity, to recall Simondon's philosophy, does not account for the medium's *margin of indetermination*—the ways in which such interfaces assign rhetorical or aesthetic functions to *vibrant matter in motion*. Here, as well as below, I play upon Jane Bennett's concept of vibrant matter (by which she attributes an intrinsic political agency to material things independent of their relation to human beings); as I will suggest, material flows also become vibrant in an acute rhetorical sense when they inhabit scenes of writing emerging with ubicomp platforms. More descriptive of such scenarios than the idea of site-specificity, the notion of techno-geographic production helps elucidate a specific class of mobile interfaces—like GPS navigation—that generate digital text by sensing and actuating the activity of certain geographical flows, as well as technologies that incorporate material flux into the very presentation of multimedia content, accommodating such flux as an integral semiotic element within the visual frame and/or the textual field.

While QR codes and geotags are more popular than ever, more robust manners of linking have emerged. In particular, radio-frequency identification (RFID) and AR technologies add further technical and rhetorical dimensions to these earlier modalities, creating conditions for a more radical interface among networked media, everyday objects, and geographical flows. N. Katherine Hayles has already written a thorough review of RFID tags, wherein she describes how these tiny, low-cost, and increasingly pervasive microchips are being embedded into a multitude of everyday products. In line with her previous work on distributed cognition, Hayles' analysis of RFID highlights "the technology's potential for shedding the burden of long-held misconceptions about cognition and moving to a more processual, relational, and accurate view of embodied human action in complex environments" (Hayles 2009, 48). When combined with appropriate sensors and receivers, RFID tags effectively create smart environments in which animate objects communicate with one another to perform a finite set of programmed tasks. In addition, RFID tags can be "read" and traced, such that any mobile entity—objects, animals, humans—can be tracked in real-time. More generally, Hayles claims, "RFID participates in a larger transition to a world where human action is coordinated with complex virtual/actual environments characterized by flow and relations between many different agents, including non-human ones, tied together through distributed cognitive networks" (53). AR technologies also contribute to this larger transition to which Hayles alludes. Attending to the layered scene of mobile AR, we can see how the most radical trajectories of ubicomp development establish links that bind media not only to a given place, but also to the vibrancy of matter in motion. Conceptually, locative media gives way to techno-geographic interfaces.

THE SCENE OF MOBILE AUGMENTED REALITY

In terms of theoretical significance and design possibilities, GPS navigators and mobile AR apps stand as the twin peaks of recent innovation in techno-geographic interface development.[8] Both of these technologies establish an unprecedented degree of interlocking among their computational operations, the media they generate, and the material flux of geographical flows. As indicated above, turn-by-turn GPS navigation devices constitute a posthuman mode of environmental knowing; they generate a running text that effectively senses drivers' non-discursive movements, translates them into data, and actuates pertinent audio/visual cues in real-time relative to precise landmarks. GPS navigation is about more than site-specificity, for it does more than simply attach texts or multimedia to a given place. Likewise, mobile AR is not only about creating multimodal texts for mobile users to access at a particular place; more crucially, it marks a techno-geographic mode of multimedia production. That is, AR enables the creation of media that incorporate geographic flows as a vital element in the composition or design process. And while GPS navigators produce *virtual* texts in response to geographical flows, mobile AR apps assign to the real-time movement of geographic singularities a rhetorical function as *indexical occupants* within the textual field.[9]

THE SCREEN IS A CAMERA

Technical explanations of AR generally define the medium in opposition to virtual reality (VR). At opposite ends of a spectrum, VR systems promote total sensory immersion in computer-generated worlds, while AR technologies generate layers of digital data relative to one's live perception of his or her extra-computational surroundings. By definition, AR technologies facilitate an awareness of the environmental surroundings that necessarily envelops the embodied act of computing. Additionally, AR initiates a radical degree of convergence between screens and cameras, or between networked digital writing and live indexical recording. In filmmaking, there is often a clear spatiotemporal distinction between camera (the activity of production) and screen (the context of reception); with mobile AR, however, the screen is a camera and the camera is a screen.

In fact, throughout the history of manuscript and print culture, most texts and images have rested on an opaque writing space. To experience digital writing and multimedia via a translucent, indexical surface is thus to depart from the textual experience associated with the static, virtuality-oriented surfaces of inscription that have dominated the history of writing and computing on account of technological necessity. Book pages, canvases, and even websites employ a fixed background color or image to establish a perceptible contrast with the inscribed text or image.[10] Conventionally, writers hope that readers will become immersed in their writing, regardless of where they are or what is going on around them. Museums typically keep gallery walls as sparse as possible, in order to emphasize the artworks inside the frames. As readers and viewers, we often try to block out the surrounding environment, regarding it as a distraction from our experience of textual field or visual frame. In short, much of our reading and writing is anchored in virtual reality. It has taken place in the library, the office, the study, and the classroom à la the logic of the studio—not *en plein air*.

When critics endeavor to understand mobile AR as a genre of locative media, they emphasize the situated quality of networked digital writing, but neglect the live indexical recording aspect inherent to AR as a translucent, real-time surface of techno-geographic inscription. For instance, Farman's engagement with mobile AR apps such as *streetmuseum* depicts the medium as little more than a digital version of text panels that provide situated commentary for museum exhibits and historical landmarks. What is most important about mobile AR, according to Farman, is the medium's capacity to display site-specific information that affects our sense of place (Farman 2012, 42). Conversely, he points out that the place in which we encounter AR information affects our experience of that information. I argue, however, that mobile AR transforms not just the way we engage with texts, but, more radically, it introduces a unique form of textuality.

Because AR assigns a rhetorical or aesthetic function to vibrant matter, environments and geographical flows do not simply affect our experience of site-specific texts. More precisely, AR textuality is fundamentally indissociable from the real-time activity of geographical flows. Once a multimedia file inhabits the scene of mobile AR, that file becomes *part* of a dynamic techno-geographic milieu. AR textuality lies in excess of the location-based media files; any attempt to isolate the media files as *the text* runs counter to this principal, and thus subjects the medium to the logic of the page or the screen. Writing and reading qua mobile AR occurs on the *screen-camera*. As such, the relationship between *text* and *context* becomes a critical rhetorical issue. Whereas particular contexts always shape a text's *meaning*, AR textuality is *shaped by context* at the level of form; the textual field is always permeable and transparent.

The multimedia files included in any given *layer*—the basic unit of AR textuality—occupy only a portion of the screen-camera, which always distributes the work of composition to other sources of agency and action: the kinetic movements of the *cinematographer* (not just a viewer) who possesses a mobile device, as well as the autonomous flux of geographical flows. Instead of imagining AR textuality on the basis of the museum text panel metaphor, multimedia files in the scene of AR might be conceived as catalysts for a real-time, singular documentary that each viewer-cinematographer lives while moving amidst geographical flows with a screen-camera. Digital rhetoric, in such cases, becomes a performance that accrues at the interstice of composed media and spontaneous happenings.

INESSENTIAL TEXTUALITY

Despite the evident novelty of AR apps, the medium—though still in its infancy—has thus far "often been deemed gimmickry" by media critics and has yet to prove compelling to a critical mass of users (Farman 2012, 40). I venture that most AR content fails to be compelling because most early adaptors have been creating it and encountering it from a *virtual reality* mindset. In departing from the opacity of conventional writing surfaces, AR also challenges some basic assumptions associated with literacy. For instance, Western metaphysics—from which many techniques of literacy were developed—has always maintained a fundamental opposition between the intelligible and the sensible. We still think by means of terms, definitions, and concepts that were formulated by ancient philosophers who abstracted "essential" properties from barrages of sensible matter. Socrates (or Plato) infamously defined "man" as a *featherless biped*; of course, this definition failed—Diogenes demonstrated that the same description could apply to a plucked chicken—but the method of defining things by deducing their "essences" clearly prevailed, most enduringly in the legacy of Aristotle. Literacy, if it can be characterized as a historical and technological way of seeing, encourages us to perceive a world divided into stable substances with essential properties—a world ripe for analysis via intelligible, verbal categories. We tend to apprehend the visibility and aurality of the sensible by reference to intelligible filters.

As indicated above, most spaces for writing, reading, and viewing texts in various media have been designed to bracket sensible plentitude, on the assumption that environmental flux is inessential, if not distracting. In cases of distracted driving, the driver immerses himself in the virtual space of texting or the so-called live feeds of social media, both of which have been designed to promote anytime-anywhere access to digital networks that, with every audiovisual notification, invite the user to neglect his immediate surroundings. For drivers immersed in the act of texting; the road, traffic signs, and other cars serve as inessential distractions from the essential textual field. More subtly, and less terrifying perhaps, one inevitably witnesses this manner of digital distraction at work in a variety of settings every day: the classroom, academic conferences, restaurants, group meeting rooms, and so forth. The techno-geographic scene of mobile AR suggests an alternative media ecology, which may potentially facilitate the development of digital cultural practices that circumvent the unintended consequences of human-centered mobile interface design.

With AR, textuality is no longer an intelligible essence distinct from the accidents of the sensible; the "accidental" flux of geographical flows actually inhabits the textual field as an essential semiotic element. In accounting for this *essential accidentality* of AR textuality, designers and content creators need to identify the consequential, autonomous geographic flows that comprise a project's margin of indetermination. On the design side, one can think of assigning aesthetic or rhetorical functions to certain geographical flows in relation to the media files she or he is composing. Locations are not just the interpretive contexts in which one experiences an AR layers; sensible plentitude pervades and (re)composes the textual field. The scene of writing becomes posthuman to the extent that vibrant matter participates in the composition and, in so doing, stimulates an awareness of precisely the environmental activity that VR writing spaces have conventionally bracketed. Understood in this manner, AR textuality hinges upon the question of how certain geographical flows function within the frame of the screen-camera, as well as what rhetorical or aesthetic effects become realized or enunciated amid singular, real-time assemblages among the networked media, the geographical flows, and the viewer-cinematographer. Moreover, from the standpoint of production, one must learn to write and design *with* the accidents of the sensible, no longer in spite of them.

(DIS)LOCATING AGENCY

Relevant rhetorical and design principals can be discerned from digital artists who, at the close of the twentieth century, created experimental narratives using early mobile media hardware such as portable music players, personal data assistants, and pocket PCs. Generally these projects sought to enrich the audience's experience of traditional narratives by turning one's bodily movement through specific environments into the principal structuring device of the story. Teri Rueb's narrative called "Itinerant," for example, "invites people to take a walk through Boston Common and surrounding neighborhoods to experience an interactive sound work that re-frames Mary Shelley's *Frankenstein*" (Rueb 2005). Hence, rather than zeroing in on the page or the screen, the audience conducts their reading of the story as they attune themselves to their embodied experience of their material surroundings. Reflecting on this project and similar ones, Rueb claims, "the narrative and its structure and meaning emerge from and are dependent on the interaction of participant, place, time, and social context" (Reub 2008, 130). When multimedia becomes techno-geographic, readers and writers come to interact with texts, images, and sounds in ways that differ from the virtual reality of cyberspace. Non-discursive actions—walking, in the case Rueb's projects—become a source of narrative agency for readers, while dynamic environmental variables exhibit agency by assuming a discursive function within the narrative frame.

Since 2010, the media platforms that Rueb and other artists had to configure on their own fifteen years ago have become readymade and as user-friendly as applications like WordPress, Google Maps, and Facebook. As a result of this increased accessibility, a wider array of writers and designers can participate in the techno-geographic production of AR layers, which can only be expected to intensify once wearable devices like the Goggle's Project Glass reach a mass market. Like early works in location-based digital storytelling, AR textuality is "shaped by all the complex indeterminacy of the participant's agency and place within an actual and discursive milieus of social, cultural, physical, and natural forces" (132). Mobile AR apps enable writers to select the location to which their (multimodal) layers will be geotagged, but writers often cannot control the composition of that location, particularly when it is a dynamic, public space in which the people who occupy it need not know that it is the place of an AR layer. The geotagged location, as a writing space, bears no sign of the marks that AR writers place upon it—only a *signal* of the sign. As such, the image-text dynamics potential to mobile AR forecast some intriguing rhetorical ramifications, particularly in terms of agency.

In his discussion of film and theater as image-text media, W. J. T. Mitchell views image-text relations as "not a merely technical question, but a site of conflict, a nexus where political, institutional, and social antagonisms play themselves out in the materiality of representation" (Mitchell 1995, 91). This play of antagonisms seems to take on an even further performative dimension in the sites of conflict mobilized through AR. Writers can caption any particular location with any image-text (and sound) of their choice—all without imposing upon the materiality of its physical space, thereby intervening into some people's experience of that place (i.e., people who have the right app), while remaining undetected by other people who occupy that place (i.e., people who don't have the app). Hence, a stream of AR layers can come to affect the discursive experience of a place without that place ever being explicitly recognized as a writing space.

On the other hand, those who (or even that which) occupy a place have the agency (even if unaware of the writer's layer) to appropriate, defuse, or otherwise redirect the composition and meaning of a writer's layer, because the material composition of the place exists autonomously from all writers' layers. If the composition of a place fluctuates (and most public places play host to constant fluctuation), then a writer's layer, which the app positions as an audiovisual overlay or textual caption to the visual scene, can be caught sleeping, so to speak. Though they clearly embody a new "conjunction of images and words" like the earlier mixed media genres Mitchell discussed, the image-text conjunctions occurring within AR platforms become inevitably disjunctive when we consider the radical dispersion of (agency over) the visual in public settings and the quasi-invisibility of the digital layers, both of which are ultimately contingent upon and imminent to the kinetic gestures of people with smartphone who are themselves in motion. Of course, writing has always been infused and surrounded by indeterminacy, especially as it is taken up in rigorous theoretical scholarship. But contemporary iterations of mobile AR make this quality of writing and textuality apparent if not inescapable, such as to make even more obvious the limits of anthropocentric, sender-receiver models of communication when applied to digital writing.

CONCLUSION

If RFID tags make manifest the theoretical underpinnings of distributed cognition, AR layers exhibit a clear model of distributed composition. Both technologies, along with GPS navigation, engage geographical flows in manners beyond the basic site-specificity addressed in studies of locative media, and, collectively, they speak to the need to rethink traditional, humanistic assumptions concerning agency and textuality. Moreover, GPS and AR foreshadow distinctly posthuman media ecologies, which aim at creating and sustaining transductive relationships among humans, multimedia, and environments. This design imperative to bond users' perceptual and cognitive experiences with digital data that is linked to their spatiotemporal *here and now* stands in sharp contrast to the diminished capacity for context-awareness associated with mobile interface designs that valorize anytime-anywhere connectivity without regard for users' present surroundings (as demonstrated in the emblematic example of distracted driving). To conclude, I will indicate three relevant concepts from related scholarship to suggest grounds for further rhetorical and aesthetic considerations of contemporary and future techno-geographic interfaces.

First, Jane Bennett's writing on thing-power and the agency of assemblages provides a robust philosophical and political framework for thinking about the ways technological and geographical bodies come to affect and be affected by one another. Her new materialist account of human beings as a "*particularly rich and complex* collection of materials," which argues that humanity cannot be positioned as an "ontological center or hierarchical apex," supplies a rich vocabulary that could carry over to descriptions of textuality and the decidedly marginal status of human agency characteristic to techno-geographic interfaces (Bennett 2010, 11). Like Bennett, rhetoric scholar Laurie Gries draws on Bruno Latour's notion of actants in order to get beyond the human-centricity bound up in conventional theories of discursive production. Proposing the notion of "rhetorical actancy," Gries foregrounds the ways in which "humans and other actants work together to construct rhetoric" and concludes that "discourse is a vital, material force" (Gries 2012, 87–88). Bennett and Gries both highlight scenarios in which discourse is shaped by encounters with material entities, and Bennett celebrates texts that "direct sensory, linguistic, and imaginative attention toward a material vitality" (Bennett 2010, 19). As I have suggested above, AR textuality seems especially primed for this task; in fact, assigning compelling aesthetic and rhetorical actancy to vibrant matter in motion is perhaps the primary challenge for writers and designers working on AR projects.

Finally, in the spirit of Umberto Eco's still influential piece "The Poetics of the Open Work," writers and designers of AR must learn to regard layers as techno-geographic processes of production instead of finished.[11] In addition to being hermeneutically unbound, the literary and musical compositions that Eco cites as open works actually distribute the act of arrangement to the initiative of the people who perform them. Rather than spelling out a well-defined arrangement of notes to be reproduced exactly, open musical compositions are more like "a construction kit handed to the performer to assemble"

(Eco 1989, 4). We might understand media files in the scene of mobile AR on this basis: as units of construction to be assembled by the autonomous actions of each viewer-cinematographer relative to the real-time movement of geographical flows. Contrary to VR systems, AR layers are *not rooted* in intelligible essences of defined objects; rather, they are *(up)rooted* in the accidents of the sensible. The landscape (or any geographical flow) is always at play within the scene of mobile AR—not only as a general setting, but more notably, as a material, semiotic assemblage of actants occupying the textual field. Geographical flows *write themselves* in the screen-camera. This ongoing material composition is as much a part of AR layers as the programmed media content that overlays it: landscapes, buildings, weather, city streets, pedestrians and nonhuman bystanders—all geographical flows—participate (albeit unconsciously, at times) in the literal and figurative make up of AR textuality.

NOTES

1. Cf., Mark Hansen, "'Realtime Synthesis' and the Différance of the Body: Technocultural Studies in the Wake of Deconstruction," *Culture Machine* 6 (2004); N. Katherine Hayles. *How We Think: Digital Media and Contemporary Technogenesis*. (Chicago: University of Chicago Press, 2012).

2. Whereas Heidegger's "The Question Concerning Technology" valorizes the windmill over and against the hydroelectric plant, Simondon and Stiegler orient their respective philosophies of technology around turbines, electronics, and other (hyper)industrial machines.

3. Informed by Simondon, Stiegler routinely defines technics as "organized inorganic matter." The term refers both to the history of fabricated objects (e.g., flint, hammers, pencils, computers) and to domain of *techne*—the techniques and practices involved in making (something with) technology.

4. In the context of (pre-digital) literacy, one can distinguish between their perception of live events and their engagement with forms of tertiary retention; granted, one's engagement with tertiary artifacts impacts, even constitutes, the criteria (i.e., secondary retention) that directs his or her lived perceptions (i.e., primary retention). Nevertheless, this impact occurs over time—it is deferred—on account of the technics of alphabetic writing.

5. Stiegler does not make this distinction on the basis of a visual/verbal divide; rather, he emphasizes the different temporal relations constituted by literate writing and audiovisual media. Clearly, written phonetic alphabets are inconceivable without their auditory and visual qualities, and verbal expression underlies much audiovisual media.

6. To clarify, I am referring to the forth chapter of Propen's book *Locating Visual-Material Rhetorics* ("Navigating the Mediated, Posthuman Body"), and not to her chapter included in this collection.

7. For a more thorough discussion of the mobile Web, see John Tinnell, "All the World's a Link: The Global Theater of Mobile World Browsers," *Enculturation* 12 (2012).

8. In this section, I wish to discuss mobile AR on a general, theoretical level. Popular AR apps at the time of writing include Layar, Junaio, Argon, and Wikitudes. For a survey of recent AR projects created by artists and cultural institutions, readers can search the Web; I recommend starting with the website of the international artists collective known as ManifestAR. Cf. "Images," ManifestAR, accessed July 28, 2012, manifestarblog.wordpress.com/Images/.

9. In distinguishing between GPS and AR, one must note that these technologies have converged in a variety of ways recently. Most AR apps utilize the GPS functionality of mobile

devices. Moreover, at the time of writing, GPS navigators with AR windshield displays have started to hit the consumer market in several countries.

10. Perhaps glass inscribed with vinyl (e.g., signage on store window fronts, car decals, etc.) constitutes one of the only popular historical examples of a transparent writing space that maintains a view of the surroundings within its borders. Of course, glass writing spaces often lack the mobility and certainly lack the networked interactivity unique to the camera view of smartphones and tablets.

11. For an example of how Eco's work has been taken up in recent design research, see Deanna Herst. "Form Follows User," *Open Design Now*, accessed July 13, 2012, opendesignnow.org/index.php/case/form-follows-user-deanna-herst/.

REFERENCES

Bennett, Jane. *Vibrant Matter: A Political Ecology of Things*. Durham, NC: Duke University Press, 2010.

Derrida, Jacques. Translated by Alan Bass, "Freud and the Scene of Writing," *Writing and Difference*. Chicago: University of Chicago Press, 1978.

Amy Devitt, Mary Jo Reiff, and Anis Bawarshi. *Scenes of Writing: Strategies for Composing with Genres*. New York: Pearson/Longman, 2003.

Eco, Umberto. Translated by Anna Cancogni. *The Open Work*. Cambridge, MA: Harvard University Press, 1989.

Farman, Jason. *Mobile Interface Theory: Embodied Space and Locative Media*. New York: Routledge, 2012.

Gries, Laurie. "Agential Matters: Tumbleweed, Women-Pens, Citizens-Hope, and Rhetorical Actancy." *Ecology, Writing Theory, and New Media: Writing Ecology*. Edited by Sidney Dobrin. New York: Routledge, 2012.

Hayles, N. Katherine. *How We Think: Digital Media and Contemporary Technogenesis*. Chicago: University of Chicago Press, 2012.

———. "RFID: Human Agency and Meaning in Information-Intensive Environments." *Theory, Culture & Society* (March/May 2009) 26.2–3: 47–72.

Hansen, Mark. "'Realtime Synthesis' and the Différance of the Body: Technocultural Studies in the Wake of Deconstruction." *Culture Machine* (2004) 6.

Herst, Deanna. "Form Follows User." *Open Design Now*. Accessed July 13, 2012. opendesignnow.org/index.php/case/form-follows-user-deanna-herst/.

ManifestAR. "Images." Accessed July 28, 2012. manifestarblog.wordpress.com/Images/.

Mitchell, W. J. T. *Picture Theory: Essays on Verbal and Visual Representation*. Chicago: University of Chicago Press, 1995.

Pedersen, Isabel. "Dehumanization, Rhetoric, and the Design of Wearable Augmented Reality Interfaces." *Small Tech: The Culture of Digital Tools*. Edited by Byron Hawk, David M. Rieder, and Ollie O. Oviedo, 166–78. Minneapolis: Minnesota UP, 2008.

Propen, Amy. *Locating Visual-Material Rhetorics: The Map, the Mill, and the GPS*. Anderson, SC: Parlor Press, 2012.

Rueb, Teri. "Shifting Subjects in Locative Media." *Small Tech: The Culture of Digital Tools*. Edited by Byron Hawk, David M. Rieder, and Ollie O. Oviedo, 166–78. Minneapolis: Minnesota UP, 2008.

———. "Itinerant." Accessed June 28, 2012. www.turbulence.org/Works/itinerant/index.htm.

Simondon, Gilbert. Translated by Ninian Mellamphy. *On the Mode of Existence of Technical Objects*. London: University of Western Ontario, 1980 [1958].

Souza e Silva, Adriana and Daniel Sutko. "Theorizing Locative Technologies Through Philosophies of the Virtual." *Communication Theory* (2011) 21: 23–42.

Stiegler, Bernard. Translated by Mark Hansen. "Memory." *Critical Terms for Media Studies*, edited by W. J. T. Mitchell and Mark Hansen, 64–87. Chicago: University of Chicago Press, 2010.

———. Translated by Steven Barker. *Technics and Time, Vol. 2: Disorientation*. Stanford, CA: Stanford University Press, 2008.

———. Translated by Steven Barker. *Technics and Time, Vol. 3: Cinematic Time and the Question of Malaise*. Stanford, CA: Stanford University Press, 2010.

Tinnell, John. "All the World's a Link: The Global Theater of Mobile World Browsers." *Enculturation* (2012): 12.

Venn, Couze, Roy Boyne, John Phillips, and Ryan Bishop. "Technics, Media, Teleology: Interview with Bernard Stiegler." *Theory, Culture and Society* (December 2007) 24: 334–341.

Weiser, Mark. "The Computer for the Twenty-first Century." *Scientific American* (September 1991) 265.3: 94–104.

II

Artifact

LEGO. Siri. Upholstery. Self-trackers. If a civilization in the distant future unearthed these artifacts in an archaeological dig, what could they surmise about the civilization that made and uses them? Archaeology habitually employs a thing-centered approach to people in the past, relishing all that we can know about these invisible, intangible beings through the tools and ornaments they made, and that we can still hold in our hands. Is archaeology a viable entrée to the human experiences and sensibilities of our own time? The question is particularly poignant given how many of the artifacts of the present are electric and digital, or mechanically mediated; it feels strange to imagine these futuristic items as relics. Nevertheless, it stands to reason that some aspects of the human are so intimately entwined with the objects we've created and used as to have become permanently embedded in those things, and that we can thus know ourselves and others through the things that we've designed, we've used, *we are*. It also stands to reason that there are aspects of these things, whether machine-made or hand-crafted, that are so alien to the human, so phenomenologically distinct, that they entirely obscure their users and makers, render use irrelevant, even bend and shape their makers to their own designs.

An archaeological approach to things does something to our attention, refocusing it the way an archaeologist studying past civilizations is driven to find fascination in a fork, a fragment of vase or farming instrument, a piece of jewelry, the trash. In "Artifact," we study in myopic detail specific things and categories of things that might otherwise remain unstudied, simply assumed and absorbed into the background because they are so commonplace.

The chapters in this section concern themselves with opposing the effects and logic of trivialization, which render minutiae invisible even when they can be, through their pervasiveness, some of the most powerfully culture-forming and person-determining objects. Unlike aesthetic critical discourses, which seek out the exceptional, we subject ordinary things, and the actions and imaginings through play and craftwork of women and children at home, to microanalysis. Through eschewing the boundary between the trivial and significant to sift for latent significance in the trivial, we find ourselves concerned with toys, a tile and textile pattern, devices that count calories or heartbeats per minute, or an electronic "secretary" made to be carried in her users' purses, backpacks, and pockets.

The artifacts in this section, and the analyses undertaken of them, are all consumer goods. LEGO and Siri represent two of the most iconic brands of our era. Self-trackers are a category encompassing multiple specific brands: FitBit, BodyMedia, RunKeeper, and others. The tile and textile pattern herein examined is not branded, but it is reproduced and sold, often embedded in the design of luxury goods like James Whistler's Victorian designer dining room and Rei Kawakubo's 1997 *couture* dresses for *Comme des Garçons*. Contemporary artifacts are embedded in the corporations that make, market, and sell them; they are shaped by the *mythos* of their brands. Yet as all four authors argue, artifacts are also always something more than their market value. Something in each exceeds its "intended use": it's the part that can't be pinned down and controlled any more than the human minds, organs, and imaginations for which the artifacts are often extraordinary, improvisatory prosthetics.

In philosophical parlance, the artifacts shift from being the *objects of* individual and corporate use patterns into *things*, the inorganic equivalent of a *creature,* which is to say—they take on characteristics of the unnamed, unpackaged, instructionless version of their own designs. Here, like *creatures* in horror movies that are seemingly powered by nothing more than their shapes, *things* do their own thing. The artifacts here examined do something interesting to the word "design," putting design in its aesthetic sense into tension with design as agency or intent. We learn that there is an extent to which design is design: style is agency. Though what "agency" might mean when possessed by a thing is another question altogether, one that these papers all take on. Jane Bennett has renamed the phenomenon of object-agency "actancy." In "Artifact," we study the actancy that accompanies specific styles.

However, no author here disregards the role or impact of humans in artifact creation and use. Even when artifacts "take over" as things and assume actant-power, people are impacted. Some even seek out this impact, seeming to want to be affected by things in specific ways; they solicit the state of *being affected* by the effects of something outside the human self but also somehow invited into it. When humans enter the artifact-picture—when they appear within the still life, as part of the still life—they are often subjects in the sense of being subjected (to things' power, and their way of being). Thus in chapter 5, "The Plastic Art of LEGO: An Essay into Material Culture," Jonathan Rey Lee examines the constitutive characteristics of a toy so pervasive as to be most almost synonymous with childhood in the west, so much so that adults often use LEGO blocks as nostalgic instruments for returning to childlike experiences and perceptions through play. Thus Lee investigates the toy's atomism—the fact that users can divide every thing made by LEGO blocks into endlessly recombinable pieces. Then their plasticity: literally, the hard material they are made of and that all things made with them must conform to. It's hard to think of attributes more alien to the human body, for which recombination is butchery and mutilation, and which is soft tissue where LEGO is hard plastic. But as Lee observes, LEGO—which means "I assemble" in Latin—always insists on the presence of a human subject. The parts are "designed for design": made to invite people to pick them up and put them together, to place themselves in the throes of the thing they're holding and follow where it leads while also impressing new possibilities upon it.

Lee contends that the LEGO subject must be accessed through creative acts and as such is elevated above nature, whose status is to be the building block to humans' building acts. Yet every LEGO game is a "non-verbal essay into material culture" in which the creative subject is unable to construct her elegant designs through mastery of the intractable plasticity of the blocks but rather must assume a relationship with the medium, mold herself to their form and work with and within it. Thus, writes Lee, "LEGO play reveals a complex performativity that condenses theory into things, enabling us to critically deconstruct consumer culture, playfully encounter the material object-other, and constructively explore ecological subject-object relationships" (96).

Emily McArthur's "The iPhone *Erfahrung*: Siri, the Auditory Unconscious, and Walter Benjamin's 'Aura,'" in contrast, investigates an auditory essay into the material; whereas LEGO exists to explore the creative potential of touch, Siri is a sound. Apple's Siri, first released in 2011 as an application for the iPhone4S and now available on all iPhones and iPads, is a Natural Language Processing system that McArthur argues activates an "auditory unconscious" in the humans who use it. The auditory unconscious is McArthur's extrapolation on Walter Benjamin's optical unconscious, which the theorist proposed was able to see and therefore record and sometimes ultimately register, with the help of technologies, aspects of culture invisible to the naked human eye. Whereas Benjamin, whom McArthur dubs modernism's new media scholar, focused on visual technologies like film, McArthur focuses on auditory technologies. The most significant thing about Siri is that, and how, we hear her.

Key in Siri's embodiment of an auditory unconscious is her errors and "malfunctions": her awkwardly mechanical voice—designed, McArthur demonstrates, specifically not to sound too thoroughly human; her inability to recognize all proper nouns but the most banal; the questions that she can't answer or gender that "she" doesn't have. By recreating through the shock of a talking mechanically-reproduced object the aura that Benjamin hypothesized was lost in the age of mechanical reproduction, Siri is just far away enough from natural to stage technology's strangeness—to make an everyday product, and everyday actions performed with her, uncanny. Siri, like any new technology whose function is not yet perfected or to which users have not yet become anesthetized by familiarity or whose designers have seen fit to program inadequacy into its structure to make people more comfortable with it (as with Siri's version of NLP technology, intentionally hobbled to keep her from sounding too human), draws humans' collected attention precisely to spaces of imperfection and therefore non-integration. It is here that human attention is redirected onto itself, where we become reflective about our interactions with machines and how those interactions "direct the course of modern life" in ways that we still have the option to choose or refuse.

Sometimes, the options we choose *not* to refuse are surprising. In "Victorian Cybernetics: Networking Technology, Disability, and Interior Design," Colbey Emmerson Reid analyzes the history and contexts of emancipatory incidences of dehumanization. Through her recovery of an ancient Roman tile pattern unearthed in an actual archaeological dig in the 1400s, called the *grottesche*, Reid likewise recovers a perspective through which human objectification was considered preferable to subjectivity. The *grottesche* pattern is characterized by the hybridization of human, plant and animal forms, which are often mutated and aborted in the design process of joining inimical shapes together. Central to the strange preference for dehumanization is familiarity with the *grottesche* pattern in the context of interior design, which is often mistaken for a grotesque form in the context of biology. Indeed, whereas grotesque shapes adhere to an organic logic according to which mutations, aberrations, and hybridizations are signs of decomposition and decay, the *grottesche* pattern adheres to the inorganic logic of design that perceives such shapes as stylish. What one framework finds disgusting the other considers *chic.*

The *grottesche* pattern, Reid argues, was far more accessible in the nineteenth-century than it is today, in part because of the prominence—arguably the dominance—of the biological paradigm. In contrast, design was the more accessible paradigm in the nineteenth-century; even biologists were inspired by it. It is in the context of design's cultural prominence that Reid claims

Victorian interior decoration was the original cybernetics; indeed, the term "cybernetic" was coined in 1819 to describe human-object combinations that were then primarily ornamental. Tiffany's broaches, made out of pieces of the first transatlantic telegraph wire, playfully positioned the humans who decorated themselves thusly as, in the medieval sense that means "ornamented," the first *gewired* people.

These days it seems we're all *gewired*—not least in moments of self- and interior decoration. A key demographic, however, can benefit from the repositioning of deformity as style: the physically, mentally, and emotionally disabled. For an example, Reid reaches back to the nineteenth-century, in an analysis of the Royal Surgeon Sir Frederick Treves' memoire on his most notorious patient, James Merrick, the "Elephant Man." Merrick, Treves contends, was able to reposition his severe and chronic congenital disabilities by considering them in the context of furniture and textile patterns, essentially "reupholstering" his identity and body as a mode of self-normalization and social integration. By considering himself and inviting others to consider him one among the many *grottesche*-style *curios* that populated nineteenth-century *boudoirs* and *salons*, Merrick became a Victorian celebrity and Don Juan, known and appreciated and desired for his extraordinary shape. Contemporary humans, mired in the *ethos* of the biological grotesque, have much to learn from the "cybernetic" Victorians who pioneered emancipatory experiments in human-thing combination.

Yoni Van Den Eede's "Extending 'Extension': A Reappraisal of the Technology-as-Extension Idea through the Case of Self-Tracking Technologies" concerns itself less with combinations than extensions of the human into the object-world. The chapter takes up a variety of brands of self-tracking devices, made to measure and record such interior biological processes as blood pressure, calorie expenditure, calorie consumption, heart beats per minute, steps taken, heat emitted, and any other facet of the body that can be statistically or algorithmically rendered. Using Marshall McLuhan's rendition of the extended mind idea, Van Den Eede investigates the form of metaphorical elaboration on the human body that self-tracking devices enact. In what ways do the devices materialize invisible aspects of human organs? Are self-trackers little pieces of our hearts, our blood, our energy—reformed through quantification? What internal tracking habits do we abjure—amputate, so to speak—when we offload them onto our devices?

The utility of McLuhan's extension idea, Van Den Eede claims, is that it refocuses human attention on technologies' connections to their users. Whereas we tend, under the thrall of a "Narcissus Narcosis" that drives us to forget the parts of ourselves that we extend into the world were ever connected to us, the extension idea reminds us that every technology prostheticizes some human function that fundamentally exceeded the confines of the original organism. The extension idea insists that we retrace lost connections, rediscover the human body's reflection in every inorganic, plastic, perfected or broken object we collectively possess. Retracing is the first step to inquiring further: what has been extended, what are we extending ourselves for, and how have we extended ourselves? What, that is, are the concrete effects of technologies on our selves, our lives, the world?

The notion that every artifact is, as McLuhan suggests, a translation of the human user into another form brings us back to the archaeological construct whereby the important parts of people, as individuals and as cultures, all eventually become things, and the counterpart of this idea: that things, even at their most opaque and distanced from human makers and users, are parts of people. *Are* people. Even when bought and sold, lost and thrown away, sometimes they are the most important parts. And when things seem most disregarded by or even severed from the human? That moment at which they are most immersed within what the videogame designer Ian Bogost has called their "alien phenomenologies"? Perhaps it's here that we are most connected to them, where our relations to and mediations by them have been most carefully mystified—here where we've self-amputated so we don't have to acknowledge some quite intimate interior as our own.

Chapter Five

The Plastic Art of LEGO

An Essay into Material Culture

Jonathan Rey Lee

Sometimes it can seem like we live in a LEGO® world, a peculiar ecology constructed from and populated with designed, assembled, modular objects. Flaunting their artificiality with unnatural colors and that distinctive plastic sheen, LEGO[1] worlds consist of constantly evolving assemblages of prefabricated parts that reflect modern spaces in the perpetual process of construction and reconstruction.[2] At the heart of LEGO media practice is a set of object-oriented sensory experiences that play with the constructedness of the world. LEGO does not merely provide an *idea* of the material world, but a tangible experience—the familiar shapes, colors, and textures, the comforting "snap" when the pieces click together, and even the pain of trying to pry apart bricks with fingers and teeth all develop a profound physical relationship between player and material. Toys matter for many reasons, not the least of which is that toys—tangible, manipulable, child-scaled objects—play with objects and object-relationships. Among toys, LEGO, which is marketed as an educational toy and touted for its role in children's spatial and kinesthetic development, is particularly suited to such object play. Unlike most toys, however, LEGO is also a medium—an open system of object play—that provides a particularly complex expression of material culture. This medium-specific[3] analysis therefore traces how the specific characteristics of the LEGO medium offer unique critical possibilities for exploring material relationships.

Living in any material culture entails complex and evolving interrelationships with things; flourishing requires somewhat more. In our often materialistic[4] Western culture, relationships between subjects and objects can be threatened by a self-perpetuating cycle of production, distribution, possession, use, consumption, and disposal of objects. Rather than propose any ideal system that could realistically replace this broken one, this chapter attempts to interject a sense of concern into the ongoing object-oriented practices of material culture. I suggest that questioning—not mastery—ought to be the attitude through which material relationships with and within the world are composed. To this end, like a LEGO instruction manual read from back to front, this chapter consists of several deconstructive stages that delve into the medium and media phenomenon of LEGO. At once brick and plastic, LEGO embodies two distinct yet related underlying logics of material culture: *atomism*, which conceives of and treats objects as assemblages, and *plasticity*, a mythos of unbounded creativity and mastery over objects. By expressing these logics in a particularly tangible way, moreover, this medium of play opens unique critical possibilities. In play, LEGO engages the basic int(e)ractability of matter as a site for encountering materiality and rebuilding subject-object relationships. The hope that such play can promote more ecological relationships can but be suggested; this chapter simply offers an exploration of the material philosophy of LEGO as something to build upon.

THE LEGO MEDIUM

For such an iconic substance, LEGO is surprisingly difficult to define, not because it cannot be characterized at all, but because it can be characterized in so many ways: object, toy, brand, company, system. From its earliest moments, LEGO was designed and marketed as more than a toy—introduced as a "System of Play," LEGO employed plastic molding technology and an innovative "stud-and-tube" design to produce interconnecting blocks that rigidly hold together yet can be endlessly assembled, disassembled, and reassembled anew. Thus, "LEGO" names a constellation of possibilities and practices that revolve around sets of objects more than it names the objects themselves. In fact, the very word "LEGO," coined in 1932 by a Danish toymaker from the phrase *leg godt*—or *play well*—indicates a philosophy rather than an object of play. Even now, both the company and the adult fans of LEGO (known as AFOLs) resist naming LEGO as an object—calling LEGO bricks "legos" remains a *faux pas*—and instead apply the name adjectivally, as in "the LEGO Group," or "LEGO bricks." Perhaps one reason for this anti-nominal heritage is the intuition that reducing LEGO to its physical bricks misses the point—more than a set of objects, LEGO is a system, activity, and culture of play. LEGO is above all a *medium*, a material possibility space for construction play designed to provide "unlimited play potential," the first of ten principles that specified its play philosophy (Wiencek 1987, 48).[5] As the medium itself is both the direct object of LEGO play and its primary product, "the medium is the message"[6] in an unusually strong sense for LEGO. At the same time, any medium-specific analysis of LEGO is complicated by its inherently dual status as both medium and *media phenomenon*, a branded cultural practice tied to the circulation of retail sets. As a media phenomenon, the medium of LEGO can never be fully extricated from the cultural contexts in which it gains significance. Yet, as a medium, LEGO is a system of possibilities whose significance always extends beyond its particular manifestations.

As a medium and a media phenomenon, LEGO is unlike most other artistic media—print, film, painting, sculpture, and so forth—in that it is closely aligned with a single company and brand (although other comparable products do exist).[7] LEGO identifies itself as a media phenomenon in a particularly elemental way: whereas a medium like paint wears its brand only on its container, the LEGO brand is literally etched on every stud. The material possibilities of LEGO are therefore always mediated by a branded play culture expressed by advertising rhetoric, product art, and explicit instructions that all interpret the medium before it is initially encountered. A retail LEGO set arrives as a set of elements (the LEGO pieces) circumscribed by a normative representation of play that shows—and thereby instructs—a single way to build and play with the set.[8] The packaging and instruction manual present only one designed and desirable play script for the countless combinations made possible by the elements: these 152 pieces make a pizzeria, those 767 make a castle. This proposed end even predetermines the selection and design of the included elements, many of which are created for specific sets. While a LEGO wheel can perform many functions within the medium, for example, its representational content encourages its use as a wheel. Thus, while a retail LEGO set enables countless play possibilities, the explicit play script presents only one.

This normalizing play script is, of course, not intended to be restrictive, but rather to thoroughly scaffold LEGO play to make it accessible and desirable to its target markets. By presenting both ideas for and images of an idealized play space, these play scripts contribute to the commercial success of the media phenomenon while simultaneously miring an otherwise more open medium deeper into consumer culture. LEGO is particularly implicated in a culture of consumption because its expressive potential depends so heavily on the amount of LEGO one possesses. That additional elements exponentially increase the play possibilities of the LEGO system provides a built-in imperative to collect more sets, as reflected in how the eighth of LEGO's ten guiding principles—"The more toys, the greater their value" leads to the ninth: "Extra sets available" (Wiencek 1987, 48). The hoarding impulse implicit in this imperative is particularly noticeable in LEGO, as journalist and participant observer Jonathan Bender describes: "That need to acquire more pieces is easy to understand. A LEGO collection is quantifiable. You can physically see how many bricks you own" (19). This drive toward accumulation is enhanced by theme sets, which group individual sets within larger, cohesive story worlds such as *Star Wars* or City. LEGO thus markets additional sets as natural extensions of an existing collection—an X-Wing generates desire for a TIE Fighter just as a police vehicle generates desire for a police station.

Despite the ways the LEGO media phenomenon circumscribes LEGO play with imperatives to consume, LEGO differs from most toys in that this closure does not simply preclude openness. Instead, LEGO encourages deviations from its provided scripts by its very nature as a participatory, interactive, recombinable medium. Thus, not only is it always possible to build off-script, it is also always possible to think "outside the box" by attending to the medium itself, as Bender (2010) describes: "I still read through the instructions, looking at building techniques and the elements mix, but I'm now looking at sets as raw materials rather than as finished products" (219). At once media phenomenon and medium, object and system, representational and abstract, product and process, simple and complex, LEGO resists definition by embodying seemingly irreconcilable opposites. In the same way, LEGO play straddles a dichotomy that French philosopher Roland Barthes (2000) identifies as the opposition between rare inventive toys, such as "a few sets of blocks, which appeal to the spirit of do-it-yourself," and common representational toys featuring a mimetic play that remains "entirely socialized, constituted by the myths or the techniques of modern adult life" (53). At once an inventive, abstract, open medium and a branded, consumerist, and socially-ritualized media phenomenon, LEGO provides an expansive play space for exploring many complexities of our material culture.

To develop a framework for understanding all these complexities warrants not just a more comprehensive study, but an entirely new field. As both media phenomenon and medium, LEGO plays out in constant tension; following the provided building instructions and free building are as different as—and parallel to—reading and writing. In the former, one retraces already trodden steps to unravel a narrative of developing ideas; in the latter, one struggles with the medium in constant pursuit of creative expression. The play scripts of LEGO retail sets are a subject for another time; this chapter emphasizes free building with basic bricks because LEGO's potential as a medium for material philosophy requires looking beyond its branded surface. In what follows, therefore, I shall explore possibilities for interrogating subject-object relationships through how—at once brick and plastic—LEGO combines the rigidity of a discrete, digital, and atomistic bricks with the continuity and fluidity of plastics.

ATOMISM

As the scientific gaze continues to fixate on the building blocks of the material world, we become more and more implicated in an *atomistic* mentality, a way of perceiving, conceiving, and treating the material world—and the things it contains—as assemblages, complexes, or compositions.[9] Materially both one and many, the LEGO System is predicated on a reflexive relationship between whole and part characteristic of atomism. Paradoxically, this underlying atomistic substructure is etched across the surface of the medium, whose seams remain visible as a network of hairline fractures that testify to its bricklike construction. In this way, LEGO materializes an otherwise theoretical conception of matter, providing a uniquely tangible, performative exploration of atomistic logic that invites hands-on material philosophy.

This investigation rests on a close parallelism between the surface experiences of LEGO and the theoretical content of atomism. Democritus, perhaps the most influential early Greek atomist, speculated that the material world could be entirely constructed from small, literally "un-cut-able" *atoms* of varied shapes. Similarly, the LEGO system is populated by a wide range of uniquely shaped, interlocking *elements* that can be combined to form unlimited unique configurations: just six classic 2x4 rectangular bricks of the same color can be combined in 915,103,765 ways (LEGO Group 2012, 23). Not only do LEGO and atomism share this defining structural principle—that objects of diverse form are composed of different configurations of smaller particles—but LEGO also expresses the four major features of Democritean atomism: indivisibles, vacuum, reductionism, and mechanism (Pyle 1995, xi):

1. *Indivisibles.* LEGO elements are distinct enough that they are "conceptually indivisible (i.e., such that one cannot conceive of their division)" and difficult enough to break that they are for all practical purposes indivisible, un-cut-able, singular particles.
2. *Vacuum.* Second, while the empty space of LEGO is typically air rather than vacuum, LEGO construction is predicated upon the relationships between presence and absence. This is no incidental relationship; beyond the relative non-presence of air, the Cartesian grid[10] built into the LEGO system and exemplified by the flat building surfaces known as "baseplates" makes the builder particularly aware of how LEGO elements occupy "purely empty space in which the atoms are free to move."
3. *Reductionism.* LEGO exhibits reductionism, an "explanation of the coming-to-be, ceasing-to-be and qualitative alteration of sensible bodies in terms of the local motion of atoms which lack many (most) of the sensible properties of those bodies" in that LEGO objects owe

their characteristics to the sum of the characteristics provided by the elements themselves.

4. *Mechanism*. LEGO displays mechanistic tendencies in which "no body is ever moved except by an external impulse from another, moving body" (Pyle 1995, xi) in that LEGO pieces are moved only by external impulses, namely the actions of a builder and gravity.[11]

Whereas the atomistic composition of most physical objects is rarely accessible, as their *post facto* atomism cannot be undone without destroying the object, the LEGO system visibly and tangibly reflects the logic of classical atomism, opening atomistic principles to both intervention and play.

While classical atomism has given way to newer models in atomic physics, an atomistic paradigm for conceiving of complex entities in terms of their component parts continues to pervade Western culture. The atomism of LEGO, therefore, draws upon the logic of classical atomism, not to anchor a new physics, but to explore and perhaps critique the underlying ethos of material culture. LEGO "materializes" atomism by making an otherwise abstract physical theory accessible, providing a way of playing with atomism that differs from and productively supplements linguistic descriptions. That LEGO presents atomism in a strikingly visual way can be seen clearly in brick artist Nathan Sawaya's *Yellow*, perhaps the world's most iconic LEGO artwork. This work (fig. 5.1), one of several in which Sawaya represents human forms dissolving into (or being constructed from) their component parts, employs only standard yellow bricks and plates to depict a human torso holding open a gaping hole at its center, from which spills a flow of loose bricks. However else this can be read—as a commentary on the constructedness or delicacy of life, or an exploration of the emotions of a fractured, deconstructed self, for example—I argue that it must also be read as a reflection on the medium itself. This figure, constructed from the most basic LEGO elements, reveals what he is made of—his innermost being *is* the atoms of which he is constructed. This uncanny exploration of atomistic logic interrogates the relationships between depth and surface, essence and appearance. Not only is this questioning about the medium, it deploys the medium to invite us to explore our own constructedness. As the expressionless face reflects back the viewers' own emotional states, this artwork—itself an object in the form of a subject formed of objects—encourages reflection on the relationship between subjects and objects.

While this questioning is rendered particularly poignant by the expressiveness of Sawaya's build, everyday engagements with the medium also enable such reflection. Such introspective art is possible because "LEGO carries with it a cultural currency that makes it intriguing," as Baichtal and Meno (2011) argue, citing an interview with Sawaya in which he states:

Figure 5.1. *Yellow*, Nathan Sawaya. Photo courtesy of brickartist.com.

> LEGO is something that I enjoy working with because people relate to my sculptures in some way. Most folks don't have large slabs of marble in their homes, but they have played with LEGO bricks at some point. They can relate to this toy and are amazed by what can be done with it. (154).

Designed to be conceptually and physically graspable, LEGO makes atomistic experiences accessible to all. Quite literally to be played with, this iconic medium has permeated cultural discourse as a common metaphor for (and metonym of) atomism. Atoms have long been described as "building blocks," and LEGO is now a powerful image of atomism in the artistic, popular,[12] and scholarly[13] imagination. Thus, LEGO represents and performs atomism in a particularly prescient way that enables the atomistic logic of modern material culture to be experienced and explored.

As a modern atomistic medium, LEGO can interject a self-reflective moment into a material culture particularly concerned with its own constructedness. Whether it reflects the atomistic foundation of physical matter, the increasing modularity of the urban landscape, or the feeling of constructedness reflected in Sawaya's art, the medium of LEGO can express and explore the sense that complex entities consist as complexes, assemblages of component parts. At the same time, the atomism of the LEGO brick does not always or exclusively present a sense of fractured, modular construction—it can, paradoxically, also represent the fluidity of modern plasticity. Similar to the surprising unity of particle and wave in modern physics, for example, *Yellow* presents a striking image of fluidity derived from an assemblage of rectangular bricks as both the flow of loose bricks and the curved yet pixelated surface of the figure express both atomism and plasticity at the heart of the medium.

PLASTICITY

LEGO is a quintessential modern phenomenon, not just because of how and when it emerged, but also because it so clearly materializes a modern ethos, playing out a preoccupation with its own constructedness. In particular, not only does LEGO perform atomism, it appeals to a modern fascination with *plasticity*, a mythos of shaping and reshaping the world symbolized by the infinitely malleable synthetic substance of plastic. Roland Barthes (2000) describes plastic as a "quick-change artistry" that produces "proliferating forms of matter" (53), capitalizing on an age-old alchemical desire to transform one substance into another. A single substance that can take on any form, plastic is a simulacrum that exemplifies the art and artifice of modernity. Thus, the medium of LEGO explores the constructedness of the material world through a central play experience simultaneously about building worlds and playing with and within those worlds. The medium of LEGO also reveals a basic fluidity, an experience of open, creative play that is particular suited to an exploration of plasticity, itself fluid and difficult to define. In expressing plasticity, LEGO participates in a material history closely aligned with modern consumer culture. Abolishing a hierarchy that requires different materials for different purposes, the fluid substances of plastic and LEGO can represent practically *anything*, offering a promise of mastery that privileges subjects over objects as their creators and owners. This representation, however, also enables a deconstruction of LEGO's plastic surface, as LEGO can unlock critical play possibilities that interrogate this mythos of plasticity.

The plasticity of LEGO can be seen in both its material history and the character of the LEGO play experience. Certainly, LEGO has embraced the aesthetic and functional capacities of plastics as, according to play scholar Maaike Lauwaert (2009), "The relatively cheap, easy to clean, durable, colorful and standardized plastic LEGO bricks epitomized the postwar vogue of indoor plastic products" (128).[14] In fact, the material history of LEGO reads like a parable of modern plasticity supplanting traditional craftsmanship: what began with carpenter Ole Kirk Christiansen making wooden toys in a small Danish town in 1932 became, less than three decades later, a major multinational corporation. This change was remarkably rapid—in 1947, LEGO acquired an expensive molding machine and "within two years, the LEGO Group's top three products were made of plastic: a fish, a sailor, and a stackable block Christiansen called the *automatic binding brick*" (Baichtal and Meno 2011, 8). But if plastic manufacture were all there is to LEGO, it would still be synonymous with fish and sailors. The plasticity of LEGO, like the brand stamped on every LEGO stud, is imprinted on the material itself through a visibly and tangibly plastic phenomenal presence that characterizes the LEGO play experience. While a fish or sailor may serve as a powerful image of plasticity, its molded form is fixed and static[15]—LEGO, in contrast, places design and construction in the hands of the builder, enabling a playful experience of plasticity within a self-contained creative environment. Thus, LEGO embodies the promise of plasticity in a way that plastics cannot: through the tangible, affective, visceral *feeling* of reshaping the material world with one's own hands.

Plasticity begins with a promise of mastery, an offer of unlimited creative expression. Similarly, LEGO's selling point is the level of creative control it offers over all stages of an object's lifecycle: design, construction, use. The idea that "For millions of people around the world LEGO bricks have a common meaning: creativity" (Bedford 2013, 1) is encouraged by the LEGO company, whose former slogan was "Just Imagine . . ." (LEGO Group 2012b) Certainly LEGO, "found in more than 75 percent of Western homes," (Baichtal and Meno 2011, 3) is a medium for the masses (as plastic was and is often considered to be). The necessary knowledge and skill for LEGO play is not preserved—like alchemical knowledge—as arcana; it is precisely calibrated to be accessible to anyone.[16] As Jonathan Bender (2010) puts it, "This is one of the main reasons that LEGO is so appealing. A novice can stack bricks alongside the professionals and find acceptance" (49). At the same time, one ought to question or at least qualify the democratic potential of any medium as heavily commodified as LEGO, which is, after all, a brand and product—and an expensive one at that.[17] Thus, the LEGO Group strives to democratize plastic's already populist and creative tendencies because it is literally invested in the results. Although the promise of fluid creativity purports to serve the self, it ultimately ensnares subjects in a cycle of consumption based on perpetual lack.

A related consequence of this promise of mastery is an anthropocentric tendency to elevate subjects over objects. While all media are anthropocentric insofar as they are designed for human perception, the participatory nature of LEGO privileges the subject to an unusual degree. Unlike film, painting, or sculpture, LEGO is more commonly accessed through creation than reception—the medium is always ready and able to be manipulated. Providing possibilities for interaction more than fully formed objects, LEGO privileges the subject through an emphasis in the individual player as creator, designer, and builder. This subjective focus is illustrated by the ironic discovery, made well after the company was named and formed, that *lego* means "I assemble" in Latin (Wiencek 1987, 44), creating a revisionist etymology that uncannily describes its atomistic structure. This alternate meaning reveals the first-person connotation of "LEGO"—the implicit "I" who initiates and controls the process of assembly. The anthropocentric mythos of plasticity likewise foregrounds the "I" in material production, privileging man-made over natural substances. Reflecting modernity's long flirtation with artificiality, this anthropocentric substance privileges creativity's figurative sense as an aesthetic, innovative, imaginative act over its literal sense as the power of reshaping matter. Unlike the elusive promise of alchemy, LEGO and plastics transmute form but not substance, offering power over representation more than over reality.

In so doing, the modern substances of plastic and LEGO elevate humanity above and beyond nature, disrupting ecological subject-object relationships. The plastic promise of the brightly-colored, straight-edged LEGO world validates a power imbalance in which the material world conforms exactly to the will of the subject.[18] This artificial plastic environment, designed to be designed, is compelling to adults precisely because it provides a simplified, accessible, childlike material reality. LEGO is almost infantile in how easily it fulfills desires, as Bender (2010) explains:

> When you become an adult, there is nobody older to tell you that you have to stop. You must be your own voice of reason, offering up a counter-voice—there's work tomorrow or chores to do now. And it's easy to see the voice of reason get drowned out, because it's never offering a fun alternative. That is how you end up with a house full of LEGO, because it's nice to feel like a kid for a while, indulging your wants in place of your needs. (103–4)

As a reaction against the alienating tendencies of an adult work culture, this touch of childhood can be incredibly generative. At the same time, LEGO's utopian promise can support a childish, self-centered misconception of the world as a passive plaything for indulging materialistic desires. Whereas plastic, artificial worlds like LEGO are quite literally made for our whims, reality is not so domesticated. To encounter and develop real, reciprocal relationships with objects, therefore, we must deconstruct this surface promise and engage the medium as a site for critical play.

CRITICAL PLAY

This essay into material culture proposes critical play as a method for mediating a material encounter upon which more ethical relationships between subjects and objects can be built. Reflecting and expressing material culture within a specialized play context,[19] LEGO provides a tangible play space for encountering materiality. In particular, the plastic art of LEGO is the material expression of what I term *int(e)ractability*: how matter is interact-able (fluid, able to be manipulated) precisely because it is intractable (atomic, resisting manipulation). Allowing limitless play possibilities while remaining fundamentally unyielding, LEGO is a medium of material int(e)ractability that simultaneously brings players into contact with the intractable hardness and inter-actable possibilities of objects. In this encounter, several responses are possible. One such response is to strive for total mastery, seeking to thoroughly domesticate the intractability of objects by completely dissociating them into constituent atoms and subsequent re-forming them into plastic molds. This consumption-driven focus de-means the holistic, contextual significances of objects into *mere* resources, undermining their genuine value (how they interact with, respond to, and enhance human agency).[20] In contrast to this problematic response, however, play may provide a viable alternative, a method for exploring and rebuilding the broken relationship between subjects and objects. I believe playfulness is vital because it is self-ironizing: play persistently questions its own performance and, perhaps, undermines the seriousness of the logics it expresses. The joy of play can demonstrate how the value of objects lies in the experiences—whether trivial or profound—they enable. Thus, read as a critical event, LEGO play reveals a complex performativity that condenses theory into thing, enabling us to critically deconstruct consumer culture, playfully encounter the material object-other, and constructively explore ecological subject-object relationships.

Despite not being in the medium *per se*, the media phenomenon of LEGO's series of adventure video games (such as LEGO *Harry Potter*, LEGO *Star Wars*, etc.) accentuates the extent to which LEGO is already driven by an attitude of consumption. In particular, these games represent the medium according to a modern worldview which, according to philosopher Martin Heidegger (1977), conceives the material world as nothing more than "standing reserve": a repository of resources, facts, and commodities that subjects are impelled to accumulate. Providing all the interactive possibilities in these games, LEGO elements represent the interface between the virtual world and minifig characters, who typically move through various scenarios in quests to collect certain items. Ironically, that these games are predicated more on accumulation than building presents the open medium of LEGO more as commodity than building material—free building is eliminated, moreover, in that the rare cases of building are entirely automated. To accumulate the basic currency of these games, one smashes LEGO objects, which explode into showers of 1x1 round "studs" (one loses studs in similar fashion when a character "dies"). In this way, the game world links atomism to destructive gameplay, encouraging players to treat the world as merely to be exploited, as no fixed number of studs are needed to win. Furthermore, reducing LEGO elements to their studs misconstrues the significance of the medium by dissociating the feature that enables LEGO's remarkable interconnectivity—the studs—from the pieces they are supposed to connect. This reductive symbology represents value through symbols which are valueless in themselves, much like virtual "coin" in video game economies or paper money in real-world economies. Representing the atomistic structure of LEGO in purely quantifiable terms, these games encourage violent dissociation of the intractable object-other by presenting plastic construction as necessarily predicated upon atomistic deconstruction.

In stark contrast to this cycle of consumption, ecological LEGO play pursues a reciprocal relationship with the medium as an end in itself. The critical potential of LEGO depends on this moment of play, the basic material encounter with the medium that delves beneath the branded surface of the media phenomenon. Genuinely innovative design in LEGO requires a complex negotiation that works *with* the rigidity of the bricks. Such creativity requires an attitude of receptivity that transcends or even values the fundamental frustrations of the medium. To love LEGO is to consider its stubborn atomistic hardness (when pieces refuse to pry apart, when the sharp corners cut your fingers, etc.) as its strength, and to find creative possibilities in moments where the promise of plastic fluidity seems a sham (when the pieces you want do not exist, when your designs drastically exceed your budget, when the Cartesian grid makes certain configurations impossible, etc.). Thus, between atomistic hardness and plastic fluidity, the creative potential of the medium relies upon a dynamic, reciprocal subject-object relationship, a state of continual discovery and development. Close, ethical relationships require such reciprocity, exemplified in the moment of *touch*, the basic physical encounter where subjects and objects inter-act, the point of mutual becoming where bodies encounter and affect each other.

In this material encounter, subject and object become united in a reciprocal ecology where play unfolds as a complex negotiation between the creative expression of the builder and the specificity of the medium. For children and adults alike, this play may manifest as an intuitive kinesthetic understanding of the medium supplemented by emotional and imaginative connections that invest meaning into the build. At its best, LEGO building becomes a kind of material poetry that gains its expressiveness through the artist-poet's inspired reinterpretation of the expressive potential latent in the medium. When there is no word for the image or feeling you want to express, for example, poetry can convey it metaphorically and stylistically by creatively redeploying everyday meanings. Similarly, the best LEGO builds express images and ideas by working with and accentuating the specificity of the medium. As builder Carl Merriam, who sculpts in media ranging from LEGO to moldable plastics, explains in an online interview, "I pretty much tailor the sculpture to the medium" (Hanlon and Kay 2013). Merriam repurposes LEGO levers and wheels to beautifully represent an ordinary hairbrush, for example, and uses magnifying glasses originally intended as minifig accessories to create a working, life-sized compound microscope (fig. 5.2 and fig. 5.3). Although such mimetic art certainly exhibits LEGO's plasticity, such elegant designs require a high degree of reciprocity between builder and medium, indicating that truly expressive building entails not mastery *over* the medium but rather a relationship *with* the medium.

Whereas a resource is something one accumulates, a toy is something one plays *with*; in play, subjects and objects interact. While the emotional bonds children may develop with their toys is not likely reciprocated, there is pro-

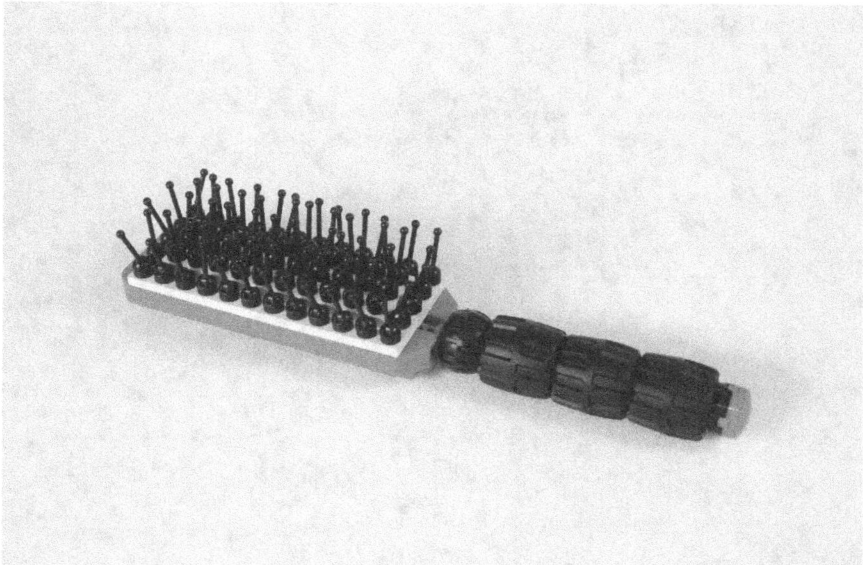

Figure 5.2. Hairbrush, Carl Merriam. Photo courtesy of Carl Merriam, bricklaboratory.com.

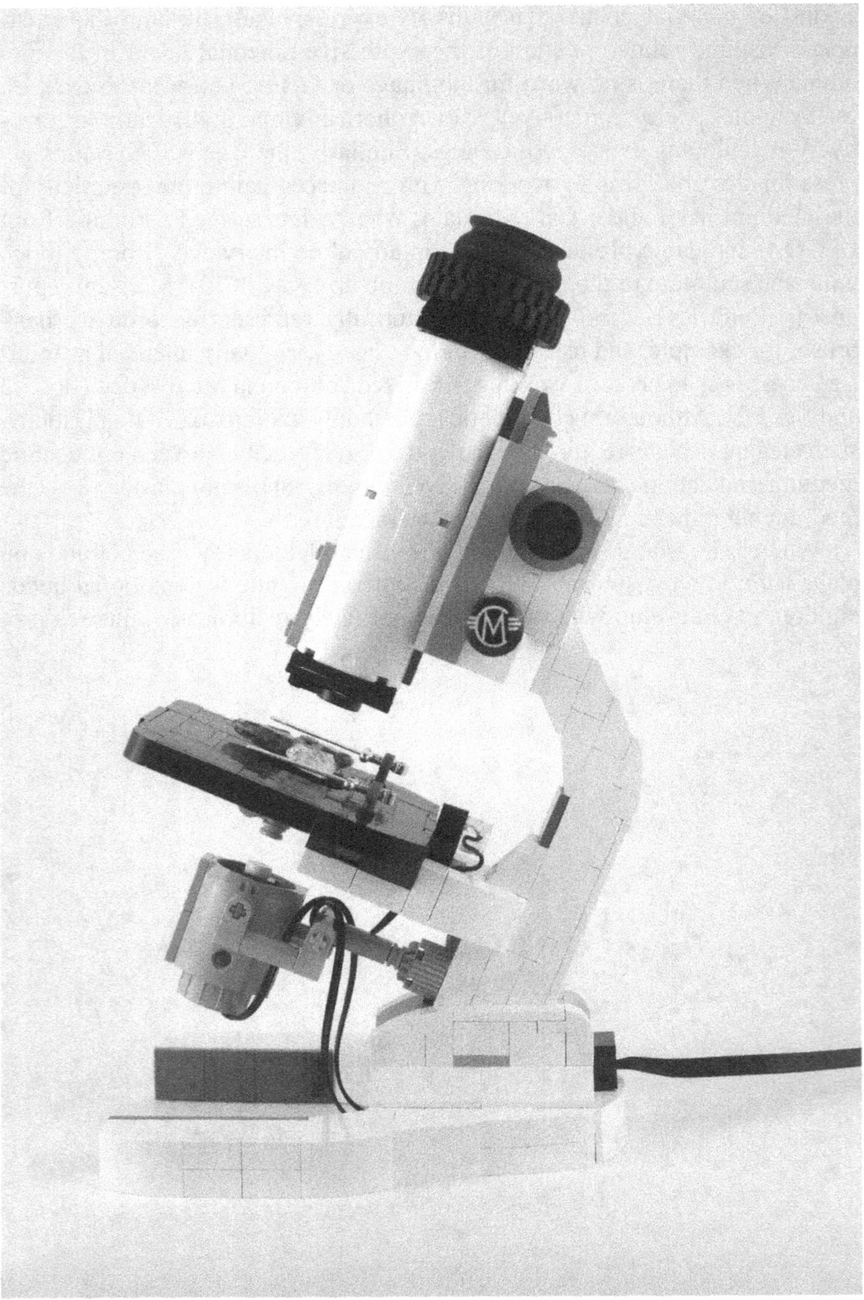

Figure 5.3. Microscope Mk II, Carl Merriam. Photo courtesy of Carl Merriam, bricklaboratory.com.

found potential in childlike attitudes that open the self to the presence of the object as other. Just as hoarding devalues toys by reducing objects of play to items in a cache, a culture of consumption reduces the value of objects to their market price or exploitable properties. In contrast, to see and care for objects as loci of interactive potential integrates them into the significances that characterize our lives, allowing objects to enrich and empower everyday activity. Whereas the cycle of consumption reduces both subjects and objects to mere means, play elevates subjects and objects to interlocutors in an ongoing, evolving relationship as, according to Maaike Lauwaert (2008), "The core of construction play lies in the process, the building and constructing of something rather than the product" (227). As process rather than product, critical play promotes a constructive ecology that recognizes the interdependence of subjects and objects, benefitting both by privileging the subject-object *relationship*.

CODA

Approaching objects as active participants in ongoing creative relationships rather than passive substances for fulfilling materialist desires reveals the possibility of *vitality* in subject-object relationships. Material philosopher Jane Bennett (2010) describes this approach as recognizing how "bodies enhance their power *in* or *as a heterogeneous assemblage*" (23), that is, how human activity is not independent from, but rather invested in relationships with objects and the environment. Seen in this light, atomism need not be reductive, violently dissociating the world into component parts, but can instead reveal the complex interrelations that characterize the material world as a network of interacting parts and wholes. Likewise, plasticity need not promote mastery, but can become an expression of the fluidity that characterizes ecological, perhaps even sustainable, subject-object relationships. At once brick and plastic, the heterogeneous assemblage of LEGO mediates possibilities for vital material relationships by unifying substance and form in a way that demonstrates the genuine value of material objects.

In the moment of material encounter, through the closeness of touch, any notion of a radical break between subject and object fades under a tangible sense of mutual interdependence. As builder and medium meet in the space of play, the basic pleasure of LEGO play lies in the simultaneous acceptance and transcendence of limits—while Bedford (2013) points out that "Learning to build within boundaries is an excellent way to improve your skills, hone your artistic eye, and maximize the enjoyment you get from your existing collection of LEGO bricks" (129), Baichtal and Meno (2011) also note that "Part of the joy of building with LEGO, of course, is to overcome the limitations of the medium" (99). This suggests that, rather than a self-contained outpouring of artistic genius, creative design invites an intimate and interdependent relationship with the medium. Approaching otherwise intractable objects through critical play therefore reveals the profound interactive potential that objects offer, not merely to be used up in the pursuit of consumerist desires, but to revolutionize our material relationships.

In a complex ecosystem of subjects and objects, the questions concerning material culture—how to flourish as who and what we are—are inseparable from that of how we relate to the material world. How one responds when one encounters the intractable others at the limits of the self, that is, determines who and what one becomes. I read LEGO as a non-verbal essay into material culture, a tangible play-space for "feeling out" subject-object relationships. In a designed, assembled, modular world, LEGO play *matters*: the material relationships engendered by this toy play with those of everyday life. While subject-object relationships can become imbalanced by the ethos of mastery implicit in consumer culture, such a critical, playful exploration may lead players to conceive and treat the material world as one that that actively participates in and productively challenges human subjectivity. The world may, therefore, be approached as consisting of *vibrant matter* instead of "mere" matter, as Bennett (2010) describes:

> Why advocate the vitality of matter? Because my hunch is that the image of dead or thoroughly instrumentalized matter feeds human hubris and our earth-destroying fantasies of conquest and consumption. It does so by preventing us from detecting (seeing, hearing, smelling, tasting, feeling) a fuller range of the nonhuman powers circulating around and within human bodies. (ix)

To be receptive to objects and environments is to divest ourselves of a too anthropocentric focus, which occludes and demeans the embodied, ecological fullness of life. Vital relationships with and within the LEGO medium thus offer possibilities for tangibly performing self-reflective yet object-oriented material play.

NOTES

1. For simplicity's sake, I use "LEGO" to indicate the classic system of LEGO bricks, setting aside products such as TECHNIC, BIONICLE, and MINDSTORMS.

2. The sense of the city as perpetually redesigning itself is particularly evident in LEGO's prominent Town and City themes. Other lines have somewhat different emphases—for example, the Castle themes emphasize fortification and combat and the Space themes imagine futuristic technologies and privilege exploration.

3. N. Katherine Hayles coined this phrase to describe analyses sensitive to how the unique characteristics of a medium influence the significance of texts and experiences in that medium.

4. This chapter is not concerned with *materialist philosophy*, which typically makes a metaphysical claim that nothing exists but matter and energy, but with *material philosophy*, a performative, exploratory method for using experiences of objects to interrogate object relations. One concern with the metaphysical claims of materialism is that, despite efforts to the

contrary, it becomes easy to hear an implicit "mere" before materialist descriptions—that things are *mere* matter and energy. That (and how) things *matter* is, to me, more fundamental than that (and how) things *are* matter, a view I share with philosophers like Kant, Berkeley, Heidegger, and Wittgenstein.

5. The list includes: "Unlimited play potential," "For girls and boys," "Fun for every age," "Year-round play," "Healthful, quiet play," "Long hours of play," "Development, imagination, creativity," "The more toys, the greater their value," "Extra sets available," and "Quality in every detail."

6. This is a central claim of Marshall McLuhan's seminal work in media theory, introduced in *Understanding Media: The Extensions of Man.*

7. Strictly speaking, the medium is not *exclusively* tied to the LEGO brand. After LEGO's initial patents expired, several companies began marketing comparable and compatible products. While the LEGO fan community often derides these brands for their inferior quality, they are clearly of the same medium, which might be called "interlocking plastic building blocks." I refer to this medium as "LEGO" not only for simplicity, but also because in popular discourse the LEGO brand does tend to stand for the medium.

8. There are some notable exceptions, such as the "Basic Sets" of standard bricks, Creator Sets, which contain three different instructions for a single set of elements, and the Architecture Studio, which teaches principles of design rather than any specific design. These sets encourage less scripted building but are never entirely devoid of socializing content.

9. Compositionality is an important idea in the philosophy of language. This atomistic view basically states that the meaning of a sentence can be built up from the meaning of each word, the syntax of the sentence, and (in some views) the temporal and spatial context the sentence was uttered in. LEGO's pixelated surface also evokes the atomistic touchstone of digitality, reflected in the popularity of LEGO video games and the naturalness of LEGO's pairing with the indie building game Minecraft (initiated by a fan through LEGO's Cuusoo program, but expanded into its own product line).

10. LEGO does include possibilities that resist the implicit Cartesian grid. The levers used in Carl Merriam's *Hairbrush* (discussed later), for example, are analog insofar as they can be bent along a continuous range of possible angles.

11. The rare motorized elements make use of non-LEGO power sources such as batteries to add motion to LEGO builds. The medium of LEGO itself is not self-moving.

12. For example, Bender (2010) quotes Duane Collicott, who states that "when the project manager of a bridge is giving a tour, they say that it went together like a bunch of LEGO bricks. People use that analogy all the time" (28–29).

13. Scholars have used LEGO as a model for economic complexity, social welfare, genetics, and—in a genuinely ironic twist—for combining polymers in papers such as: "The building blocks of economic complexity" by Hidalgo and Hausmann, "Building Blocks for a New Welfare Architecture: Is LEGOTM the model for an active society?" by Jenson and Saint-Martin, "The Lego® analogy model for teaching gene sequencing and biotechnology" by Rothhaar, Pittendrigh, and Orvis, and "Playing LEGO with Macromolecules: Design, Synthesis, and Self-Organization with Metal Complexes" by Lohmeijer and Schubert.

14. Lauwaert also notes other ways in which LEGO fit a modern ethos: "The (sub-)urban and domestic designs of the early LEGO sets reflected the sanctity of the private suburban home of the 1950s nuclear family. LEGO toys answered to and reinforced the increasing domestication of child and play and the angst-ridden relationship between the child and the 'mean world' outside the private suburban home" (128). In addition, LEGO's early emphasis on construction and town planning may reflect the state of rebuilding that characterized post-war Europe.

15. Although at this point 3D printing is not widely used in everyday contexts, this technology has the potential to make plastics exhibit plasticity more directly by bringing their production into the home.

16. LEGO also sells the DUPLO line of double-sized bricks for children as young as eighteen months. Furthermore, LEGO's wordless building instructions, like those of the popular furniture brand IKEA, eschew language to promote greater accessibility.

17. Bender (2010) notes how class divisions arise within the adult fan community, asking "*Can wealth affect your place in the social strata?* An adult version of the guy who has the most toys wins" (63).

18. This sense of mastery can even be inscribed on the human body through plastic surgery, often discussed with respect to another plastic toy, namely Mattel's Barbie.

19. In his seminal work on play, *Homo Ludens* (1938), Johan Huizinga describes the environment of play as a "magic circle," a protected world within the everyday.

20. This bears some similarity with what Marx's critique that capitalism replaces "use value" with "exchange value."

REFERENCES

Baichtal, John and Joe Meno. 2011. *The Cult of LEGO*. China: No Starch Press.

Barthes, Roland. 2000. *Mythologies*. Translated by Annette Lavers. New York: Hill and Wang.

Bedford, Allan. 2013. *The Unofficial LEGO Builder's Guide*, 2nd Edition. San Francisco: No Starch Press.

Bender, Jonathan. 2010. *LEGO: A Love Story*. Hoboken: John Wiley & Sons, Inc.

Bennett, Jane. 2010. *Vibrant Matter: A Political Ecology of Things*. Durham: Duke University Press.

Hanlon, Joshua and Matthew Kay. 2013. *Beyond the Brick* Ep. 82—Carl Merriam, YouTube video. www.youtube.com/watch?feature=player_detailpage&v=G3R7UIxXIqs#t=3910.

Heidegger, Martin. 1977. "The Question Concerning Technology." In *Basic Writings*, edited by David Farrell Krell, 284–317. New York: Harper & Row.

Lauwaert, Maaike. 2008. "Playing outside the box—on LEGO toys and the changing world of construction play." *History and Technology: An International Journal* 24 (no. 3): 221–237.

———. 2009. *The Place of Play: Toys and Digital Cultures*. Amsterdam: Amsterdam University Press.

LEGO Group. 2012a. "LEGO Company Profile." Published as PDF. aboutus.lego.com/en-us/news-room/media-assets-library/documents/.

———. 2012b. "LEGO History Timeline." Published January 9. aboutus.lego.com/en-us/lego-group/the_lego_history/.

Pyle, Andrew. 1995. *Atomism and Its Critics: Problem Areas associated with the Development of the Atomic Theory of Matter from Democritus to Newton*. Bristol: Thoemmes Press.

Wiencek, Henry. 1987. *The World of LEGO Toys*. New York: Harry N. Abrams, Inc.

Chapter Six

The iPhone *Erfahrung*

Siri, the Auditory Unconscious, and Walter Benjamin's "Aura"

Emily McArthur

In "On Some Motifs in Baudelaire," Benjamin lays out his distinction between two types of experience: *Erfahrung*—a long-term experience built over time as part of our collective consciousness—and *Erlebnis*—short-term experience that is often confined to an intense moment. We associate *Erlebnis* with the shocks of modern life: urban crowds, public transportation, and, significantly, new media technologies. Consciousness, according to Benjamin, functions as a shield against these shocks, mediating between self and world, protecting the fragile human psyche and thus guarding our collective social memory, our *Erfahrung* (1968, 194).

Despite the protective work done by human consciousness, it has become obvious that in the seventy-some years since Benjamin's death, new media technologies have entered our long-term collective *Erfahrung*. Benjamin was famously fascinated by the "new media" of his day, especially film and animation. If for Benjamin "there is no beyond or outside of technology," as Miriam Hansen claims (1999, 325), today we can hardly even imagine those positions. The interplay and overlap between humans and technology has accordingly become a site of vigorous critical conversation. Many scholars now acknowledge that things as well as humans can exercise a kind of agency and that the outline of the human must be redrawn to reflect our attachments to an ever-expanding catalog of new technologies. We call this reconstituted, expanded version of the human the "posthuman," and we now wrestle with what this concept means for our environment, our economies, and our ethics. Decades before conversations on the posthuman entered our social register, though, Benjamin anticipated these concerns. His works posit a restructuring of the human senses—and thus the human experience—due to a continuous interplay between body, consciousness, and technology.

Of course, Benjamin interpreted this reconfiguration in light of the media technologies of the 1930s. What we now refer to as "new media"—the portable, digital, democratized products of the twenty-first century—can give us a new lens through which to read Benjamin's oeuvre; at the same time, Benjamin's writings can give us crucial tools with which to interpret new media. I intend to examine Apple's Siri as a way to pursue these dual goals. Released in October 2011 as an application for the iPhone 4S (and as of November 2013 also available on the iPad and iTouch, as well as the latest versions of the iPhone), Siri is a natural language interface that processes the human voice in order to perform tasks that previously would have been performed by manipulating the device's touch-screen. Apple's website describes the program like this: "Siri lets you use your voice to send messages, schedule meetings, place phone calls, and more. Ask Siri to do things just by talking the way you talk. Siri understands what you say, knows what you mean, and even talks back. Siri is so easy to use and does so much, you'll keep finding more and more ways to use it."[1]

This interaction between person and machine using human language is called natural language processing (NLP), and programmers have long considered NLP as a kind of holy grail of human-computer interactions. According to the Center for Natural Language Processing at Syracuse University, NLP is a sub-division of artificial intelligence devoted to teaching computers to process human language, including the ability to "paraphrase an input text," "answer questions about the contents of the text," and "draw inferences from the text" (Syracuse University 2011). Research in the field has been going on since the 1940s, and while Siri represents significant advances in terms of the first two goals, the third still remains problematic.[2]

Because of Siri's NLP capabilities and the seemingly un-mediated nature of its interface, I argue that we should consider the way the program intersects with the Benjaminian idea of aura, the mystical sense of authenticity traditionally associated with an aesthetic work like a painting. By blurring the bounds between human and technology, Siri's posthuman aura gives the user a sense of power over a futuristic technological assistant and re-inscribes hierarchy by making all users into masters. Yet Benjamin differs from his colleagues in the Frankfurt School in that his stance on technology is never wholly pessimistic; he sees the potential for resistance and independence even in mass culture. Siri can help us extend Benjamin's notion of the "optical unconscious"—an "emancipatory way of seeing" fostered by new technologies (Koepnick 2009, 113)—into new territory: the auditory unconscious, an emancipatory way of hearing. The shock associated with carrying on a conversation with a disembodied, electronic voice forces us to question the relationships among technology, capitalism, and our own consciousness.

SIRI AND THE RE-CONSTRUCTED AURA

In the most basic sense, the aura is the sense of uniqueness, permanence, and authenticity surrounding a work of art. Benjamin's multiple definitions sometimes obscure rather than clarify the term's meaning, but he makes one point abundantly clear: the aura is in decay due to the forces of technological reproduction. He writes:

> Every day the need to possess the object, from the closest proximity, in a picture—or rather a copy—becomes more imperative. . . . The peeling away of the object's shell, the destruction of the aura, is the signature of a perception whose sense for all that is the same in the world has grown to the point where even the singular, the unique, is divested of its uniqueness—by means of its reproduction. (2008a, 285–86)

The notion of "the destruction of the aura" may sound ominous, but in fact Benjamin sees it as an opportunity for the further democratization of art. After all, art's aura relies on bourgeois power structures for its authority. Great art cows us—perhaps is *meant* to cow us. Because of this, we attribute to the artwork a mystical, almost spiritual quality, or what Lutz Koepnick calls a "quasi-sacred halo" around a work of art (2009, 113).

An apparatus for electronic hearing and vocalization may seem bereft of any deeper spiritual significance, but in fact, Siri reconstructs the aura in an age in which neither the human nor the divine has a monopoly on authenticity. Benjamin's original notion of the aura seems to rely on the human subject as a discrete entity, one capable of producing art and feeling appropriately reverent in its presence. Siri, however, points the way toward a posthuman conception of aura: the sense of uniqueness and authenticity granted not to a work of art but to a piece of technology that seems to dwell in the nebulous space between human and thing. This authenticity is not inherent in the device, of course. Engineers work to imbue the artifact with a sense of mystery, which in turn obscures the nature of control and agency in the human-thing relationship. Interactions with auratic technologies give the user a sense of power over those technologies but also leave him or her in awe of what they—and more importantly, the companies that produce them—can accomplish. Developers have programmed Siri to emphasize both the futuristic aspects of its technology and the human-like quality of its interactions, as I will discuss, and these coexisting emphases are the necessary prerequisites for a posthuman aura. In this sense, Siri is different than other voice-activated systems in that the engineers have attempted to build a particular "personality" and an impression of independent agency into its vocalizations. The point is not that the user literally questions whether Siri is human or even that the program's attempts at translating human speech are always successful, but that developers and users share a belief that machines can mimic human relationships, especially class relationships. In other words, just like the Benjaminian aura surrounding fine art, the posthuman aura ultimately shows that human artifacts—whether cultural or technological—can defend social hierarchies by appearing to hold a place in them.

Siri, as I have noted, is a natural language processor, and even that term—meaning a type of language developed "naturally" and used by humans, rather than "artificially" constructed computer languages—privileges the natural-ness of human language over computer code and implies an authenticity restricted only to humans. But with the ever-growing expansion and perfection of NLP, this restriction is gradually disappearing. Popular discourse frequently claims language ability as "what makes us human," but in this case, the unique human connection to language is demystified, leaving only the technology itself as exceptional. Through this demystification, Siri makes the user aware of his or her place in what Jane Bennett calls an "ecology of matter" (2004, 365), which does not necessarily privilege human over non-human actants.[3]

Unlike the decline in the aura of great art, this shift in the locus of language does not stem from Benjamin's famous "technological reproducibility." Strictly speaking, Siri reproduces nothing. Instead it participates in a complicated process of translation and synthesis that maintains a kind of auratic inscrutability. In basic terms, Siri interprets speech in the following stages:[4]

1. Human vocal chords make pressure waves in the air.
2. Pressure waves in the air vibrate a small membrane inside the iPhone's microphone.
3. The vibrating membrane creates an electrical signal that is a replica of the air-pressure wave.
4. The analog electrical signal is encoded into binary with an analog-to-digital converter.
5. Digital algorithms implement statistical methods that associate given patterns of ones and zeroes with specific words.
6. Further statistical algorithms tie these words into contextual relationships to provide the best guess at what the user is trying to communicate.

The connecting tissue between each of these steps is translation—Siri translates sound waves to an analog signal, then to a digital signal, and finally into human language. Unlike reproduction, translation implies a process of conversion; something is changed from one step to the next, which maintains the sense of magic about the process that mere technological reproducibility kills. The final steps of the process move from translation to synthesis, as Siri uses programmed algorithms to interpret spoken language or to construct human speech itself. This requires processing not just individual words but also how they work together in a context that is unique to a given syntactic/semantic moment. Ultimately, each interaction with the program participates in a distinct act of natural language creation, much like our own speech acts. Although Siri is not the first device capable of interpreting context in this way, it is nonetheless a highly significant example because it extricates the magic of NLP from the unfathomable depths of the supercomputer and places it into the pockets of the average consumer. Siri democratizes NLP, and with it the posthuman aura.

Of course, the program's complex process of translation and synthesis is hidden from the user, who sees only a small microphone image when speaking, followed by text on the screen. Communicating with Siri requires only the push of a button, much like speaking into a walkie-talkie. Too, Siri has in many ways mastered the rhythms of human speech. She uses colloquial language and calls the user by his or her first name, or even a chosen nickname. The Apple website promoting Siri urges consumers, "Talk to Siri as you would to a person. . . . It's like you're having a conversation with your iPhone."[5] Apple's claim—that its program has the capacity for authentic human conversation—both deflates the notion of a unique human relationship to language (as I've noted) and re-mystifies the program by putting a quasi-human face on new media technology. Both the visual and the auditory components of Siri's user interface work together to obscure the complex nature of digital production, thereby restoring the aura to these interactions.

The program also restores a sense of authenticity to the realm of web search, making it more like a conversation between humans than an interaction with a computer. By providing straightforward answers to a whole host of questions, Siri removes multiple layers of mediation from the search process. *Mashable* technology reporter Jennifer Van Grove differentiates between Google, a traditional search engine, and Siri, which she calls "an answer engine," citing venture capitalist Gary Morgenthaler's point that when people perform a search, they "don't want a million blue links, they want one correct answer. All the rest is noise that you'd rather have go away" (2011). Of course, while Siri can sometimes provide "one correct answer," the program frequently does resort to asking, "Would you like me to do a Google search?" But the tech world seems to see this as a short-term problem, and tech blogs abound with stories about how the answer-driven approach of Apple's Siri could eventually kill Google.[6] This notion of an answer engine is clearly a powerful one, and Morgenthaler's comment links this power to Jay David Bolter and Richard Grusin's concept of immediacy.

Humans, they argue, have a "desire for immediacy," as "the user is no longer aware of confronting a medium, but instead stands in an immediate relationship to the contents of that medium" (2000, 24). In terms of web search, that means doing away with the "million blue links"—as well as the dozens of sidebar advertisements—in order to restore a more "authentic" form of search: asking a question, and getting one answer.[7] This authenticity, though, is not entirely unthreatening. Just as Benjamin connects the aura of traditional bourgeois art with a kind of cultural control, we might ask, in a world of Siri searches, who controls the answers?

In "The Work of Art in the Age of its Technological Reproducibility," Benjamin defines "aura" as "a strange tissue of space and time: the unique apparition of a distance, however near it may be" (2008b, 23). A famous painting hanging in a gallery, for example, appears distant in terms of its historical and cultural significance, even when one stands only a few feet away. When it comes to the aura, distance cannot easily be measured in either years or meters. Siri's status as a media technology further complicates the issues of proximity and distance. Theories of the posthuman necessarily imply a sort of closeness between our technologies and our selves. Jane Bennett's ecology of matter, for example, "emphasizes the closeness, the intimacy, of humans and non-humans" (2004, 365). The user's intimacy with the iPhone is perhaps too obvious for comment; we've all seen it made manifest during subway commutes, dull parties, and long meetings. In the age of the iPhone, intimacy with technology is inescapable, since the portability of new media technologies allows us to use our devices at almost any time or place. And each time we interact with a technology, as Bruno Latour points out, the responsibility for the resulting actions is shared between the human and the machine (1994, 34).

Yet there is another node in this relationship—and in our thinking about shared responsibility—that must not be overlooked: the mediation of the engineers who designed the technology in the first place. While in some ways Siri creates sense of immediacy, its developers have also purposefully instilled a feeling of temporal distance because of the way the program plays with futuristic motifs. Rebecca Zorach of the University of Chicago's Center for the Study of Gender and Sexuality observes that Apple "seem[s] to intentionally make her speech sound artificial—they could choose to make her speech more seamless and human-like, but they choose instead to highlight the technology" (Griggs 2011). If Siri's search function disguises its mediation, Siri's voice actually draws attention to its mediated status—its hypermediacy, as Bolter and Grusin call it (2000, 31)—in order to emphasize the futuristic nature of the product. The same could be said of the way Siri answers questions about her ontological status. In response to the question, "Are you human?" Siri often remarks, "Close enough, I'd say," or "I'll leave that for you to decide." Similarly, Siri responds to questions about gender with, "I was not assigned a gender." The consumer is not supposed to imagine that he or she is talking to some friendly operator. Instead, we are to recognize the technological wonder that lives in our pockets, the type of sophisticated talking computer system that was mere science fiction just a decade ago. Thus, despite the iPhone's mass production and wide availability, this hypermediacy does not reduce the perceived aura surrounding conversations with Siri but rather reinforces their magical quality.

In his Baudelaire essay, Benjamin provides a definition of the aura that further explains the sense of magic we experience in technology: "To perceive the aura of an object we look at means to invest it with the ability to look at us in return" (1968, 188). This returned gaze acts as a means for social control, as bourgeois art enforces its values on those who look upon it. Siri makes a similarly threatening promise: to look back at the user through a vast accumulation of data and a tightly controlled search function. While Apple is notoriously closed-mouthed about its products, Jonny Evans of *Computer World* expresses an assumption made by many: that Siri is getting better at its job each time a consumer speaks. He says, "I assume it is picking up data. I'm assuming it learns as questions are asked. I'm assuming its answers become more refined as it is asked to engage in various tasks. I would expect Siri to be growing its database of questions and answers every second of every day, learning from its users" (2011). In this way, Siri—ostensibly a product we understand and manipulate—could eventually grow to understand and manipulate the consumer.

Clearly Siri's posthuman aura reflects much of what Benjamin originally theorized about the aura of art: it cows the user/viewer in an attempt to control. Given the conditions of late capitalism in the twenty-first century, though, Siri's function of social control must disguise itself by investing the user with a semblance of authority. In other words, for a commodity to be fetishized under these conditions, it must imbue the consumer with a sense of power, not powerlessness. In this vein, the consumer, who typically refuses to acknowledge the sway of commodity fetishism in his or her life, must be recreated as a bourgeois subject able to exert the kind of control over technology that previously was only exercised by the capitalist over the worker. Siri is frequently billed as a "personal assistant," accessible twenty-four hours a day to transcribe messages, set reminders, and place telephone calls. Apple's website trumpets, "Your wish is its command." Even Siri's vocalizations—feminized[8] and eternally polite—fashion a distinctly classed form of conversation. In this way, Siri gives the user power over this digital "secretary," while the program's feminine, secretarial aspect obscures its position as what Mark Hansen calls a "site for a struggle over who controls the flux of consciousness" (2006, 304). The threatening potential of Siri's aura relies not on a mythologized past, as in traditional forms of art, but on a mysterious future for which the power dynamics between human and technology are not yet fully determined.

Technology is not ethically neutral, as Peter-Paul Verbeek contends; thus, a software engineer is also a kind of moral philosopher (2006, 362–63). Technology, like art, does not exist in isolation from the forces that create and maintain it, and the posthuman aura is not an inherent attribute of technologies like Siri. Rather, engineers design programs to evoke a sense of aura. The engineer must negotiate the space between economics and aesthetics, developing a seemingly unique entity that can in fact be mass-produced, a human-like interface that is actually just computer code. The user's seeming control over the device is complicated by the fact that this appearance of mastery has been pre-programmed by teams of developers. Like the painters behind Benjamin's "work of art," these engineers recreate class relationships in a way that makes them so natural as to be almost invisible.

In thinking about Siri's interface and the aura with which we invest it, we must also consider who is harnessing the power of this artifact, and to what ends. Here, ultimate agency resides neither in the user nor in the engineer but in the requirements of our economy. In the case of the twenty-first century neoliberal American state, consumer goods must be tied to the relentless drive for corporate profitability. Apple, of course, manufactures highly fetishized goods that, perhaps more than the products of any other company, act as status symbols for the American bourgeoisie. Siri, by making its users into masters over a digitized personal assistant, takes Apple another step forward in the race for the most fetishized commodities. In Adorno and Horkheimer's famous essay "The Culture Industry: Enlightenment as Mass Deception," the authors state that the purpose of every film is "to impress the omnipotence of capital" on the watching public (2002, 98). One could say the same for Siri. It serves to "impress the omnipotence of capital" on the consumer (who marvels at the wonders of technology) and on rival corporations, which must try to compete with Siri's impression of uniqueness and authenticity.

SIRI AND THE AUDITORY UNCONSCIOUS

Yet Benjamin is in the business of redeeming, not condemning items of mass culture. While Siri does reconstruct the aura in a way that suggests capitalist domination, the potential for resistance is built into both the program's form and the overall experience of using an auratic technology. Specifically, I argue that we can read emancipatory potential into the program via the idea of the auditory unconscious. My notion of the auditory unconscious arises out of Benjamin's "optical unconscious," the hidden layers of human experience that are made visible only through the mediation of film. To appropriate the dream-language of Benjamin's early career, film gives society an impulse to awaken from its dream state, at which point the optical unconscious opens to the viewer perspectives that otherwise would remain impossible to access. As a result, "film is . . . the artwork most capable of improvement" (Benjamin 2008b, 28). This "improvement" he speaks of is the capability to recognize capitalist oppression, which the viewer can identify more readily on the screen than in the dream state of day-to-day life. The magic associated with a work of art's aura, on the other hand, has the opposite effect: that of lulling the viewer into complacency and acceptance of capitalist domination.

Why does Benjamin privilege the visual for this "improvement" rather than another sense? Adorno frequently writes about music—both jazz and the atonal music of the Second Viennese School—but Benjamin largely ignores the topic. Smell comes up several times in his *Arcades Project*, but Benjamin never develops those references into a unified statement. Some scholars have laid out theories for a revolutionary "unconscious" linked to other senses, though. Laura Marks develops the notion of an "olfactory unconscious" that recalls Proust's *memoire involuntaire*, with smell as an immediate, affective link to unconscious impressions (2008). Similarly, Robert G. Ryder constructs an "acoustical unconscious" through a study of German radio and sound film of the early twentieth century. These media, "instead of constructing meaning, disperse it among a flurry of phonemes" (2009, 16–17), thus revealing sound structures that without the help of new media would have remained in the human subconscious. Ryder uses the phrase "acoustical unconscious" to describe this phenomenon, but I will use the term "auditory unconscious," since the term "auditory" to me connotes both the projection and processing of sound.

The conditions of modern urban life, and especially our relations with technology, make a consideration of the sense of hearing especially vital. A barrage of noise now confronts us most everywhere we go, from holiday music in shopping centers to the sounds of traffic outside our homes. In most of these cases, we hear without being conscious of ourselves as hearing bodies. In this sense, sound is actually a better partner to the unconscious than sight. Ryder links this unconscious hearing to the "penetration and surroundability" of the body by sounds (2009, 13), which render us more able to take in sounds without fully processing them. Benjamin, as I've mentioned, sees consciousness as a means to shield our deeper selves. Yet we can tune out shocking sounds from our conscious mind more easily than shocking sights, enabling those sounds to more fully embed themselves in our memories. (After all, no one pays attention to the mall's Christmas carols, yet they must somehow induce us to buy more—otherwise, why play them?) Sound, as a more penetrating sense than vision, more easily escapes conscious reflection and thus can be more deeply buried in the unconscious.

According to Benjamin, the penetration of sounds has much to do with the dis-ease of modern life. In the *Arcades Project*, he quotes Georg Simmel on the topic:

> The one who sees, without hearing, is much more . . . worried than the one who hears without seeing. This principle is of great importance in understanding the sociology of the modern city. Social life in the large city . . . shows a great preponderance of occasions to *see* rather than to hear people. One explanation . . . of special significance is the development of public means of transportation. Before the appearance of omnibuses, railroads, and streetcars in the nineteenth century, men were not in a situation where, for minutes or hours at a time, they could or must look at one another without talking to one another. (1999, M8a,1)[9]

Simmel describes how the absence of hearing leads to anxiety, especially in the new nineteenth-century modes of public transportation. Yet these fellow passengers who "see, without hearing," actually do hear quite a bit from each other. They may not speak, but anyone who has spent any time at all in public places like train cars knows that these spaces are anything but silent. One overhears coughs and sneezes, fragments of conversations, and the rustles of shifting bodies. All of these sounds wind their way into our auditory unconscious. But Simmel's point here is that these contacts are dehumanizing; we see each other without directly speaking to each other, and so we have no conscious thought about the sounds around us to parry their shock value.

In the case of Siri, the situation is reversed—one actually "hears without seeing." The user interface is plain and immediate, with only the lines of dialogue and a small microphone button. Nothing suggests the face or figure of a human, but, according to Simmel at least, this should not be troublesome. However, I would posit that there *is* something shocking about communicating with the disembodied voice of Siri that is more significant than, say, talking to someone over the telephone or chatting with someone online. The shock of conversing with Siri comes in part because of Freud's notion of the uncanny, which he relates to "the idea of being robbed of one's eyes" (1955, 230). Freud goes on to state, "Uncertainty whether an object is living or inanimate, which admittedly applied to the doll Olympia, is quite irrelevant in connection with this other, more striking instance of uncanniness" (1955, 230), and he then connects the uncanny to the return of the repressed, especially castration anxiety. In the case of Siri, though, the uncanny aspect of conversation with a disembodied technological voice arises not only because of the user's lack of vision, but also because of the device's aura, or the sense of uniqueness it acquires from its liminal place at the threshold of the human. The sense of the uncanny leads to a different kind of return of the repressed: the flooding back of the shocks of modern life that have been subsumed under the auditory unconscious.

Popular discourse surrounding technology frequently asserts the idea that new media can drive a wedge between individuals by distracting us from real human relationships. To return to Simmel's illustration of a person using public transportation, we might note the common laments that "people nowadays" are so wholly immersed in their technology—listening to an iPod, playing games on a phone, and so forth—that they ignore the people just inches away from them. Given the notion of the auditory unconscious, though, we might read this deployment of technology in crowded public places as another shield—like consciousness—for protecting the human being from the shocks around him or her. But what happens when using that technology results itself in shock? What might happen when, instead of speaking to the people around one in a train, one speaks to Siri?

As it does in much of Benjamin's writings, the answer here depends on how closely the consumer scrutinizes his or her use of technology. Benjamin contends that the shock of modern technology can actually awaken us to the realities of corporate manipulation and domination. As we see with the abundant media coverage surrounding Siri's release—or even more mundane technological advances, like minor upgrades to our mobile phones or operating systems—new technologies now draw our collective social scrutiny, making the supposed immediate and transparent nature of these technologies themselves the object of concentrated attention. Reviewers and tech bloggers examine every element of design and user experience, and the technically inclined masses follow each hint and rumor of change with the fervor of a cult. Though Benjamin presents quite a different explanation for his notion of shock and subsequent awakening, it remains true that every technological advance provides a moment for society to see the medium itself rather than just see through it. We parry the shocking effects of new technologies with an intense collective consciousness of the changes that they bring.

A close examination of Siri's form and functionality reveals that the program, while giving the sense of authentic human conversation, actually emphasizes a very real interpersonal distance. One of the conveniences of mobile phones is the ability to avoid making a phone call by instead sending a text message. With Siri, however, one can actually *speak* a text message, so that rather than making a phone call to a human acquaintance, one is quite literally having a third party mediate the conversation according to the complex pattern of translations that I laid out earlier. While a face-to-face conversation between two individuals requires no translation aside from the work of the individuals' brains, a text-message conversation facilitated by Siri is mediated through multiple layers of translation and types of language and filtered through the vision of software developers and corporate leaders. Speaking a text message may seem more natural or immediate than typing it, but the appearance of closeness created by the program masks an increasing distance between user and recipient. In this way, the program mirrors the reification inherent in human relationships under late capitalism, as technologies that could serve to draw people together often perform the opposite function. Yet the very shock caused by conversation with Siri can awaken the consumer to a renewed consciousness of that reification, forcing the user to consider how conversations with Siri mimic the kinds of conversations we have with human beings.[10] If we think critically about our relationship to this piece of technology, we realize that we only speak to Siri to give orders or ask questions in the service of our own interests. We fetishize this product because it gives us a sense of control.

Siri's revelation of the auditory unconscious can further facilitate this awareness. Siri functions by translating mere phonemes—meaningless and disjointed in and of themselves—into contextually meaningful speech. In this sense, Siri closely resembles the film technology discussed by Benjamin, in which disparate images recorded by the camera are spliced together into a meaningful narrative. But Siri's acts of translation are not always successful, and the program's missteps work to break down its carefully constructed aura. Each time the program bungles a simple command, each time the server breaks down and Siri is left inoperable, each time Siri calls the wrong "John" from one's contact list, we remember that we are using a man-made and thus fallible technology. The shock inherent in the notion of a futuristic digital assistant evaporates as we come to expect Siri's anticlimactic, "I don't know what that means." The program's frequent failures draw our attention to its more problematic mediations, reducing the allure of Siri's posthuman "magic" as well as our belief in the all-encompassing power of the company that created it.

One failure stands out as particularly revelatory of the auditory unconscious: the program's difficulty interpreting the highly individualized spellings and pronunciations of human names. If you ask Siri who Walter Benjamin is, for example, the program responds with a meaningless translation of those phonemes: "Who is Walter been you mean?" In any given person's contact list, dozens of names might prove similarly problematic—a favorite Vietnamese restaurant, perhaps, or a friend who is a transplant from Hungary or Brazil. Siri has been programmed to process nearly every aspect of human language, but the realm of proper names is far too vast. It does not pay for Apple to devote countless developer hours to tracking down each possible variant. And so only the most ordinary can be included; there is no room for the uncommon or the exceptional. This is the revelation of the auditory unconscious: the intensely personal—in this case, a name—cannot be wholly conscripted in the service of capitalism. Just as film's ability to slow down and speed up time allowed the early film viewer to recognize things he had always seen but never fully understood, Siri's direct phonetic translation of speech allows us to see the limits of a capitalist vision of progress. Corporations try to peddle the magic of the posthuman aura, but the realities of profit-making mean that technology's thrilling effects are only available to those within a certain norm. And in any case, users do not always buy in to this notion of technological progress without question. We scrutinize our devices even as they "look back" at us.

CONCLUSION

Regardless of the limits of auratic magic, the number of technologies capable of "looking back" at the user has vastly expanded in recent years. These attempts include not only natural language programs like Siri but also, to give one increasingly important example, technologies classified as recommender systems. These algorithms take into account the self-generated preferences of the user in order to more carefully target him or her with products and advertisements. This technology now forms an important part of some of the internet's most highly trafficked sites: Pandora for music, Netflix for film and television, Amazon for everything imaginable. Even the venerable *New York Times* now recommends articles based on one's reading history, and it seems that almost every retail site uses recommender systems to generate suggestions for further shopping based on past views and purchases. Like Siri, these systems simulate a seemingly unique human behavior, but their magic arises from their ability to learn, not from their facility with language. Also like Siri, they develop complex power relationships with users, who feel the force of their own influence in expressing individual preferences but who remain at the whims of the system's algorithms. Even more than Siri, however, this iteration of the posthuman aura is directly linked to corporate profitability. Recommender systems hold out the promise that our devices can learn from us and know our desires better than we can ourselves, but their ability to "look back" and learn enables a sinister familiarity: mass data gathering in the service of corporate sales.

Siri, then, is hardly an isolated example of companies striving to associate their digital technologies with a sense of posthuman aura. As we see with the example of recommender systems, these efforts become increasingly important in an age in which the appearance of technological magic directly correlates with profitability. Yet Benjamin's optimism about technology remains valid in the early twenty-first century, as each corporate attempt to harness our desires is met with an equal opportunity for resistance. Auratic technologies may signal uncomfortable levels of surveillance and social control, but the perceived enchantment in new and shocking technologies also draws our collective attention, opening our eyes—and ears—to the ways that our interactions with technology both reflect and direct the course of modern life.

NOTES

I would like to thank Kimberly Hall, Susan Zieger, and the editors of this collection for their invaluable contributions to this chapter.

1. I will refer to Siri as a gender-neutral "it" throughout the course of the paper, although the English version of the program employs a female voice. If you ask Siri, "Are you a girl?" the program responds, "I do not have a gender."

2. IBM's Watson supercomputer famously used NLP to compete in—and win—a Jeopardy tournament in 2011, establishing the outer limit of what NLP has been able to accomplish to date. We might consider Watson another example of a technology granted a posthuman aura, especially as it combines an ability to interpret the nuances of English idiom with a superhuman ability to process a huge body of information.

3. I borrow the term "actants" from Bruno Latour in "On Technical Mediation—Philosophy, Sociology, Genealogy."

4. My thanks to Robert deCarvalho for helping me articulate this process.

5. The discourse of Siri's humanity was reinforced after significant outages of the application in early November, 2011. Articles with titles like "Siri, We Need to Talk," "Siri Gets Over Her Huff, Returns as Useful as Ever," and "Siri, Are You Working?" somewhat ironically position the system as capable of emotions and relationally connected to humanity.

6. Perhaps in response to this competition, Google has recently started providing its own "one correct answer" to certain questions—often a sidebar box with facts aggregated from different websites—alongside its familiar blue links. In moves like this we can trace a trend toward greater obscurity in the control and transmission of knowledge, even in the "free information" atmosphere of the digital age.

7. In some cases, Siri will provide more than one answer. If one asks Siri to find a good Mexican restaurant in Los Angeles, for example, the program will list several nearby options to choose from. But one does not see *all* the potential choices. The program is applying some sort of selection criteria, but it is difficult to know just how the results on that list are determined.

8. The voice for Siri is female in most languages, although in a few European countries Siri has a male voice.

9. Benjamin notes that he takes this quote from *Mélanges de philosophie rélativiste: Contribution à la culture philosophique*, trans. Alix Guillain (Paris, 1912), 26–27.

10. The shock is doubly emphasized because all conversation with Siri takes two forms: both spoken words and typed display.

REFERENCES

Benjamin, Walter. 1999. *The Arcades Project*. Translated by Howard Eiland and Kevin McLaughlin. Cambridge, MA: Harvard University Press.

———. "Little History of Photography." (1931) 2008a. In *The Work of Art in the Age of its Technological Reproducibility and Other Writings on Media*, edited by Michael W. Jennings, Bigid Doherty, and Thomas Y. Levin, 274–298. Translated by Edmund Jephcot and Kingsley Shorter. Cambridge, MA: Belknap.

———. "On Some Motifs in Baudelaire." (1940) 1968. In *Illuminations*, edited by Hannah Arendt, 155–200. Translated by Harry Zohn. New York: Schocken.

———. "The Work of Art in the Age of its Technological Reproducibility." (1989) 2008b. In *The Work of Art in the Age of its Technological Reproducibility and Other Writings on Media*, edited by Michael W. Jennings, Bigid Doherty, and Thomas Y. Levin, 19–55. Translated by Edmund Jephcot and Harry Zohn. Cambridge, MA: Belknap.

Bennett, Jane. 2004. "The Force of Things: Steps Toward an Ecology of Matter." *Political Theory* 32 (3): 347–372. JSTOR (4148158).

Bolter, Jay David, and Richard Grusin. 2000. *Remediation: Understanding New Media*. Cambridge, MA: MIT Press.

Evans, Jonny. 2011. "Apple, Siri, and the New Search." *ComputerWorld.com,* October 16. Accessed November 16, 2011. blogs.computerworld.com/19105/apple_siri_and_life_after_google.

Freud, Sigmund. (1919) 1955. "The Uncanny." In *The Standard Edition of the Complete Psychological Works of Sigmund Freud,* edited and translated by James Strachey, 218–252. London: Hogarth.

Griggs, Brandon. 2011. "Why Computer Voices are Mostly Female." *CNN.com*, October 21. Accessed November 6, 2011. www.cnn.com/2011/10/21/tech/innovation/female-computer-voices/index.html.

Hansen, Mark. 2006. "Media Theory." *Theory Culture Society* 23: 297–306.

Hansen, Miriam Bratu. 1999. "Benjamin and Cinema: Not a One-Way Street." *Critical Inquiry* 25 (2): 306–343. JSTOR (1344205).

Horkheimer, Max, and Theodor W. Adorno. (1947) 2002. *Dialectic of Enlightenment: Philosophical Fragments*, edited by Gunzelin Schmid Noerr. Translated by Edmund Jephcott. Palo Alto: Stanford University Press.

Koepnick, Lutz. 2009. "Benjamin in the Age of New Media." In *A Companion to the Works of Walter Benjamin*, edited by Rolf J. Goebel, 112–129. Rochester, NY: Camden House.

Latour, Bruno. 1994. "On Technical Mediation—Philosophy, Sociology, Genealogy." *Common Knowledge* 3 (2): 29–64.

Marks, Laura U. 2008. "Thinking Multisensory Culture." *Paragraph* 31 (2): 123–137. *Project Muse* (10.1353/prg.0.0019).

Ryder, Robert G. 2009. "Hearing Otherwise: Towards a Genealogy of the Acoustical Unconscious from Walter Benjamin to Alexander Kluge." PhD diss., Northwestern University, 2009. ProQuest (3386474).

Syracuse University School of Information Studies. 2011. "Center for Natural Language Processing." Accessed November 25, 2011. www.cnlp.org/.

Van Grove, Jennifer. 2011. "Siri Gives Apple a Two-Year Advantage over Google, Says VC." *Mashable.com*, October 28. Accessed November 6, 2011. mashable.com/2011/10/28/apple-google-smartphone-war/.

Verbeek, Peter-Paul. 2006 "Materializing Morality: Design Ethics and Technological Mediation." *Science, Technology, and Human Values* 31 (3): 361–380.

Chapter Seven

Victorian Cybernetics

Networking Technology, Disability, and Interior Design

Colbey Emmerson Reid

On the heels of nearly half a century's discussion in the liberal arts of public selves and civil recognition, some of the most pressing questions of the present era concern domesticity, a sphere that had been marginalized in modernity. What is, and how does one sustain, a home, a family, or a couple? These questions and a hundred more are being pushed into public discourse from within interior spaces, often by existing domestic arrangements that desire recognition. They, in turn, are creating the conditions for the other imaginative arrangements that will comprise the domesticities of our future.

The present chapter argues that the content of domestic arrangements can be productively considered using the language and concepts of interior design. As Rosalind Galt demonstrates in *Pretty: Film and the Decorative Image* (Columbia 2009), the trivialization of decorative style is the product of masculine public culture's disregard for feminine cosmetics. As the domestic interior emerges as a space of public preoccupation, however, stylistics presents itself as an abiding problem and question—and a potential space of solution. Ornamental design can provide the means for thinking about and understanding subjectivity stylistically and for taking style seriously as a mode of self- and world-transformation.

Domesticity is not only an ornamental space, however. It is also a cybernetic space, in which humans frame, make, use, and imagine technologically created and enabled minds and bodies. In fact, the human-technology hybrids today associated with futuristic laboratories and manufacturing progress have been emerging within the home for at least two centuries. Domesticity puts ornament and technology back in touch. As such, interior design becomes indispensable to producing, refining, and framing cybernetic invention. By returning to the emergence of domesticity discourses in the nineteenth-century, the material and conceptual proximity of technological invention and interior design come into focus. The present chapter considers Victorian domesticity as emerging in continuity, not conflict, with the Industrial Revolution. Home and machine co-evolved.

A consideration of nineteenth-century cybernetics particularly highlights the extent to which ornamentation reframes the strange, foreign, or outright deformed minds and bodies produced by interface with artifacts as stylish. Within the Victorian context, it becomes apparent just how entwined experiences of the deep proximity of humans and things are with episodes of disfigurement and derangement. It also becomes clear how interior design helps people to process and absorb encounters with dramatic difference by reframing them as style. What now seems like the unlikely convergence of technology innovation, disability, and interior design are historically commonplace. We have much to learn in the present from how the past negotiated their network.

NOW

In a striking scene in "Under Our Skin," Andy Abrahams Wilson's award-winning 2009 documentary, victims of Lyme disease enact an unusual brand of civil rights protest. They suffer from what the film argues is a microbial epidemic that has devastated scores of thousands of lives, while remaining unrecognized by the American medical establishment. The protesters believe that insurance companies have influenced research on the chronic illness, whose diagnosis can cost millions of dollars in long-term care. At the demonstration, signs condemn insurance discrimination and demand medical recognition using language and slogans recognizable from several genres of civil rights protest. Lyme is claimed as the salient identity of individuals whose identification with the disease preempts the visible differences of gender, race, even class. They demand and embrace diagnosis as enunciating a kind of invisible kinship that entitles them to a community, the attention of medical specialists, insurance coverage, accommodation, and public understanding when their behaviors counter social and human norms.

The Lyme community is not unique in reconfiguring their disease as a deep kinship that binds individuals "under our skin." The quests for public recognition and community identification among those with AIDS, breast cancer, multiple sclerosis, and many other illnesses all resonate with Lyme's quest for a common narrative and bond that transcends ethnic, geographic, or even demographic differences. A disease named after a rural town in Connecticut now embodies the contemporary archetype of disease as the emergent super-identity of the global millennium. Lyme's geography is in blood rather than on the map, and its population soars with the indiscriminate speed and reach of microbial infection in an era of extreme mobility.

A second recent text on this issue, Andrew Solomon's highly acclaimed *Far From the Tree: Parents, Children, and the Search for Identity* (2012), helps to explain how our understanding of illness has changed in ways that prime it for acceptance, even celebration, thus reconfiguring diseases from isolated physiological events into expansive identity categories. The book examines an emergent identity classification forged by a series of inherent and acquired traits that estrange children's identities from their parents' and position children to acquire identity from a peer group instead. Solomon calls such peer-formulated identities *horizontal*, in contrast to the *vertical* identities that children tend to inherit from their parents. A vertical identity might include ethnicity, language, or religion; specific horizontal identities that have emerged in modernity include deafness, dwarfism, autism, Down syndrome, schizophrenia, and a cluster of severe physical and cognitive congenital disabilities.

Solomon's examination of the intersection between illness and identity assumes the convergence of apparently polarized concepts and events in technology innovation, interior design, and disability studies. To highlight the role of technology in horizontal identity formations, Solomon explains that they are the combined products of information and medical technology. First, medical advancements have saved the lives of children who historically would not have survived their infancy in significant enough numbers to constitute a social group, or who would have been seen as generically ill and languished or died in the margins of medical attention. Beyond survival, horizontal identities are then cultivated by the emergence of digital and especially social media, connecting individuals who might otherwise have imagined their atypical minds and bodies as unique and experienced them in isolation. Even as technology has helped to create the conditions for horizontal identity, however, procedures like cochlear implants for the deaf and genetic screening to preclude the fetal development of autism, dwarfism, or Down's syndrome are also exterminating specific identities. Horizontal identity represents a plane of profound human-technology interface, one that is simultaneously actively creating and destroying forms of human being.

The horizontal identities Solomon explores short-circuit the common conception of technology as a perfecting instrument. Solomon's "cyborgs," from the profoundly genetically defected to the chemically and surgically altered and treated, are considered disabled. As Solomon observes, disabled minds and bodies are becoming ever more prevalent fixtures of the modern landscape: to be "saved" by technology seems to lend itself more to the manufacturing of hybridized and atypical shapes, viscera, cognitions and identities than to purity, consistency, or singularity. The mental and physical forms produced through technological intervention hardly comply with conventional images of health and wholeness. Solomon implicitly asks us to question an impulsive modern association of technology with the perfected mind and body, suggesting that the human-machine interface is instead the birthplace of disfigurement and derangement. Even apparent cures produce a range of deforming symptoms, as in the pharmacological obesity of schizophrenics, or the limbless girl to whom Solomon introduces his readers, whose healthy arms and legs were amputated to facilitate rolling and lifting, her only mobility.

Far From the Tree is both an analysis and a defense of disability, an effort to explore perimeters from within which to understand, tolerate, and even desire it. Put another way, Solomon instructs his readers in developing a taste for certain horizontal identities, endeavoring to complicate the hegemony of the vertical by teaching viewers how to see—as mothers, fathers, siblings, and disabled individuals themselves have often learned to see—horizontality on its own aesthetic and ontological terms. The phrase "far from the tree" conjures the image of a family's effort to understand and incorporate dramatically distinct identities within domestic systems typically oriented around reproduction as replication. Solomon's instruction in the perimeters of a new kind of taste for unfamiliar forms of embodiment and subjectivity enacts what Tobin Siebers has called "disability aesthetics" (2010) within homes and among families. *Far From the Tree* doubles as a manual on domestic style, explaining how to incorporate horizontal forms into the traditionally vertical family structure by learning not merely to accept but to take pleasure in the neural and physiological expressions of divergent progeny.

Solomon's selection of aesthetic terminology—"vertical," "horizontal"—to describe identity foregrounds all identity, and disability in particular, as style, a framework which expresses and embraces the adoption of an ethos of excess. Disability can be seen, he implies, as genetic excess. Solomon then frames the challenges presented by human disability as a design problem which, solved, could reframe disability as pleasure. Psychoanalysis has used the term *jouissance* to describe a pleasure-pain continuum. Though Solomon never uses the word, his depiction of disability as intensely emotionally and physically difficult but also as filled with charm, surprise, and delight of many kinds suggests it.

Solomon further addresses disability's design challenges as presented to homes—families—rather than to citizens or communities. Horizontality implies domestic divergence: difference in comparison to one's parents. This framework suggests the merging of domestic and aesthetic systems conventionally expressed through the term "interior design." Solomon, in other words, can be said to position hard questions about the construction of extremely unfamiliar forms of selfhood within the intimate context of home and family as a matter of ornament, expressed within the perimeters not of philosophy, psychology, sociology, or political science—the human and social sciences—but rather in the perimeters of a practice frequently seen as comparatively trivial and specific to the arrangement of objects, not people. To what extent does the kind of human objectification that facilitates thinking about people in the language and concepts of ornamental design foster desirable subjectivities—perhaps better renamed "objectivities"—precluded by conventional humanistic thinking? What can we all learn from buildings, carpets, paintings, clothing, pottery, automobiles, or furniture about how better, more variably, to see and be a human?

THEN

The historic design pattern known as *grottesche* illustrates the extent to which an organic or inorganic context for disfigurement can make the difference between perceiving such forms as horrific or chic. Linguistically, organic disfigurement is "grotesque," which implies an abomination of nature; inorganic disfigurement is "*grottesche,*" or ornamental. The *grottesche* pattern illustrates the extent to which perceptions of stylishness can reform the way people interact with, experience, and think about their own bodies and those of other people when they do not comply with the conventions of normalcy.

The term *grottesche,* which is the etymological origin of grotesque, originates in the Italian Renaissance and is the name of a tile pattern combining human, animal, vegetable, and mineral forms into single protracted bodies (Harpham 2007). The pattern, which can be found in Paleolithic cave drawings and medieval designs, entered Western modernity in the hands of the felicitously named Roman decorator, Fabullus. Fabullus used the images as the prototype for the ornamental décor of Nero's *Domus Aurea.* When Nero's palace was excavated in 1480, its imaginatively patterned walls found their way into the Neoclassical decorative vernacular. The pattern was wildly popular among artisans and craftsmen, at the same time that it was celebrated and even performed by visitors to the archaeological site. Renaissance archaeology was less reverent than the contemporary science. Visitors to the dig site were lowered into its grottos, ate al fresco dinners near and within them, and engaged in carnivalesque efforts to be "more bizarre than the grotesques themselves" (Zamperini 2008, 95). *Grottesche* came in time to comprehend more than merely the recovered walls of Roman ruins. The pattern design became a property of human minds and bodies.

Late nineteenth-century artists embraced a Classical resurgence, which included returning to Renaissance forms that valorized ancient values and techniques. The term "renaissance" is, in fact, a product of the nineteenth-century, coined by French art historians in 1840 and making its way into general usage by 1872. William Morris introduced the concept to artisanal culture by advocating that Victorians follow the example of domestic design set by "mediaeval or Elizabethan builders" (Morris 1911). Numerous nineteenth-century fine artists and art historians collectively advocated some version of a return to practices common within Quattrocento Italian art. The Victorian return to Renaissance conventions in color, detail, and compositional complexity included a resurrection of the *grottesche*, the design technique perhaps most associated with vibrant color, intricate detail, and complicated formal networks. The nineteenth-century artisanal revival of the form, like the original Renaissance manifestation, valorized figural monstrosities as stylish ornaments rather than as horrifying monsters. As ornaments, grotesques were invited into homes and bodies, not banished from them.

Victorian *grottesche* often incorporated all kinds of organic creatures, setting the stage for a new biological paradigm inspired by ornamental design. Inseparably entwined swans, eagles, snails, squirrels, rodents, snakes, flowers, and fruits were common motifs in the patterns. Naturalistic representations of these plants and animals were integrated into a design lexicon teeming with stylish "mutations." In the pattern, the organic world is compressed: bodies are "disfigured" by their transformation into others. The *grottesche* forms embedded in textiles and pottery models biological evolution: bodies turning into other bodies, "disfigurement" the sign of connection between species, "monstrosity" the key to variety. Some scholars have even claimed that Darwin's model of evolution was inspired by Victorian ornamental designs (Menninghaus 2009).

In *grottesche* patterns, variations in form are ornamental, "Classically" if not classically beautiful. The negative affect that clings to the grotesque is the result of a tendency to view shapes through organic paradigms that cast formal irregularity as deformity, the result of a biological "abortion, abridgment, or misdirection" (Harpham 2007, 14). The Neoclassical *grottesche*, however, did not signify within an organic paradigm. Instead, it was a feature of design culture, a mode of inorganic metamorphosis common—and highly valued—within the decorative arts.

The word "silicon" was first used in 1817 and did not refer, throughout the nineteenth century, to electrical chips but rather to a non-metallic element found in great abundance in the ground, and typically therefore also found in the sand-based compounds which comprised Victorian glass, ceramics, paint, and bricks. A substance that contemporary culture associates with modernity—and the future—emerged within pre-Victorian ornamental design. The original silicon-human combinations came from using dishes, wearing clothing. Victorian "cybernetics" was fundamentally ornamental—and culturally pervasive.

And yet, applying the term "cybernetics" to Victoriana defaces the technological sphere to which we typically refer when using the term, and disfigures the Victorian period, an era that was historically premature to and the aesthetic opposite of the sleek digital constructs commonly associated with silica. Using the phrase "Victorian cybernetics," as opposed to the less incendiary " early human-technology hybrids," is admittedly melodramatic, perhaps inappropriately hyperbolic in a critical context. However, etymologically speaking, the word "cybernetic" existed in the nineteenth-century; it was coined to describe pre-Victorian artifacts, and has been appropriated in modernity to describe our futuristic forms—*not the other way around.* Tom Standage's *The Victorian Internet: The Remarkable Story of the Telegraph and the Nineteenth Century's Online Pioneers* (2007) similarly applies modern terms like "internet" and "online" to old-fashioned technology to make the point that that technology was not so old-fashioned, or that contemporary, cutting-edge technology has an older lineage. Whereas Standage's terminology is analogic, however, in the sense that "internet" and "online" weren't nineteenth-century words, "cybernetics" is etymologically accurate. What does it mean when it defies the conventions of scholarly rigor to use a word to describe the period that invented it?

What it means is that certain cultural and scholarly conventions deny the relationship of Victoriana to forms of technology that exceed the sphere of industrial manufacturing. The same conventions bifurcate women, domesticity, and disability from technological objects, which are conventionally masculine, contemporary or futuristic, urban, and seen as perfecting instruments. The obscuring of certain technical forms like the *grottesche* in nineteenth-century culture has obscured networks in which technology is gendered female, generated within the home, embedded in the past, used ornamentally rather than to promote efficiency, and experienced as disabling or deforming, if often stylishly so.

James Abbott McNeill Whistler's "Peacock Room" (1876–1877) evinces precisely the feminized, domestic, bourgeois, ornamental, and deeply disfiguring characteristics of much of Victorian cybernetics. Whistler's room employed the concept of the *grottesche* to explore cybernetic arrangements in the Victorian dining room, the quintessential bourgeois space. The room transformed Whistler's patron Frederick Leyland's family dining room into an enormous silicon bird. Whistler's silicon peacock is one of the many Victorian cyborgs produced by ornamental culture. The room's painted leather walls form a textured feather body, while the embossed, paneled ceiling and lamps evoke an entire flock of necks and beaks.

Whistler's experiment caused a sensation even before its completion. The room was simultaneously anomalous in relation to the domestic decorative conventions of its era, and also thoroughly representative of the converging concerns and iconography of late nineteenth-century merchants, artists, and naturalists. The Peacock Room offered a stylishly monstrous amalgam of human, animal, and silicon forms as a paragon of industrial innovation and global commercial expansion, demonstrating how the latter contexts could render biological evolution through aberration a bourgeois domestic fixture.

Whistler's room was designed to showcase Leyland's collection of Kangxi-era blue-and-white porcelain. It is as a porcelain cabinet that the Peacock Room trumpets an investment in trade and technics, namely by referring to the west's discovery and capitalization on China's cutting-edge export technology: "the hidden art of porcelain" (Froula 2012, 234). Christine Froula explains that "Chinese (and later, through them, Korean and Japanese) artisans alone possessed the knowledge of this art from the seventh century to the first decade of the eighteenth, when, after long effort, European craftsmen succeeded in fabricating this elusive 'white gold.'" The blue-and-white porcelain collected by fashionable Europeans in the nineteenth-century thus signified a form of technological knowledge elusive to, but highly coveted by, the west. At the same time, "the rare products of this mysterious art became central to Eurasian trade and cultural exchange." Leyland's porcelain, which belonged to Great Britain's preeminent shipping magnate, signified the stock and trade of technological competition, commercial expansion, and cultural globalization.

Despite the masculinization of porcelain collection and display proffered by the nineteenth-century market context, however, the Peacock Room was undoubtedly also an icon of feminine domesticity. Victorian critics of the finished space fretted over such wifely concerns as how to "fathom a family saying grace in that room," what "ladies' costume . . . would not interfere with the gold and blue around and above," and whether "commonplace dining would be out of place here, where opium" or "locusts and honey" were the only conceivably fitting foods to serve (Merrill 1998, 254–255). What does one wear to dinner with—in—a cyborg? Whistler hinted at the necessity of pondering such problems to fully experiencing his creation by placing above the fireplace mantel his *Princesse du pays de la porcelain* (1865). The oil painting depicts a figure from one of the Leyland teapots, clothed in a dressing gown and standing on a rug whose patterns also mime blue china. The female body counted among the ornamental porcelains displayed on scaffolding throughout the room. The *sine qua non* of nineteenth-century British commercial expansion and technological invention doubled as a domestic ornament, framed the Western dining room as Asia, and configured British women as china and China.

Whistler's exotic hearth was considered a "chamber of horrors" (Merrill 1998, 251). Many of the artist's contemporaries saw the disturbingly chic amalgamation of women, peacocks, and porcelain as grotesque, just as a quarter-century later the figural distortions of Matisse—also inspired by textile patterns—would generate public outrage as intolerably ugly, even disgusting. Aubrey Beardsley would name the Peacock Room as one of his greatest influences in conceiving the pen and ink figural distortions for which he became known in the 1890s (307–312). Whistler too was drawn to the grotesque vernacular. His painted caricature of Leyland, which followed a violent rupture between artist and patron over the cost and creative vision of the room, depicted "a grotesque hybrid of man and beast horrifically deformed by greed" (287). The massive image, entitled "The Gold Scab," became central to Whistler's conception of the Peacock Room, which oscillated between lurid biology and ornamental wonder. The painting was one of only a handful commenced during the 1870s that Whistler was able to complete. He photographed it, hung it on the wall of his private compartment in Venice, intended according to some art historians to replace "*La Princesse*" with the original "Gold Scab" in the honored mantelpiece wall space of the Peacock Room.

The "gold scab" is consistent with the room's general conversion of organic into inorganic structures. The painting's name suggests the metamorphosis of either a beetle or clotted blood into gold. The central figure sits upon a house, feathers draped over the roof, torso hunched at a piano. Whistler hints at the easy conversion of *grottesche* to grotesque, flirts with it, invites it and condemns it as a travesty of taste. Hovering above the house, Leyland is monster. Grotesques inside the house, seen from within it and according to the home's perspective, are stylish. Interior design reframes grotesque as *grottesche.*

Whistler's focus on the peacock was not a coincidence. The bird was a notorious emblem for designers and biologists alike. In 1871, Darwin observed that male peacocks' plumage impeded their capacity for fighting, fleeing, and hiding from predators. Their feathers came to represent the paradoxical operation of sexual selection alongside natural selection, the former process instantiating a system in the wild whereby "deformities are considered ornamental." Darwin used the term "mutilation" (596) interchangeably with "ornament" (460, 560), pondering the cultural conditions and species specificity under which "unsightly deformities are considered great personal attractions" and noting, in the end, that while certain bodily modifications are "to us a most grotesque notion" it is "not more strange than that the tails of many birds should be especially decorated" (560). Many interior decorators and fashion designers of the 1870s adopted the peacock as their symbol (Neiswander 2008). Whistler brandished the peacock as a sign of the thin line between horror and decoration, and so between bodies and things.

Whistler's room illustrates his designer-sense of the inseparability of radical disfigurement from decorative style as well as techno-commercial invention: the latter was likely to produce disfigurement in the individual and national body while the former could help to assimilate it. His work emblematizes the convergence of technological, decorative, and physiological evolution in the nineteenth century. Though the concerns and endeavors of technical and decorative designers are conventionally polarized, as are both to the natural world, Whistler entertained more fluid and permeable boundaries for technology and art, ornament and nature, matter and life.

NOW AND THEN

Contemporary steampunk sets the future in the nineteenth century. The human-thing combinations usually reserved for laboratories and surgical arenas are therein discovered *already* lodged within Victorian dressing and drawing rooms. Steampunk's axiological anachronism was pioneered by the prominent cyberpunk authors William Gibson and Bruce Sterling, whose co-authored novel, *The Difference Engine* (1990), popularized the emergent literary subgenre in a fabricated Dickensian England where Charles Babbage was successful in building in 1824 the steam-powered computer with which he did in fact experiment throughout the early 1800s. Contemporary steampunk, however, has exceeded its twentieth-century literary roots to become primarily a design aesthetic for clothing, accessories, and furniture. Steampunk's insight into the cybernetic past contained within Victorian personal and domestic craft objects offers a useful theoretical reorientation of history, in which the fictional conceit of a technologically sophisticated nineteenth-century renders perspicuous the myriad experiments in mechanized humanity and "*vital materiality*" (Bennett 2010, vii) that are actually a part of the past.

In addition to extending the technological vanguard backwards steampunk significantly expands it beyond the scientific laboratory and into domestic and aesthetic spaces. Steampunk distinguishes itself from other science-fiction narratives and tropes by articulating similar motifs in the lexicon of Victorian style. The genre's clocks and watches, corsets and umbrellas, bicycles and automobiles all exaggerate, estrange, and thereby re-enchant the mechanics of everyday centuries-old human-thing collaborations. Rather than depicting an innocent, nostalgically pre- or neo-industrial nineteenth-century, steampunk elaborates an era already steeped in the aesthetics of the machine.

As Julie Anne Taddeo illustrates in "Corsets of Steel: Steampunk's Reimagining of Victorian Femininity" (Taddeo and Miller 2012), the corset is steampunk's preeminent symbol of the nineteenth-century's thing-human hybridizing tendencies. As Steampunk has become more popular, the art of corset making has undergone a modern resurgence. Googling "steampunk corset" uncovers thousands of images of the twenty-first century decorative body cages. Worn outside rather than underneath the clothing to expose instead of hide the body's inorganic compositional media, Steampunk corsets are made of a range of materials that mold and meld flesh to metal, leather, wood, and cloth. Nineteenth-century corsets shaped women's bodies using whalebone and steel, and enacted the Victorians' sense that the female body could not comply with its own nature without the mediation of technology. Corsets were prosthetics that externalized and imposed an invisible, so-called "essence."

Nineteenth-century art critics corroborated to configure domesticity and its emblem, the wife and mother, as cybernetic. John Ruskin's *Of Queens' Gardens* (1865) declares that woman's power is for domestic "arrangement": she "sees the qualities of things, their claims, and their places." Victorian womanhood was so entwined the home as to render women incapable of shedding their houses: "Wherever a true wife comes, [her] home is always round her. The stars only may be over her head . . . but home is yet wherever she is; and for a noble woman it stretches far round her, better than ceiled with cedar, or painted with vermilion, shedding its quiet light far, for those who else were homeless" (Ruskin 2002, 78). Between their houses and their corsets—and to make no mention of the thousand other objects appended to their bodies—nineteenth-century middle-class women were conceived of and experienced life as living ornaments.

Contemporary steampunk couture appropriates the corset as equipment, using the design framework as a vehicle for enjoyment. The steampunk corset therefore does not bind sexuality; it is sexually explicit, even pornographic (Taddeo 2013). Steampunk's pornographic corsets are more than merely stylish; they are icons of the design paradigm's romanticized mechanizations. Steampunk machines are neither sterile nor impersonal. They are not radically dehumanized, as in the modernist machine aesthetic. Instead they are tarnished with the marks of human use. They are often therefore unnecessarily elaborate, even baroque, eviscerating what have become sleekly invisible processes in modern technological instruments, to lay bare the material interiority of the machine; in this sense, all steampunk design is pornographic: devoted to the perspicuity of the machine's interior.

The steampunk cyborg is the opposite of Donna Haraway's seamless cultural cyborgs or Andy Clarke's wirelessly "natural born" cyborgs, both of which deemphasize mechanical form where steampunk underscores it (Haraway 1991, Clark 2004). Steampunk mechanics, by offering a non-modernist machine aesthetic more comfortable with, in its refusal to mystify, the close ontological proximity of humans to machines, broadens our ability to recognize prototypes for human-object combinations not merely in the present or in fantasies of the future, but also in the artifacts of the past. Such artifacts are unearthed through close investigation of the cybernetic styles—imagined and real—to which one can become habituated by the ethos of steampunk style and actual Victorian personal and domestic decorative arts.

Nineteenth-century design culture was almost banally cybernetic. It compromised the boundary between organic and inorganic matter simply by drawing inspiration from living nature in the ornamentation of objects. Botanical interiors were made of flowered wallpapers, carpet vineyards, and upholstered gardens. Animal interiors like Whistler's Peacock Room turned urban spaces into exotic wildlife, and gave living creatures silicon-based beaks, feathers, and pigmentation. Drawing rooms and boudoirs were the habitats of ceramic or textile bestiaries, and even, as in the case of the Rampendahl armchair (1860) in the Victoria and Albert Museum's permanent collection, antlered furniture.

Owen Jones's *Grammar of Ornament* (1856), arguably the most influential design text of the era, established the convention of treating nature as a sourcebook for pattern ideas. William Morris applied the idea most famously to wallpaper, lavishing Victorian interiors with portions of the outdoors. He used the concept of "organic Architecture" to define architecture as including furniture, and to position domestic interiors within nature rather than as barriers against it. Morris particularly elevated the home—its walls, furniture, fixtures and wares—as a space of quotidian interdependence between living and material being.

Cybernetic decorations existed in more arcane spaces as well. In 1858, Tiffany's sold commemorative broaches, fobs, canes, and other decorative pieces made from fragments of the world's first transatlantic telegraph cable, allowing human beings to ornament themselves with wire a century before we began to think of humans as "wired" in the sense of being fitted with a listening or other personal computational device. Tiffany's telegraph cable jewelry also perhaps alluded to the Old English "gewired," a decorative term used to describe something ornamented with (gold or silver) piping. Tiffany's evoked an old association between wired and decorated humans, pointing out the latent ornamental tradition embedded in cutting-edge technological invention. The fob design emphasized the connection by replicating the shape and color of the telegraph wire in the surrounding ornamental hardware, making it tough to discern where technology stopped and jewelry began (or vice versa).

William Morris envisioned, along with C. R. Ashbee, Christopher Dressler, and C. F. A. Voysey, a revival of artisanal craftsmanship in which designers and their clients would collaborate to preserve a capacity for human thought and imagination through the valorization of handmade and uniquely ornamented objects over mass-produced things (Saler 2001). Victorian designers negotiated a paradoxical anti-industrialism, however. On the one hand, the elaborate iconography of nature in the patterns of handcrafted textiles, ceramics, and books signified the privileged status of natural states and production methods. Artisanal craftsmanship was meant to protect methods of world-making that facilitated interiority and distinctiveness. On the other hand, design theorists assumed that human interiors could not transcend the flora, fauna, and minerals around them. They crafted a world where things could give shape to emotional and intellectual experiences, and where moods and mentalities expressed themselves in forms and figures (Navarette 1997).

By some accounts craftsmen were even more prone to mental mechanization than factory workers because of the intimacy of their contact with both materiality and tools. Handheld tool-use have been responsible for creating the possibility for human interiority in the first place:

> Conscious reflection in (proto)humans first emerged with the use of stone tools because the materiality of the tool acted as an external marker of a past need, as an "archive" of its function. The stone tool (its texture, color, weight), in calling attention to its projected and recollected use, produced the first hollow of reflection. (Bennett 2010, 31)

Whereas all kinds of technologies, from factory industries to television, cell phones, and video games, have traditionally been blamed for encroaching on mental life, Victorian artisans espoused a trajectory whereby technology is the prosthesis of interiority.

George MacDonald's *Phantastes* (1858: 2000) illustrates the way decorative interior objects dehumanize those who live among them. An initial scene in MacDonald's book describes a path to Fairy Land emerging within a bedroom containing patterned carpets and furniture. The human in this room becomes caught up in the *trompe l'oeil* altercation between second and third dimensionality:

> I suddenly . . . became aware of the sound of running water near me; and looking out of bed, I saw that a large green marble basin, in which I was wont to wash, and which stood on a low pedestal of the same material in the corner of my room, was like a spring; and that a stream of clear water was running over the carpet, all over the length of the room, finding its outlet I knew not where. And, stranger still, where this carpet, which I had myself designed to imitate a field of grass and daisies, bordered the course of the little stream, the grass blades and daisies seemed to wave in a tiny breeze that followed the water's flow. (MacDonald 2000, 9)

The narrator goes on to describe his encounter with a woman who emerges from a secret compartment of his secretary, as well as the transformation into a lush forest landscape of the ivy leaves carved on his wooden dresser, the clematis tendrils of its gilt drawer handles, and the branches and leaves woven into the window curtains. "I found myself," the man observes, "completing my toilet under the boughs of a great tree . . . as the cool morning wind swung it to and fro, like a sinking sea-wave" (10). The speaker has been rendered two-dimensional as he is drawn into the designs around him (carpet, furniture, curtains), but once a part of these designs they all become a three-dimensional landscape. The chapter ends as the speaker walks off into the forest; he seems to leave the interior for an exterior, but given the origins of the exterior, it is clear that he is still inside his secretary, dresser, curtains, and carpet. MacDonald observes the deep cognitive mediation between people and things with which later eras were to become increasingly preoccupied, though also in realms increasingly dissociated from bedrooms, living rooms, and dining rooms or their furniture.

Observing the carbon-silicon combinations that were fixtures in Victorian conceptions of the domestic subject applies steampunk's way of seeing nineteenth-century technology to the historical nineteenth century. Silicon was part of the chemical composition of the glass in cathedral windows, the cement in buildings, and the ceramic and vegetable products within pottery, textiles, and dye. Steampunk allows us to consider human interactions with these objects as ostentatiously cybernetic. When William Morris contended that certain forms "[contribute] toward the revolt against utilitarianism" he was exchanging the notion that humans "use" things for the notion that their minds and bodies are made by their productive and consumptive relations to things (Morris 2004, 330). The carbon-content of the human body mingles with the silicon-content of thread, ceramics, paint and glass to construct an "extended self" (Belk 1988) that would not exist independent of the combination.

THEN

The iconographic ornamental cyborg of the nineteenth century would have been James Merrick, the so-called "Elephant Man." Merrick was discovered in 1884 at a Mile End freak show by the prominent (eventually Royal) surgeon, Sir Frederick Treves. Treves introduced Merrick to England as his prodigy on two occasion that span and unite Victorian and modernist chronologies, first in the months following his discovery of Merrick at a freak show in East London, and the second some forty years later, posthumously for Merrick, within Treves' popular surgical memoire, *The Elephant Man and other Reminiscences* (1923).

Treves' discursive framing of Merrick was as pivotal to recuperating the Elephant Man from the margins of Victorian society as his surgical and philanthropic benefaction was important to Merrick's physical recuperation. Treves' tendency to consider ornamentation a biological issue made him the ideal physician to James Merrick, whose body demanded—if it was to assume a place within society—the treatment not only of a doctor but a decorator.

Merrick's prestidigitation from sideshow freak to stylish celebrity had much to do with Treves' habitual conflation of medical science and domesticity; he was thereby able to address not only Merrick's physical but also his social discomfort. The treatment was facilitated both by elements in Treves' personal history, and an epochal receptivity to the transition. Treves, as a surgeon and a writer, was positioned to think scientifically and aesthetically about his subject—essential in a patient whose complaint was every bit as much about the aesthetics as the function of his physiology.

Treves grew up in Dorset, a town whose ancestral trade was upholstery. The lingering influence of Dorset on the adult physician's sensibility is evinced in the subject of his first book, *Highways and Byways in Dorset* (1906). A surgeon by training, but an upholsterer by heritage, Treves' artisanal sensibility was explicit in his public discourse on medicine. The *Reminiscences* of his life as a surgeon describe the vagaries of private and institutional decorated interiors as much as those of human bodies. That surgical memoire concludes, rather bizarrely, with a chapter entitled "A Question of Hats." The chapter is not atypical in the volume, which otherwise brims with descriptions of the decorated interiors of the grand homes into which Treves was called as a physician. The book introduces the doctor as a domestic fixture, each house call an opportunity for interweaving medical and decorative insights. The book reminds us that nineteenth-century doctors, unlike other members of the emerging professional class, performed a masculine version of "house work."

Treves treated Merrick primarily by first rhetorically and then actually shifting his patient's context, moving his patient from biology, the reigning discourse in Treves' 1886 and 1890 *British Medical Journal* articles on Merrick, to upholstery, the discourse of choice in the 1923 *Reminiscences*. In the latter text, Treves describes Merrick's deformities using imagery reminiscent of Victorian textile patterns. His only detailed description of Merrick is based not on the man but on a "painting" (4) of him, a "life-size portrait" (1) displayed in the Mile End greengrocer where Treves first found Merrick. Treves' *ekphrasis* of the portrait, the sole physical description of Merrick in the book, figures medical treatment as repainting. Treves' restored image borrows its conventions from Cubism, which didn't exist when Treves met Merrick but did exist by the time he wrote his memoire. Cubism's formal lexicon, like *grottesche* design, appreciates physiological displacements. Treves describes, for instance, a forehead that "occluded one eye," a head whose "circumference . . . was no less than that of the man's waist," a nose "only recognizable as a nose from its position," an "enormous" and "shapeless" right arm, and a "horrible" back that "hung as far down as the middle of the thigh" (4, 5).

The surgeon doesn't simply repaint but remakes Merrick, describing him as a fresco or textile ornament. He explains how Merrick's human features are interwoven with anomalous vegetable and animal forms, emphasizing a brow whence "project[s] a bony mass like a loaf, while from the back of the head [hangs] a bag of spongy, fungous-looking skin, the surface of which [is] comparable to brown cauliflower" (4). Merrick's mouth is a "pink stump," his jaw a "rudimentary trunk or tusk," his face "a block of gnarled wood." He also possesses "cauliflower skin," a "radish" thumb, fingers of "thick, tuberous roots," and "a fin or paddle rather than a hand." From his chest hangs "a bag . . . like a dewlap suspended from the neck of a lizard" (4, 5). Merrick is displaced from the genre of biological sketching or photography to Renaissance ornamental design, or one of the Victorian textile patterns that might have formed part of the collection of designs owned by the Treves' family upholstery business.

Though Treves' discursive remaking of Merrick occurred in 1923, it was the culminating reflection on the reframing process begun in the 1880s. These changes transformed Merrick into a Victorian celebrity before his death. According to Treves, the Elephant Man became a pen pal to the Princess of Wales and future Queen of England. He was "visited by almost every lady of note in the social world" (Treves 1923, 22), and became habitually "conversant with duchesses and countesses and other ladies of high degree" (23). A passionate conversationalist despite the impediment to speech presented by his deformities, the Elephant Man became a "dandy and . . . young man about town" (27) who received fashionable women in his book-lined hospital room, attended the theater ensconced by hospital sisters in evening dress, vacationed in the country, and loved Jane Austen's *Emma*. In the solitude of his private hospital rooms, he liked to dress the part of "the Piccadilly exquisite, the young spark, the gallant, the 'nut'" (28) using props kept in a theatrical dressing-bag given him by Treves as a Christmas present.

Expressing himself in "terms any courtier would have envied" (25) and surrounded by "dainty knicknacks and pretty trifles that . . . may almost have befitted the apartment of an Adonis-like actor or . . . a famous tenor" (25–26), Merrick experienced a "glamour[ous]" (26) change in social standing during the final four years of his life. He was, in the company of his elegant friends, one among the tchotchkes that decorated his and their rooms. He was not fearsome but chic. With the help of his doctor, John Merrick experienced an "objectification" as a decorative ornament that afforded exhilarating possibilities for self-invention. He claimed his deformities as a style.

Casting "human figures . . . as decorative objects" until humans "assume the status of bric-a-brac, as though they were figurines" (Brown 2003, 140) is not here the effect of cultural and political "othering," nor is it the result of "the so-called reifying effects of a society permeated by the commodity form" (13). Instead, Treves helps Merrick to overcome "the subject/object opposition . . . [and contest] the ontological distinction between thoughts and things" (3). As an ornament, Merrick explores a "vital materialism," one that detects, studies, and imitates "the presence of *impersonal* affect" (xv, emphasis added), seeks and exploits the "vitality of matter," and does not reduce "agency to *human* agency" (xv, emphasis in original).

Merrick's upholstered subjectivity sharply contrasts with "humanity"; paradoxically, it is the latter category, dependent upon organic conceptions of the body and self that Merrick was denied because of his disfigurement, which Merrick experiences as destructive. When treated as a human at London hospital's medical college, where Treves first examined Merrick, the patient's deformed body signified the organically grotesque features of human and animal bodies distorted by the vicissitudes of reproductive mutation and disease. Within the scientific paradigm, a shape like Merrick's is a "mistake" to be studied, corrected, prevented, endured, or overcome. Treves surgical interest in Merrick at that point is in cataloguing nature's errors, against a cultural backdrop in which the new science of evolution had revealed that species distinctions were permeable and the existing shapes of biological entities no longer guaranteed. If natural forms were always changing, and slowly changing into something else, then nature was no longer "natural" in the sense of being predetermined and complete, and the grotesque body, composed of incongruous organic parts, signified the terrifying capacity of evolving forms to resemble devolving forms and suggested a latent monstrosity in every living thing. The Elephant Man, whom Treves describes as creating riots and mobs everywhere he goes without the cover of his cage or a disguise, elicited public outrage against disfigurement as an abomination of nature (Asma 2009).

Treves appealed to an alternate context in his discursive reconstruction of the Elephant Man as an upholstery pattern. Treves' *Reminiscences* carefully evoke bourgeois contexts in which Merrick's outlandishly foreign shape, the product of an African jungle in his advertisement poster, could be comfortably reassigned to the domestic sphere. As an exotic animal or misshapen human the Elephant Man might be thought to belong in a zoo, circus, or asylum. But as a sofa pattern he is welcome, and admired, in the parlor. Treves develops the conceit by recasting Merrick's indecipherable speech as a foreign language that he "learnt" (14) as though it were French, "occasionally . . . act[ing] as an interpreter" (14). Visitors "learnt his speech so that [they] could talk freely with him," casting Merrick's speech impediment as a foreign language demanding fluency by cosmopolitan socialites as eager to engage in drawing room conversations with the Elephant Man as they would be with an interesting Italian.

Merrick's skin was also recast as bric-a-brac. Anthropodermic bibliopegy, the binding of books in human skin, was an honor bequeathed to prominent dead anatomists whose skins were used as textbook bindings. For Dr. Treves, a celebrity doctor who could reasonably expect his skin to be used this way after his death, it would have been a small imaginative leap to see Merrick's skin as potential vellum. As one versed in the techniques and materials of upholstery, he would have been able to identify particular textile genres and patterns that Merrick's skin resembled. Additionally, the combination of plant and animal exteriors figure Treves' subject as the page of an illuminated manuscript, in whose ornamental borders bestiaries resembling Merrick commonly resided.

In Treves's description of Merrick, the patient is surrounded by books: his romantic sensibility the product of having become filled with books, his transformation affected by the implication that we might see his body *as* a book. The genteel prevalence of human tanning would further have connected Merrick's skin to *grottesche* patterns, since Victorian library wall coverings were commonly made of embossed leather. Treves' displacement of the Elephant Man's grotesque body from biology to the decorative arts is thus completed by the geographical displacement of his skin from the African jungle, where his advertisement poster depicted him, to the Victorian library in which he comes to dwell and which his skin might even comprise. Merrick's "reification," "understood not just as a social condition, but also as a rhetorical maneuver, and as a phenomenology" (Brown 2003, 162), describes a "self's relation to itself" in terms of "the canniness of finally settling into one's mind as though one were settling into one's study."

Yet another means through which Treves weaves Merrick's deformities into the civilized *grottesches* of the Victorian living room is the gendering of his subject according to one of the physiological displacements suggested by his body. In contrast to Merrick's uselessly deformed right arm, "the other arm was . . . not only normal but was, moreover, a delicately shaped limb covered with fine skin and provided with a beautiful hand which any woman might have envied" (5). Part of the clash of incongruous organisms suggested by Merrick's body, the left arm incorporates facets of idyllic beauty into its cacophony of shapes. The enviably delicate feature of a *grottesche* design, in this case Merrick's feminine hand, serves as a shadow underscoring the ornamental, as opposed to biological, nature of the form.

In the 1880s, however, Merrick's hand also connotes the feminization of the human taste for ornament. On the one hand, Darwin had explained that even physical deviations as dramatic as "mutilations" could be appealing to female organisms' taste for variation. On the other, the industrialization of craftsmanship shifted the association of the trade with masculine artisans to the female consumers of machine-made products. Treves' discursive realignment of Merrick with the latter is suggested to Treves by Merrick's existing left hand, a feature embellished upon until Merrick comes entirely himself to resemble the "happy woman, free from any trace of cynicism or resentment" (17) that is suggested by his left appendage.

The surgeon, in upholstering his patient, cures Merrick of his "fear of people's eyes" (19) by helping him to become something that ladies look at with approval and even desire. Merrick likewise adopts for himself the sensibility, subjectivity, and value system of one of those ladies: he is both the stylish male *and* the discriminating female.

Explains Treves: "To secure Merrick's recovery and to bring him, as it were, to life once more, it was necessary that he should make the acquaintance of men and women who would treat him as a normal and intelligent young man and not as a monster of deformity. Women I felt to be more important than men in bringing about his transformation" (20). Treves enlists a young pretty woman to enter Merrick's room shortly after his arrival at the hospital and greet him with a smile, handshake, and pleasant "good morning." Merrick is deeply moved, explaining afterwards to Treves that she is the first woman who has ever smiled at him or shaken his hand. Treves asserts that it is "from this day" that "the transformation of Merrick commenced" (22). He becomes a particular favorite of "[ladies] of note in the social world" (22), who bring him presents and "[make] his room bright with ornaments and pictures" (23). Amidst them, and the books and objects they bring him, he grows "less conscious of his unsightliness, a little disposed to think it was, after all, not so very extreme" (23).

Expanding the female self that was suggested by Merrick's delicate left hand allowed the Elephant Man to combine, within a single person, nineteenth-century female consumer desire and the objects of bourgeois acquisition. Merrick enacts female desire by "[falling] in love" (28) with each "attractive lady" he encounters, emblematizing his adoption of the persona of "Don Juan" by daily "[laying] out upon his table, with proud precision," objects a gallant lover might own: "silver brushes," "razors," a "shoe-horn" and "silver cigarette case" (28). He is able to make love to the ladies as both the owner of the objects and as one of the objects, knowing he is of the genre of articles they might desire: he is to their/his taste.

As the habitat of the *grottesche*, houses become a passion of Merrick's, their decorated interiors eliciting particular study. A "dining room with plate on the sideboard and with easy chairs into which [a] hero could 'fling himself'" (29) excite "[absurd] interest" and "untiring curiosity" (30). He longs to see a "white marble staircase," "gilded mirrors," "brocaded divans" (30) or any "modest dwelling of the Jane Austen type" (30). It is "the contemplation of all these things," Treves explains, that allows Merrick to imagine that he too is "the 'real thing'" (28).

Merrick's example illustrates the extent to which objectification, typically viewed as a form of cruelty, can facilitate empathy; indeed, in a culture in which subjectivity was formed through surrounding surfaces, it may have been the only way of establishing empathy with foreign forms and selves. (Kunin, 2010, 87–108) Merrick was twice strange in the organic context because his deformities marked him as a terrifying sign of the possibility of biological devolution, and because biological devolution was associated with so-called primitive and savage cultures. As furniture, however, he was not exotic but a domestic fixture, an intimate interior object. Reupholstering his figural deviations facilitated his reconstitution as English. Interior decoration was a way of engaging with the home as the space of cosmopolitan domesticity, and the self as a global surface. Home and self were refashioned as the world, *gewired.*

NOW AND LATER

Today, disability is having a moment. The vampires and zombies ubiquitous in recent consumer culture have been figures of infection and human-thing hybridization that have served as a collective throat clearing for the real conversation, about a category of humans that will soon contain almost anyone born in an industrialized nation, and many who aren't. As Solomon pointed out, modern medicine is manufacturing a group of people who didn't survive in large enough numbers to count for most of history. The deeper and earlier medical technologies interfere, the bigger, more inescapable, and less visible the group.

When physical disabilities are visible, however, there are those who like to show them off. Consider a turn-of-the-century catwalk. It's 1996, and the catwalk is Rei Kawakubo's for *Comme des Garçons*. The show is "Dress Meets Body, Body Meets Dress," but fashion critics call it the "Quasimodo collection"—or simply "Lumps and Bumps." Models resembling overstuffed sofas stomp and slouch in dresses equipped with what Voguepedia will later describe as "bulbous . . . padding": bustle-like protrusions from the chest, back, hips, and shoulders. But in 1996, *Vogue* just calls them "the tumor dresses." Kawakubo's medical makeovers erode the difference between sick and chic. They say: if style is divergence, then all divergence contains the potential for style. The dresses also make it hard to tell homey from exotic. "Quasimodo" reupholsters deformities, swathes them in gingham until they move from lab to living room. Dress meets body, makes furniture.

Fast forward: 2013, more running track than catwalk. Adidas launches its Springblade running shoe, equipped with sixteen polymer blades in each sole. The springy blades evoke the curved carbon fiber prosthetic legs worn by the South African Olympic sprinter Oscar Pistorius. Tipped in red so that the shoes seem to sprout like flames from a fleet foot, they also look really cool. Ads that flash between images of Springblade and Pistorius imply without irony that wearing the shoe will give all athletes the advantages in style and speed enjoyed by a cripple. The Adidas ad can exist because fashionable amputees now abound. Aimee Mullins, the double-amputee who modeled hand-carved solid ash prosthetic legs for Alexander McQueen in 1998, collects legs like shoes. In her 2010 TEDx talk, she tells the story of wearing her height-increasing "Barbie" prosthetics to a party. Her short friend pulls her aside and moans: "but Amy, that's not fair!" And it isn't, no more than that Pistorius gets to race on springy blades instead of feet.

Fashion and the popular media are rehabilitating physical and mental disabilities at a lightning pace. And they're doing it without the usual campaigns for social sympathy or civic action. Without much concern for education or scientific knowledge, designers and storytellers have advanced restorative approaches to illnesses in the face of which science and politics have been relatively helpless, and excruciatingly slow. They are doing it at the speed of fashion, whose pace is, after all, the most notorious thing about it.

Contemporary and historical examples of stylish pain and disfigurement point out design's potential to revolutionize medicine. Deaf, dwarf, and autistic communities have already advocated for experiencing symptoms as style. For them, silence, smallness, or stimming are not merely tolerable but enjoyable: vehicles of creativity, community, and sometimes, even fun. Stuart Murray, a literature scholar and the parent of an autistic son, has described his quest to understand autistic pleasure through a study of Melville's autist, Bartleby the Scrivener. Bartleby uses the language of preference, not suffering, to explain his own strangeness (Murray 2008, Murray 2011). He has a taste for it, one that Murray is trying to learn the way you might develop a taste for black coffee, dark beer, stinky cheese, or peaty scotch. These are difficult but ultimately epicurean tastes. What would it mean to apply the same aesthetic logistic to disabled forms and behaviors?

To experience physical or mental disabilities with enjoyment requires more than the insistence on civil support and social awareness that the middle-class parents of disabled children have been demanding in recent decades. Tolerance is a civil and social issue. Pleasure is an element of style. Design is its "home discipline." One of design's key insights includes its treatment of inorganic objects as deep mediators of the lives and experiences of organic entities. Light through a window can shape a plant's color and direction of growth, a shoe or skirt of a particular shape molds the human gait, a soaring ceiling creates mental and emotional sensations fundamentally different from a low one. Designers study the lives in things the way doctors study life in the body. They can teach physicians the ability to see variable physical forms under the light of the less punishing, and even often emancipatory, lamp of ornamental style.

REFERENCES

Asma, Stephen. *On Monsters: An Unnatural History of Our Worst Fears*. New York: Oxford University Press, 2009.

Belk, Russell. "Possessions and the Extended Self." *Journal of Consumer Research* 15:2 (September 1988). 139.

Bennet, Jane. *Vibrant Matter*. Durham, NC: Duke University Press, 2010.

Brown, Bill. *A Sense of Things: The Object Matter of American Literature*. Chicago: University of Chicago Press, 2003.

Clark, Andy. *Natural-Born Cyborgs: Minds, Technologies, and the Future of Human Intelligence*. New York: Oxford UP: 2004.

Darwin, Charles, *The Descent of Man and Selection in Relation to Sex*. New York: Penguin, 2004.

Froula, Christine. *Modernism/Modernity* 19:2 (April 2012), 227–254.

Haraway, Donna. "Cyborg Manifesto: Science, Technology, and Socialist-Feminism in the Late Twentieth Century." *Simians, Cyborgs, and Women: The Reinvention of Nature*. New York: Routledge, 1991. 149–181.

Harpham, Geoffrey. *On the Grotesque: Strategies of Contradiction in Art and Literature*. Aurora, Colorado: Davies Group Publishers, 2007.

Kunin, Aaron. "Decoration, Modernism, Cruelty," *Modernism/modernity* 17:1 (January 2010). 87–108.

Kurzweil, Ray. *The Singularity is Near: When Humans Transcend Biology* . New York: Penguin, 2006.

Macdonald, George. *Phantastes* (1858). Grand Rapids: William B. Eerdmans Publishing Company, 1981.

Menninghaus, Winfried. "Biology á la Mode: Charles Darwin's Aesthetics of Ornament." *The History and Philosophy of Life Science* 31:2 (2009), 263–278.

Merrill, Linda. *The Peacock Room: A Cultural Biography*. New Haven, CT: Yale University Press, 1998.

Morris, William. "Gothic Architecture." *News From Nowhere and Other Writings*. Ed. Clive Wilmer. New York: Penguin, 2004.

———. "Some Hints on Pattern Designing." *News From Nowhere and Other Writings*. Ed. Clive Wilmer. New York: Penguin, 2004.

Murray, Stuart. *Representing Autism: Culture, Narrative, Fascination*. Liverpool: Liverpool University Press, 2008.

———. *Autism*. New York: Routledge, 2011.

Navarette, Susan. *The Shape of Fear: Horror and the* Fin de Siécle *Culture of Decadence*. Louisville: Kentucky University Press, 1997.

Neiswander, Judith A. *The Cosmopolitan Interior: Liberalism and the British Home, 1870–1914*. New Haven, CT: Yale University Press, 2008.

Ruskin, John. "Of Queens' Gardens." *Sesame and Lilies: John Ruskin*. New Haven, CT: Yale University Press, 2002.

Siebers, Tobin. *Disability Aesthetics*. Ann Arbor: University of Michigan Press, 2010.

Taddeo, Julie Anne. "Corsets of Steel: Steampunk's Reimagining of Victorian Femininity." *Steaming into a Victorian Future: A Steampunk Anthology*. Taddeo, Julie Anne and Miller, Cynthia J. (Eds). New York: Rowman and Littlefield, 2012.

Treves, Sir Frederick. *The Elephant Man and Other Reminiscences*. London: Cassell and Company, 1923.

———. *Romford Garden Suburb Exhibition Guide* (1911).

Zamperini, Alessandra. *Ornament and the Grotesque: Fantastical Decoration from Antiquity to Art Noveau*. London: Thames and Hudson, 2008.

Chapter Eight

Extending "Extension"

A Reappraisal of the Technology-as-Extension Idea through the Case of Self-Tracking Technologies

Yoni Van Den Eede

"Welcome to the future, where technology isn't just around us. It's on us. It's in us. It's an extension of ourselves." This is how a recent AOL video series about innovations in wearable technology, titled "Hardwired," is introduced by its host, YouTube star iJustine (AOLOnTech 2013). In the series, iJustine sets out to explore and test-drive devices such as self-tracking technologies, brainwave reading tools, eye-tracking devices, new virtual reality headsets, and so on. What is interesting first of all are those inaugural words of promotion. The succession of sentences has a quaint rhetorical beauty about it. With apparent obviousness, the notion of "extension" tops off the use of anaphora in "It's on us. It's in us. It's an extension of ourselves." We are being persuaded that the real-life developments portrayed by these words are in any way logical and laudable. The future. Extension. Finally, we have made it.

We live in times in which technology seeps into every nook and cranny of our existence, but in which there is also a lot of talk about it. Those two levels, of objective-material conditions on the one hand and discursive interpretation on the other hand, do not necessarily correspond, let alone coincide. The scholarly study of technology is continuously challenged to articulate the difference between its object of study and the ideological discourse about it. When one hears iJustine, in her admittedly infectious style, report on groundbreaking technological advances, one should keep this split in mind. Nevertheless, at the same time there is no reason for us scholars to be pretentious: we, also, do not have an exclusive, ideologically untainted access to the object. We are part of the cultural context, marked by deep-rooted existential expectations, wishes, and needs just as much. "Fish don't know water exists till beached," Marshall McLuhan quips (1970, 191). To extend the metaphor somewhat: they may have, in their very own fish-like way, a "story" about the water. But that story is never the thing; "the map," in Alfred Korzybski's words, "is not the territory" (1994).

This is not to say that in investigating issues of technology, one should not try to relate the two in a more or less accurate manner—if only by critically comparing different stories and words. Looked at in this way, iJustine's plea for a future rife with technological extensions can be said to retrieve in just a few lines a couple of the most tenacious problems in the philosophy of technology. Where do "we" humans end and "they" technologies begin? One cannot be so sure. Even though in the last decades, so many heralds of posthuman eras have been going around proclaiming the modernist subject to be a false construction, the old humanist account, hinging as it does on the infamous subject-object dichotomy, still haunts our dreams. And if not our dreams, then certainly our pop-cultural vocabulary. How else could iJustine so swiftly put in a row the notions of the "around," the "on," and the "in," and suggest this sequence to have followed a linear, progressive timeframe? An object can only be said to be able to approach, invade, and conquer a subject in that way if the two were once clearly separable entities.

The concept of extension, then, partakes in this paradoxicality in a sublimely illustrative way. Notwithstanding its long history, it keeps eliciting debate, hovering somewhere in between humanist and posthumanist discourses. If we humans extend ourselves, upon what base can we say that extension takes place? As Ciano Aydin (2013) observes—albeit specifically in the context of a discussion about the "extended mind thesis"—the word "extended" itself already suggests an autonomous, extendable entity to be present before any extension happens. Whereas in our present scholarly climate, we would rather prefer to consistently speak in terms of networks, assemblages, ecologies, mangles, imbrications, and so forth, instead of surreptitiously confirming the old essentialist subject-object scheme once more. Is the "extension idea" in that sense not just a moot remnant of our humanist heritage, to be discarded as quickly as possible by us full-fledged postmoderns?

Apart from the particular position that the extension idea takes up nowadays within pop-cultural discourse and scholarly theory, however, there is the pertinent question as to how alive it *should* be. In what follows, this question will be explored, partly in a conceptual way, partly through a case study. In order to assess whether the extension idea stands the test of time, the concept will be framed through the prism of an artifact—again taking a hint from iJustine—that has much actual relevance: self-tracking technologies. Conversely, and concurrently, an investigation of that artifact from the perspective of the extension idea will be carried out. If it turns out that the latter is able to analytically clarify the former, then the case for the former will be fortified.

In accordance with this schema, the paper unfolds in two large parts. The first part is theoretical and relatively wide in scope, and expounds the extension idea in its historical and conceptual dimensions. First, the history of the technology-as-extension concept is (briefly) revisited, and the ongoing contemporary debate on it discussed. Then, the argument gradually zooms in, singling out a particular and crucial interpretation of the extension notion, perhaps the best-known of all, that of media theorist Marshall McLuhan. The second part of the paper, then, is more focused and specific, being in part empirical. Here, the case of self-tracking technologies is deployed. First, a description of what they are is sought after, followed by a succinct review of the research on them. Subsequently, they are scrutinized through the extension idea—in its McLuhanian version—and vice versa, the extension idea is critically investigated through them. This analysis will then be situated within the contemporary debate on the extension idea and on the human-technology relation, discussed in the first part. On the basis of these explorations, it will be argued that the extension idea, in its McLuhanian guise, can serve as an analytic heuristic, a conceptual tool for instigating critical awareness about the dynamics of technology and human-technology relations, holding a unique place within the philosophical study of technology.

THE EXTENSION IDEA: A BRIEF HISTORY

Since at least the nineteenth century up until this day, the extension notion has been fairly prevalent in analyses of technology. Indeed, "[t]he idea of extending the human being with technologies could be seen as the most obvious and widespread conceptualization of the relation between human beings and technologies," as Asle Kiran and Peter-Paul Verbeek contend (2010, 412). Although it has been referred to under different names—among them "extension theory" and "extensionism"—here the rather modest denomination "extension idea" is employed, in order to delineate the matter at hand from concepts of "extension theory" as used in mathematics and agricultural policy.

In its most rudimentary form, the extension idea could be circumscribed as follows: Technology is an extension of the human being, of human organs, body parts, senses, capabilities, and so on. That basic theme is said to go as far back as Aristotle (cf. Lister et al. 2009, 90; Mitcham 1994, 176). However, it might have been the emergence of the Industrial Revolution that truly sparked its ascent to conceptual fame. Ernst Kapp, nowadays known as the coiner of the term "philosophy of technology" (Mitcham 1994, 20), crafts a comprehensive inquiry into the nature and functioning of tools and technologies around the notion of "organ projections." He sees for example the railroad as an externalization of the circulatory system, the telegraph as an extension of the nervous system, and so forth (Kapp 1877). Around the same time, Ralph Waldo Emerson as well as Samuel Butler make references to a similar idea (Emerson 1876, 129; Butler 1872, 136).

During the first half of the twentieth century, the extension idea returns in the context of several developments within the then vigorous tradition of philosophical anthropology. This connection is interesting, for a concept first and foremost used to define technology gets deployed here within a discipline that has mostly been endeavoring to define the human being. Arnold Gehlen, for one, sees the human being as essentially an undetermined creature, in the sense of "insufficiently defined" but also "incomplete." We are a deficient being (*Mangelwesen*), but we compensate for this lack by taking action, more specifically by deploying tools and prostheses (1988; 2003). Exactly because of this deficiency and the corresponding potential for compensation, the human being is—borrowing from Max Scheler—"world open": our survival is not dependent on a particular environment. We carry our technological environment to a certain extent with us. In a more specified vein, then, Ernst Cassirer sees technologies as organic projections, although this description is meant to have a far-reaching scope: he attempts to understand technology as a form of spiritual creativity not unlike myth, language, and art (1955, 215; cf. also Curtis 1978, 69). Subsequently, Cassirer influences Jean Gebser, who defines a machine as an objectivation (*Entäusserung*), that is, an externalization (*Hinausstellung*) or "in psychological terms" a projection of human capacities (1985, 132; cf. also Curtis 1978, 70–71). The projection makes these capacities and their structure visible and conscious. Conversely, this entails that technologies make up a part of consciousness.

Such analyses already ring a familiar tone, as they move into the direction of ideas that we know today under the headings of "extended mind," "distributed cognition," and such. Before the extension idea takes on these more contemporary forms, however, it passes through an even more holistic period, so to speak. Where in the first instance it serves to delineate what technology is, then turns out to be able to shed some light on our human condition as well, it ultimately goes on to find a place within all-encompassing evolutionary frameworks, a few of them even with religious undercurrents. Pierre Teilhard de Chardin (1963) sees the human being and what he calls the "social phenomenon," that is, the extension of the organic, as part of one and the same system. The technical and the biological, in his view, converge. Together, they are continually growing and evolving toward the "Omega Point" (1964).

Teilhard de Chardin's thinking thereafter exerts a great influence on Marshall McLuhan, in whose work the extension idea perhaps finds its most famous instance. McLuhan's version deserves a special place in this overview. It, too, like Teilhard de Chardin's, has a certain all-embracing character to it, but it can be said to be unique in stylistic, conceptual, as well as methodological regard; the next section will elaborate this point more extensively. Moreover, McLuhan's reading can in a way be perceived as an axis around which the history of the extension idea rotates. On the one hand, it incorporates the foregoing approaches to a large extent, though admittedly not in a very systematic manner. On the other hand, it can be alleged to foreshadow subsequent, more contemporary accounts. For after McLuhan, the extension idea again finds other shapes, as by the 1970s and 80s another era is on its way in which the notions of the "cyborg" (Haraway 1991) and of the "posthuman" (Hayles 1999) are eventually worked out. McLuhan's extension theory anticipates those concepts, as Joost van Loon (2006, 168) and Jay David Bolter and Richard Grusin (2000, 77) argue. By extension, the same could be suggested in relation to what is perhaps the most palpable representative of the extension concept in the literature today, and that is to be located in cognitive science (cf. MacDougall 2010). The "extended mind thesis," first worked out by Andy Clark and David Chalmers (1998), proposes that cognition does not simply reside inside the brain or the "skin bag," but extends into the environment, into our bodily functioning, and into the tools we use (Clark 2003; Logan 2007). Also the related notions of "embodied cognition" (cf. Lakoff and Johnson 1999) and of "distributed intelligence" (cf. Hayles 1999) have a clear conceptual connection to the idea of technologies as extensions of the human body and of cognition.

Nevertheless, these applications are a bit further removed from the classic form of the extension idea as a general definition of technology, and thus, by implication, from McLuhan's version. In its more "generic" form, the extension idea keeps eliciting interest especially in the philosophy of technology. Larry Hickman is said to have retrieved the extension notion in working out a "naturalistic" stance on technology, based on the works of John Dewey (Garnar 2012; Hickman 2001). Philip Brey (2000) discusses Kapp's, McLuhan's, and David Rothenberg's theories and subsequently formulates his own account of the extension concept, claiming technological artifacts to extend not so much human organs (Kapp) or faculties (McLuhan), but rather humans' natural capacities—bodily and mental—for acting upon intentions. Clive Lawson (2008; 2010) further updates the extension idea significantly in elaborating an intricate model that integrates elements from actor-network theory and Andrew Feenberg's critical theory of technology, seeing extension not simply as an addition to human powers, but as an activity, that takes place in "technical and social networks of interdependencies" (2010, 217), in which material artifacts serving as extensions of human capacities must be "enrolled" in order to work.

Clearly, the extension idea is still alive and well; and some debate about the pros and contras of the concept has been going on, too. Kiran and Verbeek (2010) work out an elaborate critique of the extension idea, claiming that it harbors an intrinsic instrumentalism: it suggests that technological extensions are nothing but neutral, transparent additions to a pre-given human being. They call such a view "external": technologies and humans are seen here as independent entities, and the relation between them—the extension—as an external supplement to both. Kiran and Verbeek find this thought back in a diverse array of theorists and theories, among them Kapp, Gehlen, McLuhan, and Clark. In an ethical sense, this approach can only lead to an assessment of technology in terms of either naïve reliance or outright suspicion, a perspective that in a practical sense often takes the form of an analysis of risks. A third way is possible, Kiran and Verbeek suggest, when we take another starting point: technologies should not be seen as transparent intermediaries; they co-shape the human being. This is the "internal" view, that assumes that the relation between humans and technologies in fact constitutes both. Both are always interwoven. Technologies are not neutral instruments, they—in a Heideggerian sense—help to reveal and conceal facets of human life. And "[t]he revealing–concealing aspect of technologies, the co-shaping of the man–world relation, is lost in the extensionist idea" (Kiran and Verbeek, 418). The internal approach, then, leads to a wholly different ethical advice: instead of looking for risks, out of distrustful motives, we should realize that we are already always intertwined with technology. Confidence toward, and actively though deliberately trusting ourselves to technology are better options.

In a commentary piece, nonetheless, Richard Heersmink (2011) responds to Kiran and Verbeek's criticisms, taking up the defense of the extension idea. He distinguishes between weaker and stronger versions of "extension theory." According to Heersmink, weaker approaches, such as McLuhan's, Gehlen's, and Brey's, are close to what Kiran and Verbeek call the external view. But stronger versions, such as Clark's, exhibit more similarities with the internal approach than Kiran and Verbeek would like to admit. These perspectives, says Heersmink, also hold that human-artifact relations make for "new systemic wholes"—Clark's phrasing—in which "a strong symbiotic and interdependent relation between humans and extending technologies" (122) reigns. There is no ontological distinction between the human and the technology here. Moreover, Heersmink dismisses Kiran and Verbeek's charge that given its instrumentalist bend, the extension idea cannot accurately account for the psychological, cultural, and social impact that technology has on human beings, on their subjectivity and existence. For example, in the work of McLuhan, the study of these influences takes central stage.

Beyond Heersmink's response, it should further be remarked that the situation is even more confused than that, to the extent that the extension idea, even though Kiran and Verbeek label it as an instrumentalist stance, has actually often been associated with technological determinism. This is certainly the case with McLuhan's work. Also, when Heersmink acknowledges Kiran and Verbeek's critique that extension theory, first, does not adequately deal with the revealing-concealing—or, in postphenomenological terms, magnification-reduction (Ihde 1990)—structure of technologies, and, second, that it does not sufficiently account for the changes in our perception that extensions bring along, it should be pointed out that in fact in McLuhan's thought, these aspects have been investigated in considerable detail, albeit waywardly. That is the topic of the next section.

MCLUHAN'S RENDITION OF THE EXTENSION IDEA

The work of McLuhan has been discussed repeatedly of late (amongst others, cf. Van Den Eede 2012). Here, the focus lies specifically on McLuhan's rendition of the extension idea, in order to investigate how it holds up against recent theoretical as well as technological developments. As became clear in the short historical overview, McLuhan writes himself into an already long-existent tradition when he adopts the notion. In what follows, the idea will not be traced back to its roots in his thought or compared to some of his forerunners' interpretations (for an extensive list of the available literature, 137ff.). Surely McLuhan's version, given his typical eclectic "sampling" style—where he picks up terms and concepts from various fields and authors and, without caring too much for systematic exegesis, idiosyncratically incorporates them into his own body of thought—shares some characteristics with versions that went before. But what makes his special? Three things could be proposed: its metaphorical elaboration, its ontological foundations, and its methodological underpinnings and consequences.

First, the metaphorical elaboration. McLuhan describes media and technologies as extensions of human senses, body parts, or capabilities: script of speech, the wheel of the foot, and so forth. In the process of extension, however, all sorts of strange events occur. McLuhan uses the myth of Narcissus to clarify. Narcissus falls in love with his reflection in the water, that is, the extension of himself. In the exact same manner, we fall in love with the extensions of ourselves in technologies. But McLuhan stresses that in this context, one should not interpret the myth the way in which it is usually done, namely, presuming that Narcissus recognizes the image as himself and consequently falls in love with himself. No, Narcissus stays completely unaware of the fact that the image is an extension of him, just as we perceive technologies as foreign material, "things out there," and remain oblivious of the fact that they really hail "from us." Thus McLuhan states, responding to some of his critics: "I think of technologies as highly identifiable objects made by our own bodies. They feel that technologies are strange, alien intruders from outer space" (McLuhan and Stearn 1967, 280).

What makes and keeps us blind to the true origin of media? We are victim to a numbness, a narcosis: the Narcissus narcosis. To elaborate on this, McLuhan (2003) borrows some concepts from medical research, among them "irritation," "counter-irritant," and "autoamputation." Certain troubles that we experience in our environment—possibly brought on by older extensions—spur us into action. They function as irritation. McLuhan gives the example of written and monetary media, the introduction of which resulted in an intense acceleration of societal life and of communicative exchange. As a consequence, a counter-irritant was called into life, namely, the wheel, as an extension of the foot, in order to keep up with the sped-up pace. This new counter-irritant, in turn, by "amplifying" or "enhancing" that capacity of the human organism, made for "a new intensity of action" (McLuhan, 64), unbearable to the body. This is one of the central—unfortunately empirically not so convincing, but philosophically metaphorically still sufficiently interesting—aspects of McLuhan's extension idea. In his view, the technological extension of a human function produces a heightening of intensity within that function, body part or sense. Extending the eye, for instance, creates a kind of tension in our visual capacity that is insufferable to us. That has something to do with a sort of imbalance. In extending and thus intensifying a particular body part, we disturb the natural equilibrium between our senses—the sense ratio—and this makes for a pain of sorts. The body can only react by autoamputating and, correspondingly, numbing the concerned part or function. This, then, is the Narcissus narcosis, which, in a pragmatic sense, makes us disregard the actual origin of media extensions, namely, ourselves.

Second, then, ontological foundations. Clearly, McLuhan grounds his extension concept in a theory of perception, that sees the human organism as foundational. "[T]he starting point is always the individual," W. Terrence Gordon comments (1997, 204). The technological springs from "the" individual organism. This organism, however, should not be seen as purely or exclusively biological. It includes the wide range of all things that humans do and are able to do. All human capabilities, functions, and characteristics can be among the "body parts" potentially projected into external artifacts. The "human organism" in this sense obviously exceeds the individual body (cf. also Brey 2000, 72), and envelops for instance matters of societal, political, economic import. The point relates to McLuhan's definition of the term "medium," which is remarkably wide, as already testified by McLuhan's expansive list in *Understanding Media*, including "unusual" media such as motorcars, clocks, and weapons. Eventually, in the posthumous *Laws of Media* (1988), Marshall and Eric McLuhan wind up defining media as in fact all human-made artifacts, material as well as immaterial. Which means that, for example, an electric guitar is to be regarded—and hence studied—as a medium just as much as a political ideology is. This leads to a theory of media that is actually not so much about communication media as we know them, but about almost all the things with which we deal on a day-to-day basis, from garbage cans to scientific theories to works of art.

If the story would end there, McLuhan's media theory, though of a quite holistic nature, could simply be called organicist. Strangely enough, his thinking in fact turns out to be rather "linguistic" in essence. For in time, McLuhan begins to analyze the process of extension in terms of language. In this way, he retrieves an early interest in the history of the trivium—the set of subjects covered at medieval universities: rhetoric, grammar, and logic. Media, he argues, are linguistic entities, that "translate" one thing, that is, a human function, into another, that is, an artifact. Specifically, according to McLuhan, media are metaphors: four-part structures. "[A]ll human artefacts are extensions of man, outerings or utterings of the human body or psyche, private or corporate. That is to say, [. . .] they are translations of us, the users, from one form into another form: metaphors" (McLuhan, 116). This sounds sufficiently puzzling, but the trick here, rather than to suspect some linguistic reductionism, is to actually reconsider what language is supposed to be. In this regard, the metaphorical dynamics upon which media thrive are to be seen as the glue that binds our human reality together. The whole human-made world is hewn out of this so-called "language."

This, however, third and finally, has crucial methodological consequences. If media and technologies, that is, extensions, are linguistic-metaphorical in structure, we can deploy linguistic tools in order to study and understand them. The analysis of media then becomes a matter of close-reading, of critical interpretation, led by exegesis and etymology. McLuhan points out: "The etymology of all human technologies is to be found in the body itself: they are, as it were, prosthetic devices, mutations, metaphors of the body or its parts" (McLuhan and Powers 1989, 34). And he seeks to make this idea tangible by elaborating his four "laws of media," also known as the "tetrad." The laws correspond with the four "parts' of the metaphoric structure that all media are purported to have. McLuhan typifies the parts as follows: enhancement, obsolescence, retrieval, and reversal. Each aspect can be formulated as a question that should be asked about every medium. First, what does the medium enhance? Enhancement acts here as a synonym for extension. What aspect of human life, culture, biology, et cetera, does the medium amplify? The second law, by contrast, suggests in fact the converse process: what does the medium obsolesce? Something gets "lost" with the introduction of each new medium too. For example, the car is said by the McLuhans to enhance "privacy"—one can retreat into its solitude—while it obsolesces—makes irrelevant, outdated—the "horse-and-buggy" (McLuhan and McLuhan 1988, 148). Clearly, notwithstanding Kiran and Verbeek's aforementioned critique, this does indicate that McLuhan at least pays some attention to the revealing-concealing or amplification-reduction structure (cf. Van Den Eede 2012, 177-8). Then, third, what does the medium retrieve? Every medium is expected to retrieve an earlier medium, characteristic, or phenomenon that had previously been forgotten or made obsolescent. Thus, the car in the McLuhans' view (who often like to put things poetically) retrieves the "knight in shining armour," and also the "countryside," as it enables one to escape the city and go visit the latter more easily. Fourth, finally, all media are supposed to turn into their "opposite" when pushed through to their extreme. In a sense, this means the medium has become too successful, so to speak, and ends up effectuating exactly the reverse of what initially had been its enhancing function. In the case of the car, we are all too well acquainted with this "reversal" in the form of traffic jams (cf. McLuhan and McLuhan 1988, 148).

Now, as is well-known, McLuhan's ultimate aim is to understand the effects that media have on "us." If media and technologies are extensions of human senses, body parts, or functions into external artifacts, eventually these externalities will "feedback" upon our organism. By translating ourselves into materialities, we reach out into the environment, but this also makes it possible for the environment to reach into us. Essentially, we are a full-blown part of the environment. John Culkin phrases it, in words often attributed to McLuhan himself, like this: "We shape our tools and thereafter they shape us" (1967, 38). Media enable a reciprocal transformative process between us and them, but in a sense it is also very much impossible to distinguish between us and them.

Nevertheless, the critique by Kiran and Verbeek is still pertinent to the extent that talking in terms of effects, feedback, reaching out and in, and shaping indeed seems to imply the pre-given existence of two separate entities—the one biological-psychological-social and prior in an ontological as well as a temporal sense, that is, the human organism, the other first of all imaginary and rooting in human ingenuity and craft, and subsequently materialized into some artifactual form, that is, a technology or medium. There is indeed in McLuhan the suggestion of a boundary line between humans and technologies. If the study of media should be first and foremost an etymological endeavor, that is, acquiring an awareness, beyond the Narcissus narcosis, of the fact that technologies originate in ourselves, then this would surely presuppose that the origin from which media come forth retains a certain priority. If the individual body, despite the elusiveness of its boundaries, is the starting point, it is indispensable. Emerson already anticipates as much when he states: "Machines can only second, not supply, his unaided senses" (1876, 129). But also later versions of the extension idea, even approaches that Heersmink bookmarks as stronger variants, reserve some primacy for the body. Just as McLuhan keeps treating the individual organism as "starting point," Clark still preserves, in a highly similar way, a "cognitive core" that circles around the individual body. Although our mind is organized in "heterogeneous ensembles," he says, the biological organism continues to play an important role in their organization: "Individual cognizing [. . .] is organism centered even if it is not organism bound" (2008, 123).

Evidently, the debate is not wholly settled. In how far and in which way specifically does the extension idea still retain relevance in the contemporary context of the philosophical study of technology? The question can be investigated by way of zooming in on a specific artifact, in this case, self-tracking technologies. That is the aim of the second part of the chapter.

THE CASE OF SELF-TRACKING TECHNOLOGIES

What is self-tracking? As such, the activity is relatively easy to pinpoint, even if it is known under diverse names, among them "personal informatics," "personal analytics," and "quantified self." It entails the collection and storage of various sorts of data in or about one's body or life, often by way of a sensor-equipped device, and mostly involving a software tool that enables the aggregation and analysis of the thus acquired or constructed data streams. More and more of these tracking-related devices and tools enter the market nowadays at affordable prices. Among the best-known entry-level products of the moment are the Jawbone Up (consisting of a wristband, smartphone or tablet app, and an online dashboard), the tracking systems produced by FitBit (including a pedometer or wristband, smartphone-tablet app, and an online dashboard), BodyMedia (armband, app, and online tool), and a myriad array of standalone smartphone or tablet applications, such as the running app RunKeeper.

Most of these systems track movement by way of accelerometers, and on the basis of that information calculate—or rather perhaps, estimate—amongst others steps taken, calories burned, distance traveled, and sleeping patterns. Those are however not the only things that can be tracked. Some products enable users to manually enter food intake, water consumption, moods, or levels of productivity. This obviously requires a little bit more effort on the users' part than is the case with data collection that mostly goes on in the background. Other, often more specialized devices designed first and foremost for medical or sports purposes monitor blood pressure, heart rate, galvanic skin response, or blood glucose levels. The gathered data is then usually displayed in the accompanying apps or web tools, for the most part in the form of diagrams or other graphics.

Self-tracking is done in the first instance, as the term suggests, on an individual basis, but possibly also takes place within a community of co-trackers, online and or offline, with which one interacts and communicates about the very activity. That community may be small and coincide at least partly with existing communities—for instance, a group of Facebook friends with which one shares his or her RunKeeper results. Then there is the worldwide community known as Quantified Self (QS), that grew out of the Quantified Self Lab founded by Kevin Kelly and Gary Wolf. The QS network ("About the Quantified Self" 2013) gathers people hailing from the most diverse backgrounds—designers, researchers, or merely just interested individuals, . . .—and enables exchanges and dialogue between them, amongst others by organizing well-attended international conferences. In fact, the current "fame" of the self-tracking phenomenon may in large part be owing to the large-scale press coverage that this community has received of late. Its members endeavor to experiment with self-tracking in all sorts of ways and in that way they can really be said to pioneer the field.

Of course in a fundamental sense, some of these activities or technologies are not at all new. For example, physiologists have been able to measure blood pressure since the eighteenth century (Booth 1977). Even as an individual activity, generally speaking, self-tracking has a long history, as Natasha Dow Schüll (2012) points out: since centuries, certain people have tried to improve their lives or conduct by keeping track of them, by collecting and monitoring data, albeit in less technologically advanced forms. Nowadays, one could say, the phenomenon is receiving something of a technological boost as a rising amount of devices and software is increasingly becoming available to average users. On the one hand, measurement techniques that used to be the sole privilege of physicians or hospitals are now "individualized." On the other hand, tracking activities that could already be executed on an everyday, individual basis—monitoring weight, food intake, moods, et cetera—are now made easier and more effective through automatization and enhanced computer power. And, crisscrossing these two evolutions, new developments in sensor technology—for instance accelerometers—are starting to open up vistas onto heretofore hard-to-visualize constellations of data, such as yearlong patterns of correlation between diverse elements (e.g., miles traveled, sleep cycles).

All of this serves to illustrate that although at first sight, self-tracking seems easy to delineate, one should remain wary of all-too simple definitions. First, this mix between "old" and "new" aspects described above should alert to any all-too quick celebration of purported newness. Granted, self-tracking is a "new," emerging phenomenon, but it did not appear out of nowhere, on an empty stage. This is reflected in statistics presented by the Pew Internet Research report "Tracking for Health" (Fox and Duggan 2013). 69 percent of US adults track some health indicator, such as diet, weight, exercise routine, or a symptom. But half of them do so in their heads and one-third takes notes on paper. Only 20 percent of trackers uses a "technology" and in the case of 8 percent (of the total of tracking informants) this concerns a medical device such as a glucose meter. A mere 7 percent of all trackers employs "an app or other tool on their mobile phone or device" (Fox and Duggan, 9) Here, of course, specifically self-tracking done by way of some of these newer technologies, such as apps or sensors, is of interest. Nevertheless, it can be seen that this sort of tracking remains relatively marginal for the time being. According to some, however, this situation may soon change, if and when the right circumstances are created.

This, then, refers to another point, perhaps even more important: the phenomenon of self-tracking can be seen to branch out into many different fields. This wider context is essential in trying to understand its wider "cultural" consequences. A very brief review of some of the existing research on self-tracking or closely related areas may illustrate this. Connections have been explored between self-tracking and health or, more particularly, e-health or m-health (Swan 2009; Swan 2012a; Swan 2013a; Ranck 2012; Lupton 2012), big data (Swan 2013b), the Internet of Things (Swan 2012b), behavior design (De Maeyer and Jacobs 2013), surveillance and or biopolitics (Albrechtslund and Lauritsen 2013; Lupton 2012; Oxlund 2012; Ruckenstein 2012), and, for lack of a better word, morality (Morozov 2013). What happens, however, if one investigates self-tracking technologies and the attending activity, as proposed, from the perspective of the extension idea?

SELF-TRACKING AS EXTENSIONAL ACTIVITY: THE EXTENSION IDEA AS EXERCISE IN CRITICAL AWARENESS

Recall that the paper began with the words of iJustine, inaugurating a new era of technological extension: "Welcome to the future, where technology isn't just around us. It's on us. It's in us. It's an extension of ourselves." While she is uttering the lines in voice over, we see her showing off several kinds of fitness tracking devices. Indeed, one literally puts on these devices. Subsequently they are incorporated in one's activities, in this case, doing sports. It is hard to deny that, in such a practical sense, these tools are indeed "extensions of ourselves."

Nonetheless, as seen, the significance of McLuhan's extension idea surpasses mere common practicality. Of what would self-tracking technologies exactly be extensions? Phrased in McLuhan's pseudo-medical terminology, what is the irritation in relation to which self-tracking tools are supposed to function as counter-irritants? How would a purported autoamputation manifest itself in this context? It will be clear that in a McLuhanian framework, the notion of extension evokes a whole lot more questions than that it simplistically reduces technology to a bare mechanistic principle, as some have suspected (cf. Van Den Eede 2012, 53-54). What is striking is that in the—for the moment still relatively scarce—literature available on self-tracking technologies, precisely the latter view appears to persist, however implicitly. To be sure, the concerned researchers do not literally employ the extension idea to go about their investigations into the phenomenon. But some of the analyses fit the distinction that Kiran and Verbeek make between discourses in terms of either reliance or suspicion remarkably well.

Scholars on the "reliance" side, so to speak, conceive of self-tracking technologies as a prime potential gateway into a better future. Melanie Swan, for example, one of the most productive researchers on the topic, is hopeful that a shift toward health care models that empower the individual or "biocitizen" to a greater degree is underway, and sees self-tracking as an important element in this development. "Patient-driven" models, fueled by amongst others the use of self-quantifying devices, will not necessarily supersede, but certainly supplement and expand traditional health care systems (2009). From an exclusively "fix-it-with-a-pill mentality," we will need to evolve toward a system in which "individuals increasingly take the responsibility to self-manage health in a more empowered proactive manner" (2012a, 113).

By contrast, on the "suspicion" side, such evolution is viewed from an altogether less anticipant standpoint, as testified to by Deborah Lupton, who warns that "[b]y focusing on the individual, [. . .] these technologies reduce health problems to the micro, individual level. Such approaches do little, therefore, to identify the broader social, cultural and political dimensions of ill-health and the reasons why people may find it difficult to respond to such messages" (2012, 239). This stance is, even more forcefully, shared by Evgeny Morozov (2013) in his idiosyncratic critique of what he calls "solutionism," a worldview that endeavors to frame all human predicaments in terms of problems that should be solved, preferably by way of technological intervention. According to Morozov, the solutionist agenda dominates the mindset of ICT business people and aficionados, but he also sees it returning in the activities of self-trackers. And it has dire consequences. In Morozov's opinion, the use of self-tracking technologies implies a worldview in which health becomes totally and exclusively a matter of personal behavior. Factors of wider societal, economic and political relevance are not taken into account, for example, the organization of the food industry, the role of advertising and marketing, economic interests, and so forth. Responsibility is outsourced to the individual, and the individual offloads it in turn onto devices (cf. also Mol 2000). Over and against that, Morozov paradoxically makes a plea for *more* individual moral responsibility, but of the sort that actually requires an intense involvement with those systemic elements.

There are nevertheless also other, somewhat more nuanced accounts that strike a balance between reliance and suspicion and thus, in fact, partially converge with Kiran and Verbeek's proposed program of "trusting ourselves to technology." A couple of scholars link up the activity of self-tracking to the Foucauldian notion of biopolitics and to surveillance issues, and while one would expect any critic of biopolitical dominance to have a field day demolishing the potential naiveté of self-trackers as they gleefully disregard issues of power and ideology, these analyses in fact attain a refreshing degree of subtlety. Minna Ruckenstein (2012), following Lupton, deploys Haggerty's and Ericson's notion of the "surveillant assemblage" (2000) in order to bypass classic critiques in terms of domination, pointing out that "the self-tracking assemblage is characteristically emergent and unstable, replicating familiar biopolitical and economical aims, *but also reshaping them*" (Ruckenstein 2012, 4; my emphasis). In a similar vein, Albrechtslund and Garitsen (2013) foreground the actor-network theoretical notion of "oligoptica" to make sense of self-tracking. Oligoptica are "fragile, narrow gazes": they entail seeing only a very small bit of something, but seeing it very well. While the Foucauldian panopticon notion presumes the presence of one "big" gaze, the oligopticon concept is much more tailored to our intensely networked world in which a lot of small, very precise gazes, among them those constructed by self-tracking technologies, combine in a complex constellation. This makes that "surveillance is a multi-faceted phenomenon involving participation from many different actors" (2013, 5).

What these perspectives attempt to go beyond, is the false choice between—in McLuhanian terms—either simple instrumental enhancement or straight-out threatening obsolescence. Both aspects are present to a certain extent, and all across them, some active engagement has to be at play too, an engagement that actually grounds our relation to technologies instead of being just a supplement to it. Exactly this idea takes central stage in Kiran and Verbeek's "trust" approach. With their case study, moreover, of the Myotel system, a monitoring device for patients with neck and shoulder problems (2010, 420ff.), they offer an example very much congruent with the case of self-tracking technologies here. The Myotel measures muscle activity by way of on-the-skin electrodes, and gives patients a signal when they are too immobile and need to do exercises. On the basis of the data gathered by the device, the patient regularly consults with his or her physical therapist. In Kiran and Verbeek's view, the use of the Myotel cannot be adequately understood by way of the extension idea. Framing as they do extension theories as "externalist," they argue that seeing the monitoring device as an extension would suggest it to be merely added to an already pre-existing, present subject, with, obviously, the usual analyses in terms of either reliance or trust as a consequence. Technologies such as the Myotel do more than that; they shape lives. "[O]nce used, such devices do not merely operate as extensions; they have a profound impact on people's existence. [. . .] [O]ne's subjectivity takes shape in relation to the technology" (2010, 420).

Nevertheless, as has been suggested before, the extension idea does account for that impact—certainly McLuhan's version. McLuhan understands the human-technology relation as a cybernetic circuitry of feedback loops. This is illustrated by the vocabulary of irritation, counter-irritants, and so on. We react to the effects of media by making new media—a process Paul Levinson describes as "remediation" (1997). The extension idea in that way makes sense of the reciprocal "shaping" process going on between humans and technologies. However, if the idea would not do more than that, there would not be any special reason to keep it. What extra value does the extension concept have to offer? Its "unique selling proposition" lies exactly in the hint of "one-way traffic" it does still retain. At one elusive originary moment, media are "born" from "us," but we stay unaware of that origin. At another elusive moment (or various moments) throughout their existence, they work their effects upon us, but part of that process escapes us—in McLuhan's terms, we focus on the content of media, but are oblivious of their form, being the effects they have on us. Both these unidirectional "events," so to speak, may not be so easily discernible in practice; their description serves first and foremost as analytic heuristic. What matters is that this heuristic forces us to *expect* "something" to "happen," some impact from one to another thing to "take place," some causal chains to be set in motion, and that we do not notice it, at least not right away.

So, the valuable addition, one could say, that the extension idea, first and foremost in its McLuhanian guise, provides, is the expectation that it evokes of things unseen. As was mentioned, methodologically, McLuhan winds up studying media from a linguistic standpoint. Etymology and rhetoric are crucial tools in this endeavor. As far as etymology is concerned, on the one hand, the "functionality" or workings of technologies and media should be investigated in the light of the capabilities or functions that they extend. With regard to rhetoric, on the other hand, the forms of media should be scrutinized in their structural composition and in the way in which they work their influence upon the reader, that is, the user, namely, "us." In other words, even though one can assume the human-technology interrelation to have at base level a "two-way traffic" character, it helps to ask of which "vectors" precisely this reciprocal interrelationship is made up. This can aid us, using Brey's phrasing, in our "evaluative analyses of the role of technology in society" (Brey 2000, 59). For the extension idea urges us to inquire deeper into, first, the "why" of extension—as Brey puts it, inspired by Rothenberg, the "question of what we are extending ourselves for" (Ibid., 76)—and, second, the "how" of extension—the concrete manners in which technologies and media eventually affect our world, life, and self.

This then brings the discussion back to the McLuhanian-tetradic questions asked above in relation to self-tracking technologies. Surely, self-tracking can be seen as an activity that helps to shape our subjectivities in the sense of Kiran and Verbeek. One is reminded once more of their Myotel case. The device, they suggest, encourages patients to take responsibility for their condition and their treatment. "When using the Myotel, successful treatment requires the patient to take responsibility for tasks that otherwise would be the responsibility of healthcare professionals. Moreover, sensations of pain are directly linked to actions one can undertake oneself to prevent the pain from getting worse or even occurring" (2010, 423). Such an approach indeed evades perspectives in terms of either reliance or suspicion. Nevertheless, it is fruitful, without lapsing into either of those two extremes, to also ask the questions to which McLuhan exhorts us. For they make one attentive to the things that may escape a purely and exclusively "interrelational" approach. And even though any answer to them can perhaps only be tentative, they can assist us in inquiring into the aforementioned "why" and "how."

Thus, *of what* are self-tracking technologies extensions? They measure and track states or events in our body, in our bodily experience. As remarked, of course, we undertake such measuring already "naturally" without technological extensions. People tend to "keep track" of their physical (and mental) condition in a more or less intensive degree. We all feel, notice, and perceive things going on in and with our body. Of these general capacities for experiencing things in our "internal" environment, self-tracking technologies could be said to be extensions or enhancements. But that is not the end of the story. If they enhance our in-born tracking abilities, one can also expect them to numb those very abilities to a certain extent. Metaphorically interpreted, this does not sound too idiotic. The linear, algorithmic, statistical way in which tracking devices gather, collect, analyze, and finally display data, may perhaps partially stunt our more intuitive habits of "feeling" our way around the world. The numerically imposed coercion of the desired heartbeat rate, amount of burnt calories, or number of steps taken can lead us to disregard other signals in our body that are not—for the moment at least—trackable through sensors and algorithms.

Surely, in this regard, the wider cultural setting in which one deploys these extensions of oneself should be taken into account as well. Here, the notions of irritation and counter-irritant, however questionable in their empirical-sociological accuracy, are of utility. To what irritation do self-tracking technologies react? It is not so easy to pinpoint just one thing, but one may think of the general dictate of fitness and performance dominating our postmodern culture. Whatever the roots and dynamics of that culture are, self-tracking devices, and certainly those with fitness-related purposes, play on it and thrive on it; and at least part of their emergent success can be explained on the basis of the importance we generally seem to attach to keeping fit, healthy, and efficient.

Conversely, the use of those very devices may validate and consolidate the currently predominant cultural habits. What is more, with each enhancement, according to McLuhan, comes obsolescence and, for that matter, retrieval and reversal. The "tetrad" helps to situate a given medium within the network, the "media ecology," of other media of which it forms a part—media, of course, meant here in the broad sense outlined above. Which fitness or medical technologies, healthcare models, or even culturally ingrained ideas, say, about our physical condition, about responsibility, or about willpower will self-tracking technologies possibly obsolesce? Which older "media" do they retrieve: the once—by the installment of the welfare state—temporarily obsolesced struggle of all against all? What will they reverse into when pushed to their extremes: a situation of ubiquitous "nudging" (cf. Thaler and Sunstein 2009), of complete technologically extended self-observation, in which all health-related decisions are made "automatically" without interference of any conscious intention?

True enough, these questions have a science-fictional tincture about them (as tetrads often tend to have). Nevertheless, the authors discussed earlier in this section, on either the "reliance" or "suspicion" side or somewhere in between, can actually be seen to all deal in one way or another with them. The tetradic questions—as seen, in congruence with a methodology that is inspired by etymology, rhetoric, and grammar—illustrate the capacity of the extension idea to make us at least *expect* things of this sort to happen, even though nothing of the kind may ever take place. This expectation hinges on the assumption of a narcosis we are under. Above and beyond the choice between reliance and suspicion, one can indeed choose to "trust," as Kiran and Verbeek propose. But that trust is always based in some negotiation and mutual agreement about boundaries between "us" and "them," however relative and provisional these are. Where do "we"—humans, however hybrid we are—end and do "they"—our technological extensions, however anthropogenetic in nature they are—begin? In between these two "poles," we should expect some fog to distort our sight; a blindness we are victim to or, even more precisely, an inability to assess the "why" and the 'how' of technologies in an immediate and direct way, at a glance so to speak. The unveiling and negotiating project needs work. And so, just as self-tracking can be understood as "extensional activity"—thus superposing McLuhan's extension concept onto Kiran and Verbeek's subjectivity-shaping approach—the extension idea *itself*, in the end, turns out to be an *activity* just as much, an exercise in sharpening our critical awareness, with no predetermined outcome, and no guarantee of success.

CONCLUSION

Some final remarks are in order. Between introducing this paper by way of iJustine's proud announcement "extension!" and ending it by portraying the extension idea as an exercise in critical awareness, an ironic gap seems to run. Is this the gap between the map and the territory, between the words and the object of which mention was made at the beginning? One cannot be sure. But it is possible to put the extension idea as it has been elaborated in the service of the scrutiny of both. Across this gap, we continuously try to understand our human-technological condition, whether we do this with the frivolousness of iJustine or with the grave earnestness of McLuhan. With regard to this undertaking, the extension idea has the advantage of perceiving the human being as "the stuff to be extended." And at the risk of stating the obvious: there is a verb in that phrase. Perhaps exactly because of the inherently "actional" understanding of technology that it offers, the extension idea has withstood the test of time. From its early origins in the philosophical study of technology, via its passing through the philosophical-anthropological vocabulary, on to its incorporation in contemporary frameworks, the notion has been able to provide food for thought on the status and the dynamics of our relation to technologies. The extension idea has staying power. Hopefully, this discussion has shown that its residence permit for its stay in the realm of the philosophy of technology can at least be extended a little bit longer.

REFERENCES

"About the Quantified Self." 2013. quantifiedself.com/about/. Accessed November 29, 2013.

Albrechtslund, Anders, and Peter Lauritsen. 2013. "Spaces of Everyday Surveillance: Unfolding an Analytical Concept of Participation." *Geoforum* (May): in press. doi:10.1016/j.geoforum.2013.04.016.

Aydin, Ciano. 2013. "The Artifactual Mind: Overcoming the 'Inside–Outside' Dualism in the Extended Mind Thesis and Recognizing the Technological Dimension of Cognition." *Phenomenology and the Cognitive Sciences* (May): 1–22. doi:10.1007/s11097-013-9319-x.

Bolter, Jay David, and Richard Grusin. 2000. *Remediation: Understanding New Media*. Cambridge (MA): MIT Press.

Booth, Jeremy. 1977. "A Short History of Blood Pressure Measurement." *Proceedings of the Royal Society of Medicine* (November) 70 (11): 793–799.

Brey, Philip. 2000. "Theories of Technology as Extension of Human Faculties." In *Metaphysics, Epistemology, and Technology (Research in Philosophy and Technology Volume 19)*, edited by Carl Mitcham, 59–78. Bingley: Emerald.

Butler, Samuel. [1872] 2002. *Erewhon*. New Edition. Mineola (NY): Dover.

Cassirer, Ernst. 1955. *The Philosophy of Symbolic Forms, Volume Two: Mythical Thought*. Translated by Ralph Manheim. New Haven: Yale University Press.

Clark, Andy. 2003. *Natural-Born Cyborgs: Minds, Technologies, and the Future of Human Intelligence*. Oxford: Oxford University Press.

———. 2008. *Supersizing the Mind: Embodiment, Action, and Cognitive Extension*. Oxford: Oxford University Press.

Clark, Andy, and David Chalmers. 1998. "The Extended Mind." *Analysis* 58 (1): 7–19. doi:10.1111/1467-8284.00096.

Culkin, John. 1967. "Each Culture Develops Its Own Sense-Ratio to Meet the Demands of Its Environment." In *McLuhan Hot and Cool: A Primer for the Understanding of and a Critical Symposium with a Rebuttal by McLuhan*, edited by Gerald Emanuel Stearn, 35–44. New York: The Dial Press.

Curtis, James M. 1978. *Culture as Polyphony: An Essay on the Nature of Paradigms*. Columbia (MO): University of Missouri Press.

De Maeyer, Christel, and An Jacobs. 2013. "Sleeping with Technology — Designing for Personal Health." In *2013 AAAI Spring Symposium Series*. www.aaai.org/ocs/index.php/SSS/SSS13/paper/view/5726. Accessed November 29, 2013.

Emerson, Ralph Waldo. 1876. *Society and Solitude: Twelve Chapters*. Boston: James R. Osgood.

Fox, Susannah, and Maeve Duggan. 2013. "Tracking for Health." Pew Internet and American Life Project. Washington, DC: Pew Research Center. pewinternet.org/Reports/2013/Tracking-for-Health.aspx.

Garnar, Andrew Wells. 2012. "Hickman, Technology, and the Postmodern Condition." *Techné: Research in Philosophy and Technology* 16 (3): 231–251. doi:10.5840/techne201216316.

Gebser, Jean. 1985. *The Ever-Present Origin*. Athens: Ohio University Press.

Gehlen, Arnold. 1988. *Man: His Nature and Place in the World*. Translated by Clare McMillan and Karl Pillemer. New York: Columbia University Press.

———. 2003. "A Philosophical-Anthropological Perspective on Technology." In *Philosophy of Technology: The Technological Condition: An Anthology*, edited by Robert C. Scharff and Val Dusek, 213–220. Blackwell Philosophy Anthologies. Malden: Blackwell.

Gordon, W. Terrence. 1997. *Marshall McLuhan: Escape into Understanding*. New York: BasicBooks.

Haggerty, Kevin D., and Richard V. Ericson. 2000. "The Surveillant Assemblage." *The British Journal of Sociology* 51 (4): 605–622. doi:10.1080/00071310020015280.

Haraway, Donna. 1991. *Simians, Cyborgs, and Women: The Reinvention of Nature*. Second Edition. London: Free Association Books.

HardWired Ft. iJustine [OFFICIAL TRAILER] | Premieres 5/15! 2013. www.youtube.com/watch?v=YASspAgG2EE&feature=youtube_gdata_player. Accessed November 29, 2013.

Hayles, N. Katherine. 1999. *How We Became Posthuman: Virtual Bodies in Cybernetics, Literature, and Informatics*. Chicago: The University of Chicago Press.

Heersmink, Richard. 2011. "Defending Extension Theory: A Response to Kiran and Verbeek." *Philosophy and Technology* 25 (1): 121–128. doi:10.1007/s13347-011-0035-6.

Hickman, Larry A. 2001. *Philosophical Tools for Technological Culture: Putting Pragmatism to Work*. Bloomington: Indiana University Press.

Ihde, Don. 1990. *Technology and the Lifeworld: From Garden to Earth*. Bloomington: Indiana University Press.

Kapp, Ernst. [1877] 1978. *Grundlinien Einer Philosophie Der Technik*. Düsseldorf: Stern-Verlag Janssen.

Kellner, Douglas. 1989. "Resurrecting McLuhan? Jean Baudrillard and the Academy of Postmodernism." In *Communication: For and against Democracy*, edited by Marc Raboy and Peter A. Bruck, 131–146. Montréal: Black Rose Books.

Kiran, Asle, and Peter-Paul Verbeek. 2010. "Trusting Our Selves to Technology." *Knowledge, Technology and Policy* 23 (3): 409–427. doi:10.1007/s12130-010-9123-7.

Korzybski, Alfred. 1994. *Science and Sanity: An Introduction to Non-Aristotelian Systems and General Semantics*. Fifth Edition. New York: Institute of General Semantics.

Lakoff, George, and Mark Johnson. 1999. *Philosophy in the Flesh: The Embodied Mind and Its Challenge to Western Thought*. New York: Basic Books.

Lawson, Clive. 2008. "An Ontology of Technology: Artefacts, Relations and Functions." *Techné: Research in Philosophy and Technology* 12 (1): 48–64. doi:10.5840/techne20081211 4.

———. 2010. "Technology and the Extension of Human Capabilities." *Journal for the Theory of Social Behaviour* 40 (2): 207–223. doi:10.1111/j.1468-5914.2009.00428.x.

Levinson, Paul. 1997. *The Soft Edge: A Natural History and Future of the Information Revolution*. London: Routledge.

Lister, Martin, Jon Dovey, Seth Giddings, Iain Grant, and Kieran Kelly. 2009. *New Media: A Critical Introduction*. Abingdon (UK): Routledge.

Logan, Robert K. 2007. *The Extended Mind: The Emergence of Language, the Human Mind, and Culture*. Toronto: University of Toronto Press.

Lupton, Deborah. 2012. "M-Health and Health Promotion: The Digital Cyborg and Surveillance Society." *Social Theory and Health* 10 (3): 229–244.

MacDougall, Robert C. 2010. "The Cyborg In-Context: More Observations on Symbiotic, Parasitic and Other Quasi-Biological Relations of Some Human-Machine Couplings." Paper presented at the 11th Annual Convention of the Media Ecology Association: Media Ecology and Natural Environments, Orono (ME), June 10–13.

McLuhan, Marshall. 1970. *Culture Is Our Business*. New York: Ballantine Books.

———. 2003. *Understanding Media: The Extensions of Man*. Critical Edition. Corte Madera: Gingko Press.

McLuhan, Marshall, and Eric McLuhan. 1988. *Laws of Media: The New Science*. Toronto: University of Toronto Press.

McLuhan, Marshall, and Bruce R. Powers. 1989. *The Global Village: Transformations in World Life and Media in the 21st Century*. Oxford: Oxford University Press.

McLuhan, Marshall, and Gerald Emanuel Stearn. 1967. "A Dialogue: Q. and A." In *McLuhan Hot and Cool: A Primer for the Understanding of and a Critical Symposium with a Rebuttal by McLuhan*, edited by Gerald Emanuel Stearn, 265–302. New York: The Dial Press.

Mitcham, Carl. 1994. *Thinking through Technology: The Path between Engineering and Philosophy*. Chicago: University of Chicago Press.

Mol, Annemarie. 2000. "What Diagnostic Devices Do: The Case of Blood Sugar Measurement." *Theoretical Medicine and Bioethics* (January) 21 (1): 9–22. doi:10.1023/A:1009999119586.

Morozov, Evgeny. 2013. *To Save Everything, Click Here: Technology, Solutionism, and the Urge to Fix Problems That Don't Exist*. London: Allen Lane.

Oxlund, Bjarke. 2012. "Living by Numbers: The Dynamic Interplay of Asymptomatic Conditions and Low Cost Measurement Technologies in the Cases of Two Women in the Danish Provinces." *Suomen Antropologi: Journal of the Finnish Anthropolcgical Society* 37 (3): 42–56.

Ranck, Jody. 2012. *Connected Health: How Mobile Phones, Cloud and Big Data Will Reinvent Healthcare*. Kindle Edition. San Francisco: GigaOM Books.

Ruckenstein, Minna. 2012. "Quantifying Life: Self-Tracking and Emerging Everyday Analytics." Unpublished paper.

Schüll, Natasha Dow. 2012. "Self by Design: Personal-Tracking Applications and Life as Self-Transformation." Paper presented at the Design and Displacement—Social Studies of Science and Technology—4S/EASST Joint Conference 2012, Copenhagen, October 17–20.

Swan, Melanie. 2009. "Emerging Patient-Driven Health Care Models: An Examination of Health Social Networks, Consumer Personalized Medicine and Quantified Self-Tracking." *International Journal of Environmental Research and Public Health* (February) 6 (2): 492–525. doi:10.3390/ijerph6020492.

———. 2012a. "Health 2050: The Realization of Personalized Medicine through Crowdsourcing, the Quantified Self, and the Participatory Biocitizen." *Journal of Personalized Medicine* (September) 2 (4): 93–118. doi:10.3390/jpm2030093.

———. 2012b. "Sensor Mania! The Internet of Things, Wearable Computing, Objective Metrics, and the Quantified Self 2.0." *Journal of Sensor and Actuator Networks* (November) 1 (3): 217–253. doi:10.3390/jsan1030217.

———. 2013a. "Next-Generation Personal Genomic Studies: Extending Social Intelligence Genomics to Cognitive Performance Genomics in Quantified Creativity and Thinking Fast and Slow." In *2013 AAAI Spring Symposium Series*. www.aaai.org/ocs/index.php/SSS/SSS13/paper/view/5704. Accessed November 29, 2013.

———. 2013b. "The Quantified Self: Fundamental Disruption in Big Data Science and Biological Discovery." *Big Data* (June) 1 (2): 85–99. doi:10.1089/big.2012.0002.

Teilhard de Chardin, Pierre. 1963. "Place de La Technique Dans Une Biologie Générale de L'humanité." In *L'activation de l'énergie*, 159–169. Paris: Seuil.

———. 1964. *The Future of Man*. Translated by Norman Denny. London: Collins.

Thaler, Richard H., and Cass R. Sunstein. 2009. *Nudge: Improving Decisions About Health, Wealth and Happiness*. London: Penguin.

Van Den Eede, Yoni. 2012. *Amor Technologiae: Marshall McLuhan as Philosopher of Technology—Toward a Philosophy of Human-Media Relationships*. Brussels: VUBPRESS.

Van Loon, Joost. 2006. "McLuhan and His Influences." In *Radical Mass Media Criticism: A Cultural Genealogy*, edited by David Berry and John Theobald, 161–176. Montréal: Black Rose Books.

III

Users

In "User," the authors investigate the design of posthuman bodies and subjectivities. Forged through interface with complexly mediated objects, the User is not fully organic, not simply singular, not necessarily cohesive, not merely a subject. The affect, memory, and physiology of the User are invented anew through cybernetics, here considered not merely as a mechanical process but as a combinatory phenomenon in which the (de-centered) self is constructed through what William James has called the "relations of *withness*": with other people, things, environments, and conceptual systems. The section asks what, as one contributor puts it, is the User's "*style* of being"? What crises are incited or registered by the stylized User? While all the contributors to this section of the anthology broadly accept the turn to the nonhuman and a relational ontology, they seek to critically challenge the terms of this turn and raise some evocative questions. Does, for instance, the posthuman represent a recent turn in culture owing to the development of networked and digital technologies or ought we to trace the networked user back further into our technological past? Is the posthuman gendered and what are the implications of gender for an analysis of technological mediation? How does our understanding of mediation and the posthuman change as we vary the objects human beings enter into relation with, from bicycles to videogames to the very planet itself?

Fleckenstein and Mehler's contribution to this volume, "Mobility Regimes and the Constitution of the Nineteenth-Century Posthuman Body," challenges what we might think of as the prototypical image of the twenty-first-century posthuman digital citizen: a male body passively immobilized in front of a computer screen as he "jacks in." Building on and yet importantly extending Katherine Hayles's account of the posthuman in *How We Became Posthuman*, the authors effectively question each element of this prototypical image: the figure's gender, their immobility, and the focus on digital technologies. Seeking a more capacious definition of technology with which to generate, and gender, the posthuman, Fleckenstein and Mehler ask whether other forms of technology might give birth to the posthuman. They turn to two nineteenth-century technologies, the tourist guidebook and the safety bicycle, and argue that both give rise to networked posthuman identities. Furthermore, their analyses focus on a dimension they argue is largely missing in contemporary discussions of the posthuman: mobility. The tourist guidebook and the safety bicycle add an element of mobility to our understanding of the posthuman and in doing so highlight the gendered nature of posthumanity. Nineteenth-century tourist guidebooks and the safety bicycle, Fleckenstein and Mehler contend, gave rise respectively to the figure of the "sporting gentleman" and the "New Woman," networked, posthuman subjectivities that exist at the intersections of a distributed cognitive system. As the authors note in their essay, "Mobility and non-digital technologies thus represent important characteristics of posthuman bodies, underscoring new aspects of the interface among the machine, discourse, and the organic, as well as revealing the importance of mobility regimes in this process" (191).

If Fleckenstein and Mehler turn to the nineteenth century for insights into twenty-first century posthumanity, Matthew Levy reaches further back, to Diogenes and the ancient Cynic tradition, to suggest that we look backward to recover insights into how to live moving forward into a life increasingly mediated by complex technologies. In "Living Deliberately, Less or More: Affirmative Cynicism and Radical Design," Levy identifies a strain of thought that runs from the early Cynics, especially Diogenes, through to Henry David Thoreau which he seeks to recover and revise as part of a process of coming to terms with the deeply technologically textured shape of our lives. Levy suggests that Cynicism is best understood as a living tradition that evolves to meet the needs of the present, but he notes that there are competing ways of understanding this tradition and he articulates an affirmative cynicism that ritualizes experimentation and evaluation while problematizing the Cynics' distinction between the natural and the unnatural, which Levy argues hampers our affirmative relationship to technology. In reflecting on prosthetics, for instance, Levy questions whether technological design must remain overly beholden to the natural form of the user and affirms a more experimental, Cynical, approach to design. Concurring with a line of thought that one finds in the work of Bernard Stiegler and Mark Hansen, among others, Levy suggests that we recognize that technology is natural to the human user. As he notes, "we became a technological species before we had a chance to consider whether 'being technological' was a choice. Better to recognize that human beings are always already imbricated with technology—this is our natural state."

Technological imbrications, especially of the media variety, are the focus of the remaining three chapters in this section. Each, in dramatically different ways, focuses on a media text to complicate our understanding of users, mediation, and technology. In "Seduced by the Machine: Human-Technology Relations and Sociable Robots," Weiss too looks backward in an attempt to gain a foothold in understanding our possible future relationships with sociable robots and relational artifacts. Beginning with an analysis of an episode of *The Twilight Zone* in which the central character falls in love with a robot in the form of a woman, Weiss seeks to develop a framework in which to interrogate our relationship to Furbies, Tamagotchis, and the many varieties of robots currently on the scene. These are machines that display behaviors that make people feel as though they are dealing with sentient creatures that care about their presence, relational artifacts built with psychologies and needs of their own. Weiss is interested in asking what happens when our object world starts talking back to us and engaging us emotionally and relationally. How ought we to respond to the growing recognition that our social lives are increasingly mediated by technical artifacts and that the boundary between user and artifact, human and thing, is disappearing? Weiss turns to the work of Sherry Turkle and Peter-Paul Verbeek to begin to define a philosophical framework in which to address these questions. While Weiss argues that there are virtues in both Turkle's and Verbeek's approach to technological mediation, he maintains that we still lack a comprehensive enough framework in which to understand and evaluate the complexities of human-technology relations. Developing such a framework, he argues, will ultimately require a clearer view of both human beings and technology and how both are situated within a human community constituted by caring relations. In the concluding section of his essay, he suggests that we can discern the outlines of such a framework in feminist ethicists' emphasis on a relational ontology that begins not with human-technology relations but with human relations in which technology clearly plays a significant if not foundational role. Weiss implicitly questions whether the nonhuman turn should not itself be situated in a more foundational turn to human caring relations.

Brendan Keogh's contribution to this anthology, "'You Really Are You, Right?': Cybernetic Memory and the Construction of Identity in Videogames," as its title suggests, also begins with a media artifact, a close, textual analysis of the videogame *Final Fantasy VII*. Through his evocative reading of the engagement between the game player and the primary playable character of *Final Fantasy VII*, Cloud Strife, Keogh demonstrates the complex interplay between multiple, complex memories, giving rise to what he refers to as cybernetic memory. Using Alison Landsberg's account of prosthetic memory as a jumping off point and challenging her distinction between lived and prostethic memory, Keogh argues "memory always emerges from and actively mediates posthuman couplings of bodies and their worlds" (237). Keogh's account of cybernetic memory suggests that technologies are not mere extensions of the human being but that through their mediation, we actively construct new memories, new subjectivities, new identities. According to Keogh, we construct our identities within a constant circuit of mediated memories that are simultaneously bodily and technological, lived and displaced, actual and virtual. Videogames and Keogh's account of cybernetic memory point to the limits of the liberal humanist subject. As he notes, our worlds and our bodies are penetrated by increasingly ubiquitous and pervasive technologies, and as our memories are increasingly generated and mediated through engagements with computers, the identities we construct for ourselves are less and less able to be contained within the notion of an autonomous, liberal humanist subject. Keogh's analysis exposes the need for a conceptual framework of memory that comprehends it as equally posthuman, as cybernetic.

The final chapter in this anthology also gestures toward the limits of liberal humanism and anthropocentrism in its exploration of the total destruction of the world. Nicole Merola's essay "Mediating Anthropocene Planetary Attachments: Lars von Trier's *Melancholia*" explores how filmic technology, in particular von Trier's meditation on the end of the world in *Melancholia*, can contribute to human engagement with contemporary socioecological crises and how technology and materiality co-constitute our experiences of self and environment. Merola contends that *Melancholia* functions as an ideal technology for phenomenologically and imaginatively engaging with the very thing we can never experience, the annihilation of our planet, and offers an exemplary case study for examining the contours of environmental mourning. Like many of the other authors included in this anthology, Merola challenges distinctions between the human and nonhuman and advocates a more relational ontology. Her approach, though, begins with the affect of melancholy, which she suggests represents a different way of being a "user" of the planet. Merola questions how to make visible and feel planetary disaster and she finds the answer in cinematic technology and particularly in von Trier's film. Much in the same way that the earlier essays in this section have argued for more expansive accounts of the posthuman or Cynicism or memory, Merola's close reading of *Melancholia* serves to expand Bennett's notion of thing-power to the media, to the planet, and to its destructive, as well as constructive, possibilities.

Chapter Nine

Mobility Regimes and the Constitution of the Nineteenth-Century Posthuman Body

Kristie S. Fleckenstein and Josh Mehler

In her prologue to *How We Became Posthuman*, N. Katherine Hayles narrates the concatenation of forces that give rise to a posthuman. This formative moment, for Hayles, is initiated via Alan Turing's famous Turing Test, proposed in his 1950 article "Computing Machinery and Intelligence." In this test, an "interrogator" conducts a conversation via a terminal screen with two physically absent entities. This interrogator is charged with determining which invisible interlocutor is the machine and which is the human. However, Hayles returns to this challenge to tease out the paradoxes of Turing's *first test*: an "imitation game" that attempts to distinguish not which invisible interlocutor is the human and the machine, but which is the man and the woman. By posing the question of identity in terms of gender, Hayles (1999, xiii) contends that Turing implicitly emphasizes the necessary negotiation between the "enacted body" and the "represented body" via the computer screen, transforming the "subject into a cyborg" or into a posthuman. Furthermore, the test-taker is not immune from this transformation; that person is similarly reincarnated into a posthuman the moment he or she logs on to the terminal. Ultimately, both individuals are "spliced" into the distributed cognitive system where enacted bodies and represented bodies are fused through a machine interface. Thus, all discrete identities disappear, becoming, instead, posthuman identities networked among the three integrated components of machines, semiosis, and enaction, thus forming a thinking system. As the Turing Test demonstrates, the presence of technology—specifically a proto-computer—is crucial to Hayles's depiction of a posthuman birth, and it is this technology that plays a rhetorical role in persuading the interrogator that it is, in fact, the human. However, there are two implicit assumptions in both Turing's tests and Hayles reading of the "imitation game" that require exploration. First, the implication is that the posthuman subject relies on *digital* technologies for its generation.[1] This, in turn, limits potential exploration into other forms of technology—more broadly defined—that might give birth to the posthuman. Also implicit to Hayles's narrative is the inherent immobility of participants: the test-taker, albeit embodied, is situated in front of a terminal screen, as is the absent entity. Both are stationary, fused, or—as William Gibson's protagonist Case experiences it—*plugged into* the matrix. In fact, they become part of the machine through that physical contact which renders them immobile in what Gibson (1984, 6), in the canonical *Neuromancer,* calls the "meat" of the body. The only movement is that of the "flickering signifiers" leaping across the terminal screen and flowing through information pathways between test-taker and absent entity via the machine-body linkage. This, similarly, limits exploration into how more capacious definitions of technology generate the posthuman body by setting it into motion. Thus, both digitality and immobility are seemingly prerequisites and outcomes of the posthuman encounter.

In this chapter, we address each point: *digital* technologies as the tipping point for the emergence of a posthuman and *immobility* as a requirement for immersion in the machine-semiosis-enaction network constituting the thinking system. We extend Hayles's concept of the posthuman to make two connected arguments: first, we emphasize the value of addressing posthuman moments in eras preceding the digital computer. This historical perspective enriches our historical understanding of the distributed cognition from which posthumanity emerges while simultaneously extending our definitions of what constitutes technology and the posthuman. Second, we contend that focusing on mobility, especially on "machines" that invite mobility as a criterion of the user-machine fusion, adds depth to our understanding of posthumanity, especially in terms of a gendered posthumanity. To address both points, we analyze two nineteenth-century technologies—the tourist guidebook and safety bicycle—contending that they manifest a distributed cognition by which a particular gendered posthuman identity is defined. Specifically, these two technologies delineate this gendered posthuman identity via particular configurations of mobility we describe as "mobility regimes." We begin by conceiving technology as rhetorical, historical, and mobile. We then offer two examples of non-digital thinking systems: first, *The Gentleman's Companion*, published in 1870, which promulgates hypermasculinity through both physical and moral mobility regimes; second, the safety bicycle, which offers a new mobility regime for women, transforming them—as Frances E. Willard's memoir illustrates—both physically and mentally. Each distributed cognitive network—comprised of interconnections among machine, semiosis, and enaction—highlights the value of viewing the posthuman through a historical lens as a phenomenon inextricable from mobility.

TECHNOLOGIES AS RHETORICAL, HISTORICAL, AND MOBILE

By seeing technology as rhetorical and historical, we underscore that it is "articulated in language" and positioned "at the very heart of rhetoric," and, therefore, presents an interface with the organic that invents and advocates for certain configurations of human being (Bazerman 1998, 383). These configurations of human being, we argue, also include the schemas for human mobility that formulate mobility regimes. In this section, we first argue that technology is rhetorical and plays a formative role in these regimes of mobility. Second, we emphasize that technology is historical, manifesting the varied and different ways in which human and technology merge in eras prior to the current digital moment. Digital technology, then, is a sufficient but not necessary condition for the posthuman birth. Reading technology in these two ways allows greater leeway for looking at technologies that instigate posthuman birth via mobility and demonstrates that the three components of the distributed network that define the posthuman particularly initiate movement. Charles Bazerman (383) argues that technology should be envisioned as a "human-made object, [that] has always been part of human needs, desires, values, and evaluation," grounded in language and operating rhetorically. However, despite this grounding in semiosis through rhetoric, technology, Bazerman notes, has a "seemingly physical obduracy" since, although technology does embody discursive "intentions and plans and perceived uses," in the end, it "circulates objects and material processes." Ultimately, Bazerman concludes, "it is the material object that conveys the primary rhetoric and not the language that went into forming the technology and conceiving its uses and meanings." For Bazerman (385), this represents a "dialectic between rhetoric and the material design" that makes technology "fit into the world and makes the world fit with technology." Thus, all technology primarily represents a *topoi* for rhetorical invention that is grounded in the dialectical exchange between language and materiality. Further, since Bazerman's materialist dialectic emphasizes how technology's materiality engages with and extends the semiotic constructs that prescribe particular schemes of usage, it suggests a posthuman consideration of enaction, or a rhetoric of usage: how the dialectic is put into play through the ways in which users employ technologies to simultaneously challenge and align with the semiotic and material prescriptions of acceptable usage. Maurice Merleau-Ponty (1998, 453) envisions the world as "already constituted, but also never completely constituted" in two ways at once: "in the first case we are acted upon, in the second we are open to an infinite number of possibilities." The simultaneous semiosis and materiality of technology, then, plays a role in this constitution of the world, populated by the human and channeled specifically through the human body.

If Bazerman provides insight into the formative rhetorical function of technology in conjunction with semiosis and enaction, Geoffery B. Pingree and Lisa Gitelman provide insight into the importance of the historicity of the technology and human interface. Like Bazerman, Pingree and Gitelman (2003, xx) similarly argue that media and technology are designed to "fit," highlighting that "if media fit humans, humans adjust themselves in various ways to fit media, knowingly and not." Similarly, their collection configures technology, much like Bazerman's dialectic, as "evolving, contingent, discursive frames" that foreground "the relationship between material and idea" (xvii). Ultimately, this "fit" requires both the incorporation of the body and the technologies, a move that aligns with posthumanity's emphasis on the configuration of the human as "seamlessly articulated with intelligent machines" (Hayles 1999, 3). Further, Pingree and Gitelman argue for expanding our notions of what new media technologies entail, suggesting that "new media studies" often represents what they see as a "narrow devotion" to present technologies that is heavily influenced by "experiences of digital networks and the professional protocols of the social science of communication" (Pingree and Gitelman 2003, xi–xii). They aim to "encourage thinking about what 'newness' means in the relationship among media and societies" in different historical moments (xii). Pingree and Gitelman (xiv) suggest that due to the imposition of a technological narrative of "linear progress unthinkingly ascribed to modern technology," our contemporary media themselves tend "anachronistically, to mediate our understanding of the past." Ultimately, by exploring emergent media technologies in different historical moments, Pingree and Gitelman expand our definitions of what media, and by extension, what all technologies entail. At the same time, they invite us to consider how that historical technology-semiosis-human interface invents posthuman identities, particularly in terms of a more capacious understanding of mobility.

By expanding our notion of technology to include the non-digital, we open ourselves up to considering multiple forms of mobility, particularly those grounded in the physical world. Adrian MacKenzie (2010, 19) notes that the mobility often tied to wireless technologies is, more often than not, "indexed to stable locations." Similarly, Teddi Fishman and Kathleen Blake Yancey (2009, 39) understand that to be wireless is not to be "untethered" from physical locations; rather, it is always "embedded in a physical context and even more so when the physical contexts are multiplied." Further, beyond such physical proximities, they argue, other kinds of "proximities"—such as "proximities of discourse, proximities of perspective, proximities of knowledge"—are required, operating as kinds of "power sources." Fishman and Yancey (41), first, emphasize that "human actions are mediated by artifacts that are instrumental in the pursuit of performative objectives" and, second, remind us of the "extent to which tools and users affect each other mutually and continually." To be wireless and portable, then, highlights once again Bazerman's dialectic between material technologies, and physical and semiotic contexts. If, as Michel Foucault (2009, 165) writes, "space and language were born together" and, as Michel de Certeau argues (1984, 115), "every story is a travel story—a spatial practice," it is clear that this dialectic between the semiosis and materiality also intersects with space and movement. A wide array of technologies, then, including the bicycle or the tour book, particularly emphasizes this intersection. In this way, the two nineteenth-century artifacts that we offer as examples highlight an expanded notion of technology that moves beyond this contemporary, digital bias implicit in posthumanism and aims to emphasize the different kinds of movement such technologies afford.

A MOBILE POSTHUMAN MAN: THE GENTLEMAN'S COMPANION

In his preface to *Principles of Literary Criticism*, I. A. Richards (2001, vii) opens with the claim that "a book is a machine to think with" proposing that it might be best compared to the loom. In this way, Richards implicitly suggests that the connection between the technology of the book and the human reader creates a thinking system. Although Richards's innovative view of the book considers this "technology" in the context of literary interpretation, others have also explicitly considered the book as a technology that demonstrates the posthuman. Andy Clark and David Chalmers (1998, 10–23) argue that "external factors," such as features of an environment, contribute significantly to cognitive processes, and they offer the cases of Inga and Otto seeking to visit an exhibition at the MoMA to illustrate this claim. Inga recalls that the MoMA is on 53rd street and walks to that location. Otto, on the other hand, consults his notebook, in which he religiously records information, which tells him that the MoMa is on 53rd street. Like Inga, he successfully walks to the correct location, although the book, for him, Clark and Chalmers argue, "plays the role usually played by a biological memory." However, despite utilizing the technology of a book to illustrate their argument, Clark and Chalmers do not consider mobility in this process. Drawing from both Richards's and Clark and Chalmers's concept of the book as a "machine," we argue that the nineteenth-century guidebook *The Gentleman's Companion: New York in 1870* demonstrates a distributed cognitive network or an intelligent weaving across technology, semiosis, and enaction that gives birth to the posthuman as it initiates movement. Published anonymously in 1870, *The Gentleman's Companion* documents the location of houses of prostitution in lower Manhattan. The majority of *The Gentleman's Companion* is composed of lists, including addresses, of "boarding houses," "houses of assignation," and "brothels" with quick evaluations of each location. For example, Miss Jennie Creagh at 17 Amity Street is said to have "splendidly furnished her palace . . . from top to bottom, sparing neither expense nor labor." As a result, the anonymous authors claim, "This emporium of love and beauty is one of the finest in the city" (1870, 28). The house kept by Hattie Taylor at 111 Spring Street, however, is "patronized by roughs and rowdies" and by "gentlemen who turn their shirts wrong side out when the other side is dirty" (19). In this section, we explore how this prostitution guidebook represents a distributed cognitive system that generates a posthuman masculine identity for mobile men, first, by considering the guidebook as a technology emerging from and developing in concert with a burgeoning nineteenth-century information environment; second, by investigating, via semiosis, how this technology "jacked" into, supported, and challenged prevalent nineteenth-century discourses on gender and sexuality, particularly the "cult of true womanhood"; and, third, by looking at enaction to discover how the circuit—the connection between individual components of a system—linking the materiality of technology and semiosis is played out via the guidebook, particularly through mobility.

First, this book, like the computer of Turing's test, constitutes a "machine" plugged into an informational matrix, and, therefore, represents the first component of distributed cognition. David Michalski (2004, 202) describes the mid-nineteenth century as "a world of taxonomies, both scientific and popular." He writes that "[i]n the nineteenth century, both intellectuals and laypersons engaged in the pursuit of a kind of knowledge management, attempting to use their known technologies and conceptual apparatuses to describe and control the burgeoning information society" that was growing around them (201). In particular, Michalski emphasizes that growing American cities like San Francisco, Chicago and New York represented "complex information environment[s]" that required the "technology" of the guidebook as a means to organize this information and to assist in moving through this often tumultuous space (199). Indeed, throughout the nineteenth century, the guidebook genre proliferated and transformed alongside these cities. Paul Alan Marx (1983) writes that "[b]eginning in the 1830s, guidebook literature changed significantly, reflecting the changed nature of the city itself." He continues: "As the simple republican town developed into a complex metropolis, the uniform vision of the city as resource center was no longer sufficient and the new experiential genre of guidebooks evolved." In short, this evolving guidebook machine plugged a reader into a pre-packaged information environment; these arrangements of space, ultimately held consequences for a reader's potential for mobility. Like a map, Michalski (2004, 188) writes, a guidebook determines the potentials for mobility since it can "open and close roads . . . [and] both shield and reveal possibilities"; thus, the guidebook represents a technology that "fuses" with the human body, one that both determines which spaces it can see as well as how it can move. In this way, the guidebook initiates the posthuman; however, in contrast to the posthuman birth described by Hayles, mobility is one of the prerequisites for this birth.

Furthermore, not only was nineteenth-century urban space "reorganized to form stable collections of cosmopolitan images" through these guidebooks, but these arrangements of space also "catered to, and constructed certain identities" (207). Which roads are visible to some and which are shielded from others is, ultimately, dependent upon which bodies are jacked into the guidebook. Indeed, "[n]ot only does the guidebook record, observe, or recommend, it applies force on what it aims to interpret"; in addition, the guidebook not only makes assumptions about the mobility of various bodies, but also applies force upon them to encourage and initiate movement (188).

In many of these cities, in sync with their exploding urban populations, prostitution progressively became more profitable, widely commercialized, and publicly visible, standing as a brazen challenge to middle-class notions of sexual morality. By 1872, the reporter Edward Crapsey (1872, 142) observed in *The Nether Side of New York* that in every American city he has visited, "the walking of streets after nightfall by prostitutes has become an alarming evil; but New York is entitled, I am afraid, to preeminence in this respect." As prostitution became more prominent in New York and elsewhere, the familiar technology of the guidebook was adapted to provide information on the locations of prostitutes and brothels. These guidebooks, such as Butt Ender's 1839 *Prostitution Exposed* and Charles DeKock's 1855 *Guide to the Harems*, became increasingly available, appearing in local bookstores and on corner newsstands (Gilfoyle 1992, 146). Like other nineteenth-century guidebooks, the prostitution guidebook represented an intelligent machine that inserted the user within a very particular information environment. However, the prostitution guidebook specifically constructed and catered to a particular male identity that was intimately connected to dominant cultural narratives about male and female sexuality. Ultimately, then, this guidebook sub-genre participated in the development of a mobility regime that determined appropriate and inappropriate movements for men and women participating in prostitution both in Manhattan and in other urban centers.

Not surprisingly then, the highly designed information technology of the *Companion* mediates and circulates prevalent cultural discourses about the nineteenth-century body that link cultural representations of male and female bodies with enacted performances of male and female sexuality. In particular, semiosis, the second element that constitutes distributed cognitive system, constructs a particular mobility regime that makes specific arguments about morality and immorality by demarcating what constitutes acceptable movements for both men and women. Leslie Kanes Weisman (1994, 87) writes that "the Victorian wife and mother, endowed with love and gentleness, not energy and power, was banished from the public arena and confined to the domestic hearth where she could remain unsullied by the events of public life." By the middle of the nineteenth century, the private house and the work world, she explains, had become "two distinct and isolated spheres." The house was conceived as "a place of sanctuary, repose, and renewal" and the wife served as "guardian angel of the house" while being "locked out from economic productivity" (86–87). Barbara Welter (1985, 21), in her examination of the attributes of the dominant turn of the century "cult of true womanhood" writes that a nineteenth-century woman was judged "by her husband, her neighbors and society" according to four cardinal virtues: piety, purity, submissiveness and domesticity. Ultimately, as Welter notes, these four virtues focused on maintaining the woman in the domestic sphere, a site of sanctuary and immobility, and the "cult" enlisted innumerable women's magazines and conduct manuals to enforce them. Thus, "the cult of true womanhood" represented a particular mobility regime that connected female "piety, purity, submissiveness and domesticity" to fixity and immobility; to be mobile was to violate the moral code of the true woman ideology.

However, the vastly changing labor market forced many women to seek work beyond the home, often to get a wage for basic sustenance. Yet, despite these dramatic social and economic changes, a body of "gay literature" arose and served to prop up these discourses about women and female sexuality even as the force of the cult's tenets began to wane. Gilfoyle (1992, 145) writes that "[a]s increasing numbers of young urban females entered the labor market, moralists and novelists alike treated seamstresses, factory girls, and female migrants as major subjects. The fallen woman was an integral part of this new antebellum canon." Ultimately, such women were depicted in this way because their behavior flouted the dominant mobility regime; their required mobility within this expanding labor market stood in contrast to that of the "moral" woman immobilized within the home. Further, according to the cult, the absence of purity was seen as "unnatural and unfeminine" and that without purity a woman was deemed a "fallen woman" or even "no woman at all, but a member of some lower order" (Welter 1985, 23). A woman that did not follow prescribed social conventions of acceptable mobility and publicness was presumed to be both impure, fallen, and unnatural.[2]

Of course, the prostitute represented a challenge to the regulatory power of the cult's tenets both at its height and through its subsequent literature. As Gilfoyle (1992, 84) notes, "At a time when gender roles were in the initial stage of redefinition and the 'cult of domesticity' was gaining ground, the behavior of some prostitutes served as a vivid counterpoise." In particular, this counterpoise was demonstrated through mobility: her position outside the traditional home and her circulation within the economic sphere considered the exclusive domain of men challenged traditional discourses of domestic immobility; and her sexual intercourse outside of marriage and, even worse, in potentially visible public locations, challenged discourses of feminine purity and submissiveness. However, it is interesting to note that this dominant mobility regime replayed itself even in these spaces, since the most successful prostitutes and madams were themselves ensconced within another home space, the brothel, while those less valued and often the targets of attacks were the more public and mobile streetwalkers. In addition, the *Companion* and other such guides similarly challenge the cult's restrictions on women's economic and physical mobility, frequently characterizing prostitutes in positive economic terms, deeming them "enterprising young ladies" (1870, 49) and noting the intelligence and wit of skilled prostitutes. As a whole, then, the *Companion* constitutes distributed cognition since it constructs a particular mobility regime that makes moral arguments by demarcating what constitutes acceptable movements for men and women. However, the *Companion* also represents a site of semiosis that both aligns with and diverges from dominant discourses about acceptable movement and behavior, and so represents a technology that selects a particular slice of reality that best supports the interests and movements of its presumed users: hypersexualized males.

The Gentleman's Companion highlights the third element in the distributed cognitive system—enaction—in which the dialectic between the machine, the body, and nineteenth-century cultural discourses about gender and sexuality is played out. Nineteenth-century prostitution guidebooks participated in the constitution of a particular world that did not involve simply the management of information; these guides implicitly, by virtue of the fact that they encouraged male readers to move within and explore such illicit spaces, cultivated a mobile, hypersexual masculine identity known as the "sporting man." Sporting male culture, according to Gilfoyle (1992, 99), was "organized around various forms of gaming—horse racing, gambling, cockfighting, pugilism, and other 'blood' sports" and both "defended and promoted male sexual aggressiveness and promiscuity."[3] In contrast to "respectable" notions of domestic male sexuality, which the sporting male saw as "a female preserve organized to control and restrain his sexual freedom," "self-indulgence, not self-sacrifice, meant freedom" for this hyper-masculine male. Ultimately, "unregulated sex was the categorical imperative for the sporting male." Indeed, the technology of the guidebook presumes that the reader, also implicitly understood to be male, is moving without any regulation or restriction through the spaces described in the text. For example, the authors orient their readers with language such as, "Further on we arrive at Houston street" or "As you pass down Houston street, faro banks abound" (13–14). In this way, the male mobility taken for granted in the text aligns with the dominant discourse that sees the public, and increasingly capitalistic, sphere as the domain of men; however, the text also aligns with the sporting male subculture, which ultimately takes this traditional view to frightening extremes. Moving in the ways advocated by the guide—movement unrestricted by any physical and moral constraints—constructed a mobility regime that authorized men to participate in shocking, chaotic violence toward prostitutes. "Sprees," alcohol-fueled binges by groups of men who moved freely through these spaces, frequently led to violent attacks that left prostitutes publically humiliated, maimed, or blinded (Gilfoyle 1992, 79–81). Ultimately, in this rapidly changing information environment, Gilfoyle notes (115), sporting male sexuality "frequently determined a portion of one's identity"; in particular, it served to distance men from what they saw as "women's 'feminine weakness' and emotional attachments" while simultaneously promoting gender solidarity among sporting males.

Whereas traditional male sexuality was rationalized as the natural outgrowth of masculinity that was best expressed through marriage and kept within the private home, "sporting culture"—mediated through the technology of the prostitution guidebook—took this to extremes, challenging and undermining traditional morality openly. Gilfoyle (99) writes that sporting-male culture believed that "prostitution, sexual display, and erotic entertainment brought excitement" to a banal and difficult urban existence. Included in this public and highly visible sexual freedom was the assumption that these bachelors could move amongst illicit neighborhoods without any moral or physical restriction. Ultimately, it is *The Gentleman's Companion*, an intelligent machine linked into the dominant discourses of male and female sexuality, and contributing to the emergence of both mobile and immobile bodies, that initiates such a violent posthuman birth.

A MOBILE POSTHUMAN WOMAN: THE SAFETY BICYCLE AND RESCHOOLING THE BODY

Fewer than two decades after *The Gentleman's Companion* led the sporting man on a peripatetic, and frequently brutal, journey through Manhattan's redlight district, generating a particular hypermasculine posthuman identity, the safety bicycle challenged the restrictive mobility regimes of Victorian women, especially white, middle-class women. The distributed cognitive system of the "steed of steel," laudatory cultural discourses, and enaction transformed protocols of mobility for women and, in the process, made "deservedly unpopular the sickly heroine of less than a generation ago" (Benedict 1897, 473). The technology-human interface regendered female identity as acceptable in public, especially the public byways, rather than immobilized within the confines of domesticity and the parlor. Out of the reciprocal network of bicycle, semiosis, and enaction arose the emancipated woman whose body was reschooled for mobility and equality. Frances E. Willard (1895, 41) optimistically boasts that women's skill and ingenuity "in the manipulation of the swift steed . . . diminishes the sense of superiority in men." Indeed, other writers also claimed that the bicycle was "the great leveler," equalizing the rich and the poor, the male and the female ("Topics: Reign" 1894, 306). In this section, we explore the emergence of this emancipated posthuman female identity by first describing the technology—the safety bicycle—and its social impact. We then briefly review the popular cultural discourses, or semiosis, representing (and thus mediating) the bicycle as a commendable pastime for men and women. Finally, we turn to enaction, or the experience of the safety bicycle that oscillates between semiosis and embodiment, drawing particularly on Frances E. Willard's memoir. In her fifty-third year, a time marked by the loss of her mother and the onset of what she called "nerve-wear," the long-time president of the Woman's Christian Temperance Union learned to ride the safety bicycle. In the process of training her body to conquer what she calls a "new implement of power," Willard (1895, 73) celebrates a mobility that, she claims, transforms her mentally and physically.

The first element in the distributed cognitive system fostering the emergence of a mobile posthuman woman is technology, or, in this case, the circulation of the safety bicycle and its social impact. Although the bicycle is not itself an "intelligent" machine in the conventional sense of the phrase, it still participates in a distributed network constituting the thinking system. As Hayles (1999, xiii) notes, technology "mediates" enacted and represented body in a "contingent production" wherein technology "becomes so entwined with the production of identity that it can no longer meaningfully be separated from the human subject." The safety bicycle constitutes just such an entwined technology, for, according to various nineteenth-century authors, "[t]he bicycle has, in fact, become a necessary part of modern life, and could not be abandoned without turning the social progress of the world backward" ("Topics: Reign" 1894, 306). Writing in 1896, an anonymous commentator concurs, claiming that "it really begins to be debatable whether anything has happened to the human race since the first locomotive drew the first train of cars that will affect it so materially as the bicycle" ("Point of View" 1896, 783). Such hyperbole stems from the "bicycle passion" ("Topics: Reign" 1894, 306) that struck the United States in the early 1890s with the invention and distribution of the safety bicycle. Unlike previous incarnations of the bicycle, such as the ordinary which consisted of an immense front wheel coupled with a small stabilizing rear wheel suitable only for adventuresome athletes, the safety bicycle featured two wheels of equal diameter, air-filled rubber tires, and a gear system with an infinity chain for propulsion. Within less than a decade of its introduction in 1887, the number of bicycles sold increased from the single digits to more than 400,000, with upwards of a million riders populating the public roads (Hubert 1895, 694). Cyclists were so ubiquitous in New York City that "they swarm[ed] like flies upon all the parkways and other thoroughfares having asphalt or macadam pavement" ("Topics: Bicycle" 1895, 474). Given the widespread popularity of bicycles as a tool of mobility for pleasure and business, it is no surprise that it elicited such economia. "As a revolutionary force in the social world," an anonymous author writes in 1895, "the bicycle has had no equal in modern times" ("Topics: Bicycle" 1895, 474).

The second element in the distributed cognitive system constituting posthuman identity is semiosis, the represented body that, mediated by technology, overlaps with the enacted body. The cultural discourses generated by the production of the safety bicycle and its immense popularity offer insight into the representation of bodies, especially as the bodies in conjunction with bicycles formed a new mobility regime. Two categories of semiosis are key: that which focused on the bicycle per se, regardless of the gender of the user, and that which focused on the bicycle in conjunction with gender, especially women. To begin, popular reports published in late nineteenth-century periodicals highlight the mobility offered by the bicycle for all people: social elites, day laborers, pleasure seekers, travelers, and business owners. As one anonymous author rhapsodizes, the bicycle provides the rider with "the nearest approach to flying at present possible," an experience intrinsic to the "exhilarating joy of riding" ("Topics: Reign" 1894, 306). The bicycle quite simply exemplified freedom. The nickel-clad horse liberated all riders from the slow pace and physical restriction of pedestrian travel; it unfettered clerks, shop girls, and other working people from living within a narrow geographical areas surrounding their work place; it released men and women into nature; and it freed everyone from the ill health resulting from a sedentary lifestyle and the confines of crowded housing. Via the bicycle, riders could travel fifty miles a day in the country, across the country, or out of the country as numerous travelogues testified. Cyclists were limited only by the availability of time and good roads. Nor were women immune from bicycle madness, a madness that was gaining widespread social approval, as Philip G. Hubert, Jr., observes. He writes in 1895 that "the woman who objects to bicycling will soon likely be looked upon as more eccentric than her sister who skims along the road in bloomers" (694). A. L. Benedict (1897, 472) reinforces Hubert's insight, claiming the bicycling especially benefits the lonely spinster: "The bicycle is more than a vehicle: it is almost as much of a companion as a horse or dog, while the exhilaration of rapid motion, the accessibility of charming bits of nature, the mastery of time and space, afforded by this steed of steel, more than atone for social companionship which depends on no deep-seated affection." Thus, an array of cultural discourses represents cyclists, both male and female, as partaking jointly in the mobility offered by the safety bicycle. For women, however, long confined to the domestic realm and accorded only limited access to the public, the safety bicycle presented an unprecedented degree of mobility, emancipating them from the restrictions of the Victorian parlor as well as, eventually, from the restrictions of corsets and long, smothering skirts.[4]

Linkages between bicycling and emancipation in the popular press and popular fiction underscore the degree to which this new mobility bestowed on women, who comprised 25 to 30 percent of the bicycling market in the 1890s, a freedom similar to that of their male counterparts (Smith 1972, 25). Women cycling on city and country roads became a quotidian feature of *fin de siècle* life, indicating the degree to which women bicyclists marked the public thoroughfares as their own. Madelyne Bridges in 1893 poetically celebrates the change in women's mobility regimes: "The maiden with her wheel of old/Sat by the fire to spin, / . . . And now the maiden with her wheel/Goes spinning round the world" (76). Another writer ("Editor's Study" 1894, 315) directly associates the emancipated woman and the mobility symbolized by the bicycle, commenting that the "the long-time growing emancipation of woman" is inseparable from "her perch upon the bicycle," which typifies her "going fast and far." In addition, the spirit of the New Woman, a phrase that entered the lexicon in 1894, was readily evident in the 1890s, with the bicycle providing an apt symbol of that change. Young and pretty, the New Woman stereotype in popular fiction was "invariably a keen bicyclist . . . single and well-educated, with strong views on women's rights" (Willis 2001, 53). Thus, the everyday mobility afforded to women riding bicycles as well as the everyday recoding of public space via the popular press and popular fiction indicates the degree to which the "bicycling woman" became a gender identity integral to the discourses and images of the American cultural scene. Close attention to the actual experience of bicycle riding—women's enaction—provides insight into the regendering affected by the new mobility regime.

The third element of the distributed cognitive system that yields a posthuman identity is enactment. As Hayles points out, the critical moment in the emergence of a posthuman comes when the individual is "spliced" into a distributed cognitive system where enacted bodies and represented bodies join. At this point, it is the system that thinks, a system that, by incorporating machines, representation, and enaction, gives rise to the posthuman. Frances E. Willard, perhaps the most politically powerful woman in the *fin de siècle* United States, provides a rich source for tracing enaction, for she writes glowingly about her cycling experiences in an engaging memoir *A Wheel Within a Wheel. How I Learned to Ride the Bicycle, with Some Reflections*

Along the Way. Willard's memoir provides a record of the technology-user interface in action, a process that reschools the female body into a new mobility regime and, thus, regenders her identity. The emergence of both can be traced in three ways: through Willard's celebration of the speed and motion of the bicycle, her record of the changes in her body arising from her mastery of the silent steed, and her reflections on the subsequent changes in mind and spirit.

An especially provocative index of the newly mobile female identity emerging from this a distributed cognitive system is Willard's appreciation of speed, for she is drawn to the bicycle by its "swiftness of motion." This speed, she believes, is "perhaps the most fascinating feature of material life" and an experience long barred from women (1895, 11). She sets up a contrast between the mobility of the bicycle and a heartbreaking point in her life when she had to give up "run[ning] wild" as an adolescent and accept the immobility of adult womanhood. At sixteen, Willard mourns, she became hampered by "long skirts" and their "accompanying corset and high heels"; even her hair, like her body, was restrained, "clubbed up in pins," obedient to the "limitations thus imposed" on her mobility and physical freedom (10). Then, at fifty-three, bereaved by her mother's death and physically ailing, Willard determined to again frolic like a colt in a "pleasant pasture" by learning to ride "the silent steed so swift and blithesome" (13). Nor is she the only woman to find in the bicycle's speed and freedom a palliative for soul sickness. In 1895, Marguerite Merington (1895, 703) advocates a similar cure: "Now and again a complaint arises of the narrowness of woman's sphere. For such disorder of the soul the sufferer can do no better than flatten her sphere to a circle, mount it, and take to the road." Such activity, Merington's claims, transforms the woman into a cyclist, "belonging not to a picture"—and thus stationary—"but to a moving panorama," "mounted and in flight." In a similar spirit, Willard names her steel steed Gladys, having in mind, she explains, "the exhilarating motion of the machine, and the gladdening effect of its acquaintance and use on my health and disposition" (Willard 1894, 53).

However, to embrace the mobility promised by the bicycle, Willard had to retrain her body, an element that manifests the necessary role of embodiment in a posthuman identity, a key point for Hayles. Hayles notes: "Embodiment makes clear that thought is a much broader cognitive function depending for its specificities on the embodied form enacting it," (1999, xiv) and enacting thought requires retraining the body in new mobility regimes. For Willard, this reschooling was an act of re-embodiment, achieving a "new intelligence in the muscles" in relationship to Gladys (23). Willard notes at the beginning of her learning experience that her bicycle required her to overcome her "sedentary lifestyle," (19) to learn "the laws of balance and exploitation," and to adjust herself to the "regulations of gravity" (16–17).

This corporeal education in a new mobility regime included teaching herself new physical habits: to look up at the path, rather than down at the handlebars, and to get "forehead and foot into line" (17). It involved discovering how to sit, pedal, turn, and dismount without risking a fall (22). She disciplined her muscles in this new mobility, and its attendant identity, in precisely the same way she learned her alphabet—through repetition and use—equating her education in cycling with her education in truth-telling (53). Both types of knowledge—each operating at the intersection of thought and action—were lodged in her body.

Finally, reschooling the body results in reschooling the mind and spirit, a move that completes the system of technology, representation, and enaction. "Indeed," Willard says in her memoir, "I found a whole philosophy of life in the wooing and the winning of my bicycle" (25). Cycling became not only favorite pastime, but also a metaphor for living life. A painful fall becomes an illustration of the necessity of overcoming the obstacles—the "darkness and despondency" of life—without "quit[ting] the crowded race-course" (24). In other words, a fall in cycling or in life requires one to get up, remount, and ride again. Furthermore, the necessity of acquiring the mechanical knowledge integral to maintaining the bicycle—learning "the location of every screw and spring, spoke and tire, and every beam and bearing"—becomes a demand for a similar painstaking study of life and of loved ones (25). Without such intense attention to our environments, physical and emotional, we cannot achieve an "equilibration of thought and action by which we conquer the universe in conquering ourselves" (26). Eventually, Willard says, "I began to feel that myself plus the bicycle equaled myself plus the world, upon whose spinning-wheel we must all learn to ride, or fall into the sluiceways of oblivion and despair" (27). She who masters a steed like Gladys, Willard concludes, will also master life, "by exactly the same methods and characteristics," (28) highlighting the emergence of a posthuman identity from the technological mediation of represented and enacted bodies.

CONCLUSION

In this chapter, we have argued that the birth of a posthuman identity encompasses the construction of mobility regimes that are complicit with new subjectivities and perspectives, suggesting the importance of adding movement and non-digital technologies to our understanding of the posthuman. Hayles (1999, 3) notes that posthumanism "configures human being so that it can be seamlessly articulated with intelligent machines," a move Donna Haraway gestures to in "Cyborg Manifesto." Further, both Hayles and Haraway emphasize that the posthuman fundamentally plays a role in "the construction of subjectivity" (4) and a playful and "perverse shift of perspective" (Haraway

2004, 13). By reading technology historically and rhetorically, we offer a new consideration of intelligent machines and thus a new consideration of the posthuman body. Our analyses of the "sporting gentleman" and the "New Woman" demonstrate that these two nineteenth-century figures rely on the distributed cognitive system comprising technology, semiosis, and enaction in much the same way as we see with posthuman figures in the twenty-first. The two artifacts that we offer as examples, in particular, highlight an expanded notion of technology that moves beyond the contemporary digital bias implicit in posthumanism while emphasizing the posthuman body as mobile. Mobility and non-digital technologies thus represent important characteristics of posthuman bodies, underscoring new aspects of the interface among the machine, discourse, and the organic, as well as revealing the importance of mobility regimes in this process. Further, the tourist guidebook and safety bicycle both blur the organic-machine-semiotic boundary and provide a matrix for the emergence of new gender identities: the hypermasculine male and the emancipated female. Thus, the emergence of a late nineteenth-century posthuman body encompasses the construction of mobility regimes that are complicit with new subjectivities and perspectives, suggesting the efficacy of considering movement and non-digital technologies in our examination of the posthuman.

At the same time that our analysis traces the materialization of a posthuman subjectivity from the distributed cognitive network comprised of technology, semiosis, and enaction, it also points to the instability of that network and its emergent identity. Since a posthuman subjectivity, including a posthuman gendered subjectivity, maintains its stable identity only through the mutual transaction among its constituent components, our focus on nineteenth-century technologies invites a consideration of the mutability of any thinking system. In fact, this chapter demonstrates that these constituent components themselves evolve through reciprocal exchanges within the whole. Neither technology, nor semiosis, nor enaction operates in isolation, nor do they operate deterministically. Each exists as a "mutual stimulation" of the others (Fuller, 2004 1); each exists as "an explosion, a passion or capacity settled temporarily into what passes for a stable state" (Fuller 2004, 1). Shifts in stimulation bring forward latent potentialities that have the power of transforming the posthuman body. More crucially, as an active process, this mutable thinking system is inherently volatile, and our rhetorical and historical reading of late nineteenth-century technologies reveals that this volatility can manifest itself in radically different ways, from emancipation to brutal violence.

Expanding the scope of posthuman to include non-digital technologies, especially those from different historical periods, enriches our understanding of the posthuman by directing our attention to the important role that mobility regimes play in the constitution of all posthuman bodies, including twenty-first century digitally-inspired posthuman bodies. At the same time, it returns our attention to gender as the (un)stable product of a complex thinking system, one subject to the dynamic transactions among technology, semiosis, and enaction. It highlights that any posthuman body is itself an outgrowth of and dependent on the mobility—the mutual stimulation—of a thinking system. Such insights are well worth pursuing in an era of rapid technological change, as digital technology continues to ingratiate itself into our daily lives.

NOTES

1. This implication is made even more explicit in Donna Haraway's concept of the posthuman cyborg as articulated in *A Manifesto for Cyborgs: Science, Technology, and Socialist Feminism in the 1980's* (New York: Routledge, 2004). She notes how "late-twentieth-century" machines have distinguished themselves from earlier machines, arguing that they have particularly "made thoroughly ambiguous the difference between natural and artificial" that characterizes the cyborg (11). Further, she notes that the contemporary machines that generate such an ontological breakdown are "quintessentially microelectronic devices" that are both miniaturized and portable. For Haraway, it is the silicon chip itself that serves as "a surface for writing" upon which the image of the cyborg is etched.

2. Of course, what constituted "acceptable" mobility and publicness for women were also dependent upon social class. See David Scobey, "Anatomy of the Promenade: The Politics of Bourgeois Sociability in Nineteenth-Century New York," *Social History* 17:2 (May 1992): 203–227. Scobey describes the complex and tightly regulated social rite of the promenade for the New York elite, in which upper class men and women strolled up and down a fashionable street in their finest clothing, encountering and acknowledging other members of their class. Acceptable spaces for such public mobility were strictly prescribed and public encounters between bourgeois men and women were closely policed and highly ritualized. Ultimately, this kind of public mobility was designed to both establish and display the position of elites for all members of society to see.

3. Although the term "sporting" suggests participation in athletics, it is unclear the degree to which sporting males actually participated in such blood sports, particularly pugilism. In fact, most indications suggest that sporting males were generally spectators to such sporting events, although that was often dependent on class. In the *Gentleman's Companion*, the authors make a brief reference to a famous pugilist in the sporting male tradition, Hezekiah Orville "Awful" Gardner, a Bowery thug who was regularly in the newspapers for his fights in the ring, but seemingly more often for his barroom brawls and assaults. "Sport," instead, specifically references a kind of masculine solidarity through physicality that was perceived to stand in opposition to "'feminized' family and the trappings of domestic life" that compromised male sexuality (Gilfoyle 106).

4. However, the bicycle as technology becomes liberatory not because of a set of intrinsic material features of the steed of steel but because of the ways in which those features mediate and are mediated by specific cultural discourses and enaction. It is because of this mediation that the discourses praising the therapeutic effects of women's cycling configured the bicycle as a new liberatory mobility regime for white, middle-class woman. At the same time, these cultural discourses lauding the bicycle were coupled with cultural discourses condemning cycling as threatening women's physical and moral health. See Robert A. Smith, *A Social History of the Bicycle: Its Early Life and Times in America* (New York: American Heritage Press, 1972), 63–82. Even when bicycling for women was advocated, bicycling extremes, such as "scorching" (riding fast), were discouraged. See Isaac B. Potter, "The Bicycle Outlook." *The Century* 52, no. 2 (Sept. 1896): 786. moa.library.cornell.edu/. Willard gestures to the power of these negative representations, citing and refuting the popular medical discourses that de-

nounced female cycling (1895 54–57). In so doing, she establishes that both discourses—those of praise and blame—exist simultaneously. When censorious discourses mediate technology and enaction, a different mobility regime emerges, one characterized by restriction rather than liberation. Within this thinking system, a woman's public movement and identity are curtailed, rather than liberated, by the protocols of the True Woman wherein cycling's emancipatory potential is rigidly controlled by chaperones, chaste apparel, demure behavior, sedate speeds, and approved venues.

REFERENCES

Bazerman, Charles. 1998. "The Production of Technology and the Production of Human Meaning." *Journal of Business and Technical Communication* 12.3: 381–387.

Benedict, A. L. 1897. "Open Letter: Dangers and Benefits of the Bicycle." *The Century* 54.3: 471–473. moa.library.cornell.edu/.

de Certeau, Michel. 1984. *The Practice of Everyday Life*. Translated by Steven Rendall. Berkeley: University of California Press.

Clark, Andy and David Chalmers.1998. "The Extended Mind." *Analysis* 58: 10–23.

Crapsey, Edward. 1872. *The Nether Side of New York or, The Vice, Crime, and Poverty of Great Metropolis.* New York: Sheldon and Company.

"Editor's Study." 1894. *Harper's New Monthly Magazine*. 89.530: 311–316. moa.library.cornell.edu/.

Fishman, Teddi and Kathleen Blake Yancey. 2009. "Learning Unplugged." In *Going Wireless*. Edited by Amy C. Kimme Hea, 35–49. Cresskill: Hampton.

Foucault, Michel. 2009. "The Language of Space." In *Space, Knowledge and Power: Foucault and Geography,* translated by Gerald Moore, edited by Jeremy W.Crampton and Stuart Elden. Burlington: Ashgate.

Fuller, Matthew. 2005. *Media Ecologies: Materialist Energies in Art and Technoculture*. Cambridge, MA: The MIT Press.

The Gentleman's Companion: New York in 1870. 1870. New York: n.p.

Gibson, William. 1984. *Neuromancer*. New York: Ace.

Gilfoyle, Timothy J. 1992. *City of Eros: New York City, Prostitution, and the Commercialization of Sex, 1790–1920.* New York: W.W. Norton & Company.

Haraway, Donna. 2004. *A Manifesto for Cyborgs: Science, Technology, and Socialist Feminism in the 1980's.* New York: Routledge.

Hayles, Katherine N. 1999. *How We Became Posthuman: Virtual Bodies in Cybernetics, Literature, and Informatics.* Chicago: The University of Chicago Press.

Hubert, Philip G., Jr. 1895. "The Bicycle. The Wheel of To-Day." *Scribner's Magazine* 17.6: 692–702. mao.library.cornell.edu/.

MacKenzie, Adrian. *Wirelessness: Radical Empiricism in Network Cultures*. Cambridge: The MIT Press.

Marx, Paul Alan. 1983. "*This is the City: An Examination of Changing Attitudes Toward New York as Reflected in Its Guidebook Literature, 1807–1860.*" PhD diss., Harvard University.

Merington, Marguerite. 1895. "The Bicycle. Woman And The Bicycle." *Scribner's Magazine* 17.6: 702–704. moa.library.cornell.edu/.

Merleau-Ponty, Maurice. 1998. *The Phenomenology of Perception*. Translated by Colin Smith. London: Routledge.

Michalski, David. 2004. "Portals to Metropolis: 19th-Century Guidebooks and the Assemblage of Urban Experience." *Tourist Studies* 4.3: 187–215.

Pingree, Geoffery B. and Lisa Gitelman. 2003. *New Media: 1740–1915.* Cambridge: The MIT Press.

"The Point of View," *Scribner's Magazine* 19, no. 6 (June 1896): 781–784. moa.library.cornell.edu/.

Potter, Isaac B. 1896. "The Bicycle Outlook." *The Century* 52.5: 785–791. moa.library.cornell.edu/.

Richards, I.A. 2001. *Principles of Literary Criticism.* London: Routledge.

Scobey, David. 1992. "Anatomy of the Promenade: The Politics of Bourgeois Sociability in Nineteenth-Century New York." *Social History.* 17.2: 203–227.

Smith, Robert A. 1972. *A Social History of the Bicycle: Its Early Life and Times in America.* New York: American Heritage Press.

"Topics of the Time: Bicycle Problems and Benefits" 1895. *The Century* 50.3 474. moa.library.cornell.edu/.

"Topics of the Time: The Reign of the Bicycle." 1894. *The Century* 49.2: 306–307. moa.library.cornell.edu/.

Weisman, Leslie Kanes. 1994. *Discrimination by Design: A Feminist Critique of the Man-Made Environment.* Champaign: University of Illinois Press.

Welter, Barbara. 1985. *Dimity Convictions: The American Woman in the Nineteenth Century.* Athens: Ohio University Press.

Willard, Frances Elizabeth. 1895. *A Wheel Within a Wheel. How I Learned to Ride the Bicycle, with Some Reflections by the Way.* New York: Fleming H. Revell Company.

Willis, Chris. 2001. "'Heaven defend me from political or highly-educated women!': Packaging the New Woman for Mass Consumption." In T*he New Woman in Fiction and Fact: Fin de siècle Feminisms*, edited by Angelique Richardson and Chris Willis. London: Palgrave.

Chapter Ten

Living Deliberately, Less or More

Affirmative Cynicism and Radical Design

Matthew A. Levy

An early attempt to understand how technologies mediate our experience of the world and affect our behaviors was made by the ancient Cynics, for whom a major preoccupation was how material comforts become an obstacle to self-reliance. Cynicism is embodied in the legend of Plato's contemporary, Diogenes, and recast into an American context by the work of Henry David Thoreau. The Cynics responded to the problem of dependence on things through experiments in ascetic disimbrication, attempts to disentangle themselves from the clutter of contemporary expectations. This chapter considers the challenge wrought by technological change to this traditional cynicism. Through rigorous questioning and material practices against the grain of contemporary expectations, cynics sought distinctions between wants and needs, between the just and the unjust, and between the natural and the unnatural. Of these three oppositions, I direct my interrogation at the third. The need to simplify our ways of life and pursue justice remain pressing and related concerns. In fact, their urgency increases due to technological speed-up and ecological disaster. If rhetorical appeals to "the natural world" enable the preservation of wild spaces, I see no problem. Efforts to distinguish between "the natural" and "the unnatural" within civilization, however, produce little more than projection. For instance, when we call one animal behavior "natural" and another animal behavior "unnatural" (such as violence, polygamy, heterosexuality), we rely on reified notions of the nature of "Nature" and the nature of "the good." This trope of "the natural" paints technology with a broad, negative brush, making it more rather than less difficult to make good choices about things.[1]

Inventions create a kind of hard ideology, the Law poured into molds, the culture cast. Or, put another way, the environment—conceived broadly to include fabricated things and networks—forms the concrete grammar of our lives. Each "thing" has rules you must follow (more or less) to make it work—or in the case of a speed bump, just so you can get past it without spilling your coffee. Like grammar, technology can be used consciously or unconsciously. It is fixed and rule-bound in one moment and poetically flexible the next. And to be sure, like grammar, technologies distribute privilege unevenly. The degree of plasticity offered by a particular technology varies according to the level of a user's access and ability: the architects, the engineers, the designers, the programmers, and, of course, the practitioner-philosophers (cynical rhetoricians, hackers, and geeks). When thinking about agency in our technological reality, one cannot seriously consider being *free from* "things," but must consider which thing-environments (a hyphenation intended to point to the immersive nature of tools) bring about the best self and society. To make these evaluations, we need access to and familiarity with the many options, as well as an affirmative cynical tradition that ritualizes experimentation and evaluation. While specific widgets have always been and will continue to be abandoned or rejected, without an affirmative relationship to technology-in-general, cynicism can no longer relevantly pursue its quest to connect personal experiences to public ethical and political considerations.

While I do seek to persuade in this chapter, my work here—about and by design—is as much narrative as argument. In order to convince of the relevance of Cynicism and how imploding its definition of Nature can make it even more pertinent, I will share anecdotes about how Cynicism has developed—in anecdotes. While the name "Cynic" became overwhelmingly pejorative after the fall of Rome, the tradition that began in Athens with Diogenes, a sort of philosophical action hero, maintained its influence not only through how its critical and democratic traditions have been written into legal institutions, but also through stories.[2] The trope of the natural loomed large in these stories, frequently serving democratic and experimental ends. Technology's advance, however, has broken down the explanatory and evaluative power of "the natural," as my concluding examples demonstrate. In order to offer a relevant and ethical critique of civilization and "things," I argue that Cynics must not retreat into the so-called "natural" but must engage with or, better, participate in technological design.

GET BACK TO WHERE YOU NEVER BELONGED

Whether the appellation came from their gym (Cynosarges) or from how they embraced their animal natures, the original Cynics were called dogs (Dudley

1937, 5). The original dog, Diogenes of Sinope, lived in Athens in the fourth century BCE during a time of great intellectual activity and technological change, as well as a time of profound uncertainty. The Athenian empire had recently suffered failure in the Peloponnesian War. The actual facts of Diogenes's life cannot be reliably extricated from the apocrypha. Almost all of the information we have is from anecdotes (Laertes 1953); there are a lot of them, because Diogenes was a favorite character for chreia, brief pedagogical and humorous anecdotes based upon the reputations of famous people. Diogenes's reputation starts wild and gets wilder. He flees to Athens when his father has ostensibly been accused of destroying currency. From this experience, Diogenes takes his motto, "to deface the common currency," to signify his purpose of debunking mainstream values, beliefs, and practices. This is a wild, unnaturally natural move for an outsider among the xenophobic Athenians. Diogenes quips, "I am a citizen of the world" (Laertes 1953). Either because his domicile in Athens isn't ready or because he thinks it is more in keeping with his philosophy, Diogenes takes up residence outside in a tub. He gives up all possessions other than one piece of clothing, a wallet, a staff, and a cup. Then, after seeing a boy drinking with his hands, he castigates himself for keeping an extraneous object and throws away his cup.

Stories about Diogenes often pit him against Plato, who called Diogenes a "Socrates gone mad." Diogenes answers Plato's discourse on spiritual love by masturbating and saying, "This is love." Diogenes demonstrated thus a pithy materialism. Diogenes rejects Plato's notion that the physical is an imperfect copy of the ideal real. The physical is real, and it is our ideas and practices that are out of keeping with Nature. Diogenes mocks Plato's definition of the human being as a featherless biped by bringing a plucked chicken to the agora. This is not just a joke on this one nonsensical result of Plato's method but on his whole project of institutional, abstract philosophy, insofar as it places higher importance on technical linguistic precision than on ethics and quickly becomes a rationalization for the status quo. Whereas Plato turns Socrates into a character for his speculative pedagogical dialogues, Diogenes puts Socratic ideas into practice. Plato's Socrates thinks freely but then reminds his students to obey Athenian conventions. For Diogenes, though, the Socratic good life is not about hypothetical reflection. His self-examination is a material practice. If an action does not improve the physical body and the body politic, it is false. Hypocrisy is obscene, and obscenity is put to use to expose human dishonesty. Diogenes is commonly pictured with a lamp, making a performance of searching in vain for an honest man. When a boy farts, Diogenes instructs him to save the next one to make a point. One should not just fart; one should fart on greed and pretension.

In the Athenian empire, men have grown accustomed to a certain way of life, subjugating their neighbors and women, and living off the spoils of war and oppression. But now, Sparta is bringing an end to Athenian exceptionalism, and Diogenes offers a timely message (then and now). Self-reliance does not mean having power over another person; it means self-discipline. Those so-called luxuries enslave; more tools means more toils. It may feel powerful to possess status and luxurious objects, but this desire actually enslaves one to their upkeep and falsifies one's relationships. Along with cosmopolitanism and frankness, simplicity is a core Cynical virtue. For Diogenes, simplicity yields freedom. Diogenes comes to be revered not just as a philosopher but as a cultural hero. One frequently recounted tale has Alexander the Great visiting Athens. The citizens have tripped over each other to greet him, but the infamous Diogenes has not. Alexander goes out to find the infamous Diogenes in his tub, sprawled on his back, munching on wild onions and sunning himself. Alexander greets him and says he would like to offer him a gift: "I can give you anything the world has to offer. Is there any favor I might grant you?" With his characteristic rudeness, Diogenes only says, "Yes, you can get out of my light." Alexander understands the import of Diogenes's retort and later remarks, "If I were not Alexander, I should like to be Diogenes." The images of this encounter between the man who has everything and the man who has nothing enshrine the cultural value of speaking truth to power. Within the Greek tradition of parrhesia or "straight talk," truth was already linked to risk-taking (Foucault 2001, 15–20); however, the chreia tradition gave Diogenes a lot of the best lines, as in this scene with Alexander. The Cynics thereby put their trademark on frankness, and thus were given an important place in the free speech tradition.

Donald Dudley argues that Diogenes's reputation is shaped in part by the eagerness of the Stoics to trace their lineage back to Socrates by way of Antisthenes (Dudley 1937, 1–17). Though calling the embodied materialism of Diogenes "seminal" is irresistible, the Diogenes character we have may be revisionist history (of the *vulgar* sort). The idea that these stories are apocryphal and have Stoic origins helps us to recognize another important meaning in the Alexander story. Alexander doesn't just like Diogenes; he admires him more than he does Plato (or, rather, the Stoic inventor of this anecdote did).

Scholars have focused more on the conflict between Plato and the Sophists than on that between Plato and Diogenes. Platonism, Sophistry, and Cynicism are all concerned with a vital question for academics, *What counts as truth?* Sophistry supposedly turns its back on the truth because flattering the audience is more effective. Plato puts his stake in rigorous method. Plato is institutional; the Sophists are sophisticated; both offer civilized values that support the project of educational institutions. The conflict between Diogenes and Plato folds on a different axis. Diogenes and Plato have similarities: While not above fawning over someone for a handout, should circumstances demand it, Diogenes usually seems as hostile to flattery as Plato and also puts his stake in method. But Diogenes has a far more countercultural ethos than Plato. Diogenes is not against civilization, per se—he chooses to live in Athens—but he questions more radically what is passing for civilization in his time. And he looks to Nature for help in making that evaluation. Plato sees truth as knowledge, recognizes it by its consistency, and pursues it by ferreting out contradiction through his dialectical method. Diogenes, by contrast, sees truth as freedom, recognizes it in self-sufficiency, and pursues it by attempting to live in greater accord with Nature by examining actual daily practices, however mundane.

The Cynical ethos of simplicity reemerges in American rhetoric most famously in *Walden.* Thoreau recorded his experiences doing a Diogenean experiment which he called "living deliberately." He wrote,

> Most of the luxuries, and many of the so-called comforts of life, are not only not indispensable, but positive hindrances to the elevation of mankind. With respect to luxuries and comforts, the wisest have ever lived a more simple and meagre life than the poor. The ancient philosophers, Chinese, Hindoo, Persian, and Greek, were a class than which none has been poorer in outward riches, none so rich in inward. We know not much about them. It is remarkable that we know so much of them as we do. The same is true of the more modern reformers and benefactors of their race. None can be an impartial or wise observer of human life but from the vantage ground of what we should call voluntary poverty. (Thoreau 1845)

In the writing of Thoreau, getting rid of convenient devices becomes a rhetorical device: methodically breaking down the arrangements of life and rebuilding it with fewer things. One chair for solitude, two for friendship, three for society—and that is enough. Living deliberately throws into relief the roles that objects play in mediating human relationships and in camouflaging other kinds of relationships in the world.

Thoreau's discussion of clothes begins with a modest conjecture that "perhaps we are led oftener by the love of novelty and a regard for the opinions of men, in procuring it, than by a true utility. Let him who has work to do recollect that the object of clothing is, first, to retain the vital heat, and secondly, in this state of society, to cover nakedness" (Thoreau 1845). Royalty who wear a suit for only one day, he tells us, never get the joy of wearing clothes that have been worn in to take on "the impress of the wearer's character" (Thoreau 1845). Well-worn garments are so loved that we never want to give them up; we apply more mending and patching to them than we would bother to do with our own bodies, he jokes in good humor. Then Thoreau turns to the real object of his satire, those who slavishly follow French fashion and would rather walk to town with a broken leg than with a patch on their trousers. What we need, Thoreau quips, are not new clothes, but new people. New clothes are like a consolation prize for those who don't renew themselves through practice and experience. The woman who sells him clothes tells him what "they" are and aren't currently making. *Who is this they*, he asks. *What is their relationship to me?* The answer is no different than one we might get when we ask who makes our phone cases and "skins."

> I cannot believe that our factory system is the best mode by which men may get clothing. The condition of the operatives is becoming every day more like that of the English; and it cannot be wondered at, since, as far as I have heard or observed, the principal object is, not that mankind may be well and honestly clad, but, unquestionably, that corporations may be enriched. (Thoreau 1845)

According to Thoreau, then, overemphasis on and overconsumption of clothing deserves contempt because of the false dignity it provides the consumer at the expense of spiritual health, as well the horrible toil to which it condemns the worker at the expense of physical health.

In the story of Diogenes giving up his cup and in Thoreau's extensive travels around Concord and Walden Pond, we see a reexamination of human needs and wants and especially the exposure of false needs established by civilization. The human animal needs food, water, shelter, and love, but the specific conventions each culture establishes for satisfying those needs may be arbitrary or even decadent. Thoreau is bold enough to refuse paying taxes to support slavery and an unnecessary war, but his experiment has an ephemeral quality. Two years at Walden in a tiny cabin shows impressive commitment; it helped Thoreau develop his writing and, thereby, helped define an American ethos of self-reliance for himself and for many future countercultural seekers. It was, though, temporary—a pragmatic experiment. Even Diogenes's experiment in self-reliance, though presented as a lasting choice in most stories, is a compromise. He is not a survivalist getting as far from society as possible; rather, he engages Athenian society against the grain. Many of the Diogenes anecdotes have him taking a negative stance toward conventional mores. When invited to a rich man's home, he spits in the man's face, saying, "I'm sorry; every other place in your house was unsuitable for spitting."

Diogenes's living circumstances and attitude toward certain rich men in the anecdotes may suggest someone who has arrived securely at a morality that embraces poverty and condemns the rich; however, Bracht Branham (1994) argues that asceticism is not a consistent feature of Diogenes anecdotes. He sees Diogenes as a pragmatist who invented his way of living because that is what was produced when he applied his personal gifts to his personal circumstances. Branham boldly suggests that the ascetic Cynics, philosophical beggars, and Roman stoics who followed Diogenes's example mistook his vicissitudes of fortune and impressive rhetorical improvisation for a set of philosophical principles. As I described above, Dudley presents an equally compelling explanation, which has our Diogenes anecdotes written later according to later needs. Revisionism might also explain the contradictory anecdotes that Branham feels point to improvisation. The narratives of Dudley and Branham, in any case, both support my contention that Cynicism is a living tradition that evolves to meet the needs of the present; embracing it will not and could not require us to accept every Cynical practice or belief about the nature of Nature.

Contemporary society owes a great debt to the critical values that survived with the Cynical tradition—whether the real Diogenes was devoted to them or not. We all seem to agree that civilization is out of keeping with Nature but disagree on the needed correction. Not everyone takes the same lessons from Diogenes and Thoreau. The "liberal" Cynic—my example will be the architect of the film version of *12 Angry Men* (1957)—embraces enlightened suspicion and values "self-reliance" for its relationship to egalitarianism. The "reactionary" Cynic—my example will be the architect of the novel *Fountainhead* (Rand 1971)—hates taxation and interprets "self-reliance" as inconsistent with the welfare state. In either case, if "self-reliance" means complete independence, it is impossible. Human beings are not solitary animals; we depend on a network of relations between humans, other animals, and things. Cynical experiments in disimbrication have helped us see the nature of the network of relations and how they are more than human, including not only the physical things that we have fixed as tools but also to a broader universe of interdependent entities. They have helped us see what it means to make a tool, which is all the more important as tools get more powerful. The enlightenment continues to demystify the world, and technology is making the world an object of our creation in an increasingly literal sense. We gain the capacity to take and make whatever we want, but the maturity to distinguish those wants and fantasies from needs and benefits lags behind.

CYNICAL BULLDOZERS AT TWENTY PACES

Just as Diogenes makes a good philosophical hero for anecdotes, architects make good dramatic heroes for books and films. Architects are strong like action heroes because they know how to harness technology, but they can manage better dialogue. They are entrusted with great responsibilities: the costs of building, the effects of their designs on our cityscapes and living spaces, on health and safety. Like the Cynics, architects are greatly concerned with the connection between practical and ethical issues, like those Thoreau reflected upon in the construction of his cabin. They not only make vital assumptions about the needs and behaviors of the people who will use their buildings, but also greatly influence the actual behavior of those people. If a story seeks to explore the ethical implications of the environment human beings have built for ourselves—architectural, institutional, technological—who better to put at the center than an architect, someone who understands and shapes our thing-environment? *12 Angry Men* and *The Fountainhead* both use Cynical-architect-protagonists, men who transform the situation rather than bending to it. Such stories speak to a longstanding anxiety about the idea of the autonomous individual, exacerbated by our increasingly technologized environment and our dependence on "things." While they are both figures like Diogenes in the Cynical tradition, these two fictional architects take that tradition in different directions in terms of their ethical force.

12 Angry Men contains almost all of its action in a hot jury room. At the start of the film, eleven jurors stand ready to convict a young man for murdering his father based on the seemingly airtight case presented by the prosecution and the weak defense presented by the lawyer of "the boy." Juror #8, played by Henry Fonda, holds up the verdict until he finally persuades the rest that reasonable doubt exists, and the boy is exonerated. Affirmed is the idea that one individual can make a difference. Our hero is an architect; the significance of his career to this idea of personal agency is never explained, but the exchange in which we discover it gives special focus to the vocation of architecture and the relationship between institutional design and devices (in the sense of motives, ends, purposes).

The most impatient juror—a salesman who wants to get this capital murder case over with so he can go to a baseball game—asks, "Hey, you a salesman or something? You sure got that soft sale down. Me, I'm really aggressive, with jokes, tricks, stories" (*12 Angry Men* 1957).

"No, I'm an architect," Fonda's character replies. Architectural "statements" may be grand and imposing; however, their persuasive power does not draw attention to themselves as such. They are built into their character, into their design. In many episodes throughout the film, we are shown that the salesman just wants to get the trial over with—to close the deal (an American attitude)—so that he can get to the baseball game (the American pastime). Who couldn't imagine feeling the impatience of a man wanting to get out of what seems to be a pointless trial with a predetermined outcome in order to spend time with his family? This architect, however, offers an alternate American archetype: a designer of the future who has a specific capacity to recognize the wisdom behind the design in our justice system, which gives twelve peers the power of decision and, in such important cases, requires a unanimous verdict.

The other jurors trust the *authority* of the system rather than its process, failing to question the evidence as presented by the prosecution and left unquestioned by the inept defending attorney. They are encouraged by the seeming confidence of the judge, who declares that no punishment less than the death penalty will be applied if "the boy" is found guilty. Our architect sees the injustices that will occur if its jurors do not own their authority. The process of justice has the architect's respect because the tradition of Cynicism—or enlightened suspicion—is built into it through the mechanism of "reasonable doubt." The architect knows the power of design to encourage certain behavior and even instill certain virtues is considerable but not deterministic; if it is to work, the process relies on the cooperation of individuals willing and able *to think for themselves*. This architect is a textbook liberal in the enlightenment tradition.

Fonda's character provides a civics lesson about the importance of reasonable doubt and the power of the individual that Supreme Court Justice Sonya Sotomayor cites as influencing her decision to go into law. Another fictional architect looms large in US political culture, Howard Roark, the main character of Rand's first literary success, which packaged in pulp her case for radical individualism and against any altruistic or community-based moralism. Her philosophy of objectivism has inspired libertarians and social conservatives alike, despite her frequent bitter diatribes against religion. Her central tenants are (1) that human thought can directly access objective reality and (2) that following one's own individual self-interest (though it may be against that of the group) is the only true morality. Rand's architect, Roark, has a powerful influence on other architects even as he suffers in anonymity; however, in Rand's allegory, the impact of the architect on the world is not to be interpreted as the power of the mind to create reality, but as the power of the mind to discover objective reality, in this case an architectural principle that form should follow function. Once again, the character of the architect becomes a way for a narrative to engage the question of the relationship between social organization and individual motives.

In *12 Angry Men*, an important turning point in the case comes when the architect brings a piece of physical evidence into question. He presents a knife identical to one the prosecution had presented as being exceptionally rare. For the liberal architect, the meaning of the artifact is open to interpretation and that interpretation is motivated by compassion. He has nothing to gain beyond the satisfaction of taking part in the renewal of civic justice. By contrast, the important turning point in *The Fountainhead* is when, by giving a rousing speech on the importance of ego and unique vision, Rourk is able to win through jury nullification—by undermining democratic principles. He is rewarded with the right to build a gigantic phallic symbol, the tallest building in the world, as "a monument to that spirit which is yours" (Rand 1971, 692). Rand's Cynicism is not a Cynicism of "reasonable doubt" grounded in a shared recognition of human fallibility, but a Cynicism that rejects compassion in favor of unrestrained assertiveness on the part of the individual, seen as the only true reflection of Man. Both the liberal Cynicism of reasonable doubt and the objectivist Cynicism of anti-altruism have been imitated and ritualized enough to be recognized as influential contemporary examples in the long tradition of Cynicism. Clearly, the various strains of Cynicism that have developed over time and have been institutionalized in various ways can have radically different moral content. I prefer the Cynicism of Henry Fonda's character over that of Howard Roark's, which I regard to be the product of an intellectual sociopath. For my present purpose, however, it is more important to demonstrate that Roark's architectural principle does not, in fact, point to objective reality. It represents, rather, an example of the natural/unnatural opposition that undermines the ability of Cynicism to inform useful assessments of technology.

"NATURE" AND THE METAPHYSICS OF FORM AND FUNCTION

Rand's architect-hero was inspired by Frank Lloyd Wright's mentor, Louis Sullivan, one of the modern architects who embraced newly available materials and began the movement away from adornments and embellishments. In 1896, Sullivan wrote in "The Tall Office Building Artistically Considered":

> Whether it be the sweeping eagle in his light or the open apple-blossom, the toiling work horse, the blithe swan, the branching oak, the winding stream at its base, the drifting clouds—over all the coursing sun, *form ever follows function*, and this is the law. Where function does not change, form does not change. The granite rocks, the ever brooding hills, remain for ages; the lightning lives, comes into shape, and dies, in a twinkling.
>
> It is the pervading law of all things organic and inorganic, of all things physical and metaphysical, of all things human and all things superhuman—of all true manifestations of the head, of the heart, of the soul—that the life is recognizable in its expression, that form ever follows function. *This is the law.*
>
> Shall we, then, violate daily this law in our art? (Sullivan 1896, his emphasis)

It stands to reason that such a functionalist motto should come from an aesthetic treatise written about office buildings. But Sullivan here breaks into prose poetry to make a larger philosophical point. Sullivan believed that to subordinate form to function was to live more in keeping with Nature, which, as we will remember, was one of the principles of the Cynics that led them to embrace simplicity.

Sullivan's writing contains a significant logical failure, however. When he asserts that "form follows function" is a natural law, he supports it by saying it is an automatic and unavoidable fact, something more akin to a law of physics than to a moral law or aesthetic principle. Yet, like a sort of original sin, human beings seem to be uniquely capable of breaking this law "in our art." If it is true that "form follows function" is a law because it is true everywhere, then human beings wouldn't have to work to follow it any more than we have to work to breathe. In truth, there is much ornament and play beyond the human sphere. Furthermore, pulled down by gravity, water conforms to the shape of the land; rivers do not naturally take their shape in pursuit of a functional use-value (for instance, to get where the animals need water). In other words, form may follow function insofar as form helps support the function of reproduction, but there are many places in the organic and inorganic world where function follows form as well. The great appeal of Sullivan's "form follows function" formula is its seeming simplicity. Yet, it lacks descriptive agency.

Sullivan's more specific point was that, in his view, architects were building nine out of ten buildings badly because they were racked with fear that if they made improvements, they might be deemed "out of form" by the critics and lose work. The result was that the newly possible tall buildings were needlessly constrained by the conventions of short buildings. This is an admirable insight in and of itself. Sullivan uses the trope of the natural in order to make the point that it will be necessary to embrace new technologies in order to throw off dysfunctional old ones. I admire Sullivan's positive attitude toward invention and, even more so, his desire to understand human society as something continuous with Nature. Nevertheless, though it was a big idea and well intended, his desire to universalize "form follows function" into a natural law was arrogant and had unintended consequences. What started as a needed call to reconsider design features in relation to purpose grew into an epistemological menace at the moment it was mistaken for an objective principle of nature. If anyone were to doubt Sullivan's influence on Rand, they need only consider this passage from "The Tall Office Building":

> The man who designs in this spirit and with this sense of responsibility to the generation he lives in must be no coward, no denier, no bookworm, no dilettante. He must live of his life and for his life, in the fullest, most consummate sense. He must realize at once, and with the grasp of inspiration, that the problem of the tall of office building is one of the most stupendous, one of the most magnificent opportunities that the Lord of Nature in his beneficence has ever offered to the proud spirit of man.
>
> That this has not been perceived—indeed, has been flatly denied—is an exhibition of human perversity that must give us pause. (Sullivan 1896)

Sullivan's exuberance about the tall building gave Rand her novel's conclusion. His arrogance about the importance of his insight may have had equal influence on the egotistic character of Roark. Yet, unlike Rand, Sullivan seems to be a (small d) democrat.

When form is allowed to follow function, Sullivan holds forth, the free expression of architects will yield a new American form of architecture that embraces the way of Nature, and then "it may be proclaimed that we are on the high road to a natural and satisfying art, an architecture that will soon become a fine art in the true, the best sense of the word, an art that will live because it will be of the people, for the people, and by the people" (Sullivan 1896). When the individual American architect frees himself from the orthodoxies of the critics, he will truly be living his life, Sullivan argues, because the result will be democratic. In Sullivan's treatise, we hear echoes of Walt Whitman: his celebration of his self is one with a celebration of the other. Sullivan undertakes a radical reconsideration of the building not altogether unlike Diogenes's interrogation of the rules of the agora or Thoreau's meditations on the construction of his home. As a thinker with a Cynical bent, Sullivan knew that the artifacts of our life reflect and shape our lives.

Yet, as his slogan took hold, some forgot that function also follows form, that design has consequences beyond the intentions of the maker. "Function" has sometimes come to mean "purpose" or "end" in a severely restricted, instrumentalistic sense. And the end came to justify the means, without the benefit of the understanding that each "means" has its own effects apart from its relation to the motives of the designer. For instance, if the purpose is to provide low-income housing cheaply, inadequate regard might be given to the quality of life of the inhabitants and means might be employed that degrade that quality of life—this despite the fact that improved quality of life should obviously be regarded as a crucial function of housing. "Form follows function" can serve as a welcome rejection of pointless architectural conservatism on the one hand or a quasi-aesthetic justification for ugly, rapacious development on the other; the latter is a sad outcome for an approach ostensibly grounded in Nature. Even Roark took into account the benefits his architecture would have on its users, unlike his hack imitators. Rand's followers seem to have lost sight of this much broader understanding of function in their worship of self, to the detriment of our ecologies (media, political, physical).

DOING WHAT COMES NATURALLY (BUT DIDN'T YESTERDAY)

When I began graduate school, MOOs and MUDs, primitive precursors to today's Massively Multiplayer Online Games, were state-of-the-art virtual spaces. These interactive spaces were built using object-oriented programming. Everything, including one's own avatar within the space, was programmed as an object with properties. When you "entered" a room, you would be greeted with a description of the place and a list of "who" and "what" was contained within. In this text-based reality, you could meet others and communicate, build on to existing structures, create objects to place within those structures, and give those objects various properties/capacities. I created classrooms, and, for a few semesters, took my students occasionally. There was a sense that many of the people who regularly visited these puppet motels, as Laurie Anderson (1994) called them, were active participants in their creation. Experimentation in the form of the space was one of its primary functions. That was true of the Internet as a whole in those days. Suddenly, it seemed everyone was an architect and a designer. Our computers were no longer tools but spaces that we lived in—an increasing number of minutes each day. We were building a building in which everyone, it seemed, would eventually work and live. We knew life was changing but we didn't know exactly how or how fast.

Virtual space, with its perfect plasticity, may be escaping its electronic prison. Just as one can currently format a document and print it out on the surface of paper, those with 3–D printers are already printing out circuit boards, fabrics, and forms of almost any kind. A 3–D printer can now be bought for under a thousand dollars; designers share files to print online. Anthony Atala, a surgeon, describes an early stage experiment in which specialized 3–D printers are being used to print organs for transplant (2011). Behrokh Khoshnevis, a professor of industrial and systems engineering at the University of Southern California, describes a large 3–D printer under development that he expects will be able to print an entire house in a process he calls Contour Crafting. Instead of having specialists of various kinds, the printer will do the concrete work, the plumbing, the electrical work, the basic construction, and even the tile work according to CAD designs. Round shapes can be used rather than rectilinear walls, which are the least structurally sound, with no increase in cost. The cost, time to construct, and ecological impact will be greatly reduced; and every house can potentially be unique, because you only have to change the program. While he admits this will remove many job opportunities, he believes it will open up many other less dangerous jobs, while greatly reducing the housing shortage in developing countries (Khoshnevis 2012).

As Sullivan would have it, form ever follows function. The explosion of technological capacity we see in developments like 3–D printers remove obstacles to bringing the form of our designs closer to the purposes we have for them. However, in a certain real sense, function also follows form. In *The Medium is the Massage*, Marshall McLuhan asserted that "when faced with a totally new situation, we tend always to attach ourselves to the objects, to the flavor of the most recent past. We look at the present through a rear-view mirror. We march backwards into the future . . ." (1967, 74–5). Gradually, however, the qualities of a new medium assert themselves and the new capacities it offers lead to new functions. The printer was originally designed to put ink images on paper, but its form gradually suggested new functions that we had not previously imagined. These new capabilities and functions will be radically disruptive to our current ways of life. Downloading and printing products at home will do to many other brick and mortar companies what Amazon.com has done to many book stores. Printing human parts will transform our sense of self and bring down the cost of transplantation; building homes in a completely automated fashion will change the job market and create new kinds of living spaces. Upon hearing of these changes, Thoreau might have retreated again into the woods and Diogenes might have turned his tub so the opening faced down; however, an affirmative Cynic will recognize and insist that the current arrangements simply are not meeting the needs of enough people. Change, when it involves elegant new solutions made possible by radical design, will doubtless create new problems, but the problems we have cannot be allowed to continue unaddressed.

My friend from high school, Jon Kuniholm, lost his right arm in Iraq on the first day of 2005. I hadn't yet seen him since it happened when he came to stay for a couple of nights while I was visiting my folks in Wilmington, NC. I had seen a feature on Bob Dole that said that Dole carries a pen in his right hand as a visual cue for people to shake his left hand instead. So, when I saw Jon, I made a point of offering my left hand. For his own reasons, my father insisted on shaking the prosthetic with his right hand. Each of us had given advance consideration to this potentially awkward moment. Jon brought in a pile of arms and dumped them in the corner as if to announce, "This is not a taboo subject." There was an arm for playing the guitar (no fingers but a pic holder), an arm that he prefers wearing for pictures, and at least one other he regards as useless. Most of the time, Jon wears the same arm that Vietnam vets prefer, the same design used in WWII but with better materials. It has a claw that opens and closes when he adjusts the tension on a cable by moving "the forearm" in relation to his shoulder. Its dependable simplicity has not yet been surpassed by an electronic model—at least not in an affordable arm (Kuniholm 2009b). Jon's choice of the body-powered hook (the Hosmer Dorrance 5x) was not only practical, but Cynical. It made a legible critical statement about the lack of investment and advancement in prosthetics, especially for the upper body.

(Why the upper body? Jon explains that diabetes affects the legs, so there are more patients and more incentive for companies to develop mechanical legs. Form and function trope together toward profit, like plants to the sun.)

About a year after Jon was wounded by an improvised explosive device, DARPA launched the Revolutionizing Prosthetics program. The goal of the project is to produce a prosthetic arm that weighs the same and provides the same functionality as a regular arm (Adee 2009) so "service members with arm loss may one day have the option of choosing to return to duty" (DARPA 2012). Because Jon is both a veteran and an engineer, he has been sought out to test new designs such as those being developed by the consortium of universities funded by DARPA for this purpose. You can find videos online, some featuring Jon, which seem to suggest that we are months away from replacing the human arm with a fully-functional prosthetic. To put this in perspective, Jon cites a 1965 *New York Times* article whose title announces,

"New Process Will Help Amputee To Control Limb With Thought" and which says that "practical use [is] still a few years away" (Feron 1965). He calls this longstanding and ongoing over-optimism "The Six Million Dollar Meme." Jon believes the technological bar of feedback is much higher than many engineers, and certainly the news media, recognize (Kuniholm 2010). Yet, by using his traditional hook, Jon can pick grapes without crushing them and throw them into his mouth, and he doesn't need a computer or an outlet to do it.

According to Jon, DARPA has spent over one hundred fifty million dollars. Kuniholm points out, however, that one fighter jet costs more, and, to this date, the project has not made a single functional arm available to the public. Jon wishes they had invested the money differently. There should be, Jon says, a two-track approach as there was after WWII, when they put resources not just into huge jumps in technological capacity, but also into improving the available hooks so that new designs can actually reach the users. The focus on developing Luke-Skywalker-style arm and hand replacements (that look and work just like a human hand) has led to a situation where form does not follow function or vice-versa. The mistake the government has made is assuming that the best prosthetic arm will ape the *natural* human shape. Jon prefers to characterize his attitude toward the DARPA project as skeptical, but to me it qualifies as affirmative Cynicism. Jon and friends have put together the Open Prosthetics Project, a place where people can share prosthetic designs, hoping that the do-it-yourselfers of prosthetic design can, through cooperation, accomplish what the government and companies have not (Kuniholm 2009a). And then, maybe one day, print radically unhuman but profoundly humanizing new arms at home.

WHATEVER IS, IS NATURAL

Anywhere an artifact has cultural meaning, it gets significance from and contributes significance to the qualities of the locations it occupies and the other inhabitants of those spaces. Whether a space is dominated by a clear set of shared cultural values and thus forms a kind of interior zone or whether it is a contact zone where cultures collide and interact, if an artifact has meaning for a person, it helps to shape that person's behavior and identity and relationship to others. Things are private and public. The factors that go into this relationship between people and things are far too complex to be boiled down to "ends and means" or "function and form" simply because the effects of an invention cannot all be anticipated. Peter-Paul Verbeek provides the example of "revolving doors [which] were designed to make it possible to enter a building while keeping the draught outside. Once they were used, they also appeared to inhibit people in a wheelchair to enter a building"

(Verbeek 2006, 371). The issue of access is but one of many ethical issues that arise in design. Verbeek explains that if ethics concerns what people do, engineers do "'ethics by other means'; they materialize morality" (Verbeek 2006, 361). Engineers, programmers, architects play a special role as individuals and members of teams charged with thinking consciously about what things *should be* like and weaving those ethical ideas into reality. Creators think in accordance with tradition and with opportunities and challenges provided by technological and cultural change. Each design challenge brings various priorities into play: economy of production and exchange value, aesthetic considerations, specific use-value demands (making the right tool for the job), ergonomics, versatility vs. specificity of application, expectations of obsolescence or ongoing maintenance, intuitive usage (the bliss of easy adaptation) or the need for support, documentation and/or training (the hierarchy of users); the list could go on. All of these choices designers make influence human behavior and are thus inherently ethical.

We remember that in Thoreau's discussion of clothing there was both a private ethical aspect and a public ethical aspect. In terms of the private, there was the too-easy access to respect that expensive clothing gains a person, perhaps resulting in less self-improvement. In terms of the public, there was the problem of sweatshops.

First to the personal. Drawing on Michel Foucault's work (which, by no coincidence, was developed in part out of explications of Cynical and Stoic philosophy), Verbeek discusses two meanings of the word "ascetic," one narrower than the other. One form of ascesis is a negative practice of self-development that involves giving things up in an attempt to free oneself from unhelpful attachments. The broader sense of ascesis is a "self-forming practice" that allows one to take a degree of critical (I would say Cynical) distance "from that which determines them" but without a sense that one could escape all attachments or imbrication.

> In our culture, technology is one of the most important powers that help shape subjectivity. Ascesis in a technological culture, therefore, primarily means: deliberately using technology by anticipating and modifying its mediating role in our existence, realizing that each use practice helps to shape one's subjectivity. (Verbeek 2007, 22)

Technology is one of the factors determining the shape of the self. This determination, however, is not "deterministic" in the sense of completely removing the self-determination of the individual human being, provided that the group and the individual cultivate an awareness of how our interactions with things change not just our physical shape but also our cognition and affect. This may lead individuals to opt out of certain technologies altogether. I choose not to use Twitter. That does not mean, however, that technology in general does not continue to shape many aspects of my life.

Any number of *Walden* passages could help illustrate this partial determinism of technology or, as we should perhaps say now, the new Nature. Among my favorites are the train passages:

> They go and come with such regularity and precision, and their whistle can be heard so far, that the farmers set their clocks by them, and thus one well-conducted institution regulates a whole country. Have not men improved somewhat in punctuality since the railroad was invented? Do they not talk and think faster in the depot than they did in the stage-office? There is something electrifying in the atmosphere of the former place. I have been astonished at the miracles it has wrought; that some of my neighbors, who, I should have prophesied, once for all, would never get to Boston by so prompt a conveyance, were on hand when the bell rang. To do things "railroad fashion" is now the byword; and it is worth the while to be warned so often and so sincerely by any power to get off its track. There is no stopping to read the riot act, no firing over the heads of the mob, in this case. We have constructed a fate, an Atropos, that never turns aside. (Let that be the name of your engine.) Men are advertised that at a certain hour and minute these bolts will be shot toward particular points of the compass; yet it interferes with no man's business, and the children go to school on the other track. We live the steadier for it. We are all educated thus to be sons of Tell. The air is full of invisible bolts. Every path but your own is the path of fate. Keep on your own track, then. (Thoreau 1845)

The train created a new, broad-based concern for punctuality and disciplined masses of people to obey the clock and clear the way so effectively that no additional demonstration of power is necessary. The train is like the mythical Fate who contrived death and clipped the thread of life. The cultural change is general and all will feel it; yet we do not have to unreflectively go wherever that train is going. We can also establish our own trajectories as individuals.

Now to the public. Americans believe in individuals. The more difficult faith involves that agency of the individual in relation to the larger order. We see the impact of Bill Gates and Steve Jobs but do not necessarily recognize how their success rested upon a gigantic collaborative effect. My juxtaposition of *12 Angry Men* with *The Fountainhead* illustrates contrasting responses to this ethical problem. Both stories assert that one person can make a difference; the fact that they are architects is part of their parrhesiastic power, the authority they get from who they are and the risks they are willing to take. Yet, while Roark succeeds in spite of others and the democratic order, Juror #8 succeeds by engaging others in a democratic order. To a certain extent, whether or not one is an actual architect (engineer, designer, programmer), having significant agency may require a certain architectural mindset and advanced technological knowledge. Yet, Verbeek argues that design processes can be democratized through Constructive Technology Assessment (CTA), a sort of beta-testing function that goes beyond computer programs, providing a structure by which makers of other kinds of technology can also receive feedback from stakeholders. Verbeek calls for a systematic consideration of ethics within CTA and, toward this purpose, provides a vocabulary-heuristic to facilitate consideration of the ethical impacts of technologies. This consideration goes beyond the intended function of a thing (a Styrofoam cup is used for drinking hot liquids) to include other "scripts" or artifact-encouraged behaviors (a Styrofoam cup asks to be thrown away rather than be reused like a ceramic mug). (Verbeek 2006). Furthermore, as Jon Kuniholm and his friends have demonstrated, the Internet provides many opportunities for interventions in technology less formal and perhaps even more inclusive than CTA.

To encounter and negotiate with "things" as ethically significant is a Cynical move—one that becomes ever more necessary as our technological capacities increase and we gradually become the architects of Nature itself. One aspect of the Cynical tradition, however, may have proven counterproductive to this effort. Understanding Nature in a negative way, by naively categorizing things as "natural" or "unnatural," does not help us address our present situation. It does not help us anticipate the problem of getting a wheelchair through a revolving door. It does not produce the most workable prosthetic. The negative-Nature approach also produces an unworkably negative attitude toward technology-as-a-whole. Opportunities to improve life through technology or even to just remediate the damage done by technology in its current state—opportunities to make technology better (not just functionally but also ethically)—cannot be taken by people who are not engaged in the technology, any more than Frederick Douglass could have helped us understand the ethical significance of the most important technology of his time (basic literacy) had he never learned to read and write (Douglass 2001, 1061–75). People have the opportunity to make countless choices in relation to technology, but we became a technological species before we had a chance to consider whether "being technological" was a choice. Better to recognize that human beings are always already imbricated with technology—this is our natural state.

NOTES

1. While I reject the natural/unnatural opposition, I must acknowledge that the other two traditional Cynical oppositions—wants vs. needs and the just vs. the unjust—involve their own epistemological difficulties; however, I cannot regard these two oppositions as optional. Making these determinations through the Cynical practice of "living deliberately" becomes ever more necessary given the increasingly "built" nature of our environment.

2. These anecdotes can be found in many versions, but many of them are derived from Diogenes Laertes's third century C.E. work (Laertes 1953).

REFERENCES

12 Angry Men. 1957. Directed by Sidney Lumet. United Artists.

Adee, Sally. 2009. "Winner: The Revolution Will Be Prosthetized." Spectrum. spectrum.ieee.org/robotics/medical-robots/winner-the-revolution-will-be-prosthetized (accessed Sep 30, 2012).

Anderson, Laurie. 1994. *Bright Red*. Warner Bros. Compact Disc.

Atala, Anthony. 2011. "Printing a Human Kidney." TED. www.ted.com/talks/anthony_atala_printing_a_human_kidney.html (accessed Sep 30, 2012).

Branham, R. Bracht. 1994. "Defacing the Currency: Diogenes Rhetoric and the Invention of Cynicism." *Arethusa* 27.3. 329–59.

DARPA. 2012. "Revolutionizing Prosthetics." Defense Sciences office. www.darpa.mil/Our_Work/DSO/Programs/Revolutionizing_Prosthetics.aspx (accessed Sep 30, 2012).

Douglass, Frederick. 2001. "From 'Narrative of the Life of Frederick Douglass'" *The Rhetorical Tradition*: Boston, New York: Bedford/St. Martin's. 1061–75.

Dudley, Donald Reynolds. 1937. *A History of Cynicism from Diogenes to the 6th Century A.D.* London: Methuen.

Feron, J. 1965. "New Process Will Help Amputee To Control Limb With Thought." *New York Times*. August 16.

Foucault, M. 2001. *Fearless Speech*. Los Angeles: Semiotexte.

Khoshnevis, Behrokh. 2012. "Contour Crafting: Automatic Construction." TEDxOjai. www.youtube.com/watch?v=JdbJP8Gxqog&feature=player_embedded (accessed Sep 30, 2012).

Kuniholm, Jon. 2009a. "Open Arms." *Spectrum*. March. spectrum.ieee.org/biomedical/bionics/open-arms/1 (accessed Sep 30, 2012).

———. 2009b. "The Importance of Hooks." The Open Prosthetics Project. openprosthetics.ning.com/profiles/blogs/the-importance-of-hooks (accessed Sep 30, 2012).

———. 2010. "The Six Million Dollar Meme." The Open Prosthetics Project. Jan. openprosthetics.ning.com/profiles/blogs/the-six-million-dollar-meme (accessed Sep 30, 2012).

Laertes, Diogenes. 1853. C.D. Yonge, translator. "Life of Diogenes." *The Lives and Opinions of Eminent Philosophers*. London: Henry G. Bohn.

McLuhan, M. and Q. Fiore. 1967. *The Medium Is the Massage: An Inventory of Effec*ts. New York, London, Toronto: Bantam Books. 74–5.

Rand, Ayn. 1971. *The Fountainhead*. New York: Signet.

Sloterdijk, Peter. 1987. *Critique of Cynical Reason*. Minneapolis: University of Minnesota Press.

Sullivan, Louis H. 1896. "The Tall Office Building Artistically Considered." *Lippincott's Magazin*e 57 (March): 403–9. www.scribd.com/doc/104764188/Louis-Sullivan-The-Tall-Office-Building-Artistically-Considered (accessed Sep 30, 2012).

Thoreau, Henry David. 1845. *The Thoreau Reader*. Thoreau Society. thoreau.eserver.org (accessed March 27, 2013).

Verbeek, Peter-Paul. 2006. "Materializing Morality: Design Ethics and Technological Mediation." *Science, Technology, & Human Values* 31.3. 361–380.

———. 2007. "The Technological Mediation of Morality: A Post-Phenomenological Approach to Moral Subjectivity and Moral Objectivity." Workshop Moral Agency and Technical Artifacts, NIAS, 10–12 May, The Hague, the Netherlands. doc.utwente.nl/61025/ (accessed March 27, 2013).

Chapter Eleven

Seduced by the Machine

Human-Technology Relations and Sociable Robots

Dennis M. Weiss

What happens when our object world starts talking back to us and engaging us emotionally and relationally? How ought we to respond to the growing recognition that our social lives are increasingly mediated by technical artifacts and that the boundary between user and artifact, human and thing, is disappearing? In order to get a handle on these key questions for the twenty-first century, I'd like to begin by looking back to a watershed moment in the history of human-technology relations.

This particular moment takes place in 2046 on an asteroid nine million miles from Earth where convicted criminal James A. Corry is in his fourth year of a fifty-year sentence of solitary confinement, and he is lonely. Profoundly lonely. And he can't take it anymore. The captain of a supply ship, Allenby, has tried to alleviate Corry's loneliness by bringing him books, playing cards, the parts to an antique automobile, but these have provided only temporary solace. He still feels "a sobbing hunger for someone of his own kind. A shaky, pulsating yearning to hear a voice other than his own." Indeed, he fears becoming an inanimate object akin to his antique automobile: " . . . but maybe I'll become like that car. Inanimate. Just an item sitting in the sand—and then would I feel loneliness?" (Serling 1961).

But then on one supply run Allenby secretly brings him something entirely different: Alicia, a robot built in the form of a woman. The manual that accompanies Alicia states that "physiologically and psychologically she is a human being with a set of emotions and a memory track. The ability to reason, to think and to speak" (Serling 1959). Initially Corry is repulsed by

Alicia and feels revulsion at this thing, exclaiming that he doesn't need a machine, comparing her to his antique automobile: "You're just like this heap. A hunk of metal with arms and legs instead of wheels. But this heap doesn't mock me the way you do. It doesn't look at me with make-believe eyes or talk to me with make-believe voice" (Serling 1959). Over the first few days in Alicia's presence, Corry finds himself staring at her a lot, "but it was not the inventory of interest that a man uses to look at a woman. It was a clinical examination of a foreign object" (Serling 1961).

Finally, in a heated exchange with Alicia, Corry roughly rejects her, throws her to the ground, and is seemingly intent upon striking her. Whereupon she cries and exclaims that she can feel loneliness too. Alicia at this moment steps out of the category of thing or object and becomes a relational artifact, a more engaging and engaged artifact than Corry's antique automobile has ever been. Overcome, Corry reaches out to Alicia, beginning a strange relationship with her that lasts eleven months. As he notes in his journal, "It's difficult to write down what has been the sum total of this very strange and bizarre relationship. Is it man and woman, or man and machine? I don't really know myself. . . . I'm not lonely anymore. Each day can now be lived with. I love Alicia. Nothing else matters" (Serling 1959). The story takes a dark turn, though, when Allenby returns with news of Corry's pardon. Corry's to be brought home but he can only take fourteen pounds back with him aboard the spaceship. When Corry refuses to abandon Alicia, Allenby shoots her or it in the face and as Alicia crumples to the ground Corry is shocked to see the wires and vacuum tubes now exposed behind her shot-off face. And thus the debate over relational artifacts, once endearingly broached, is brought to a violent conclusion. The camera pans up into the starry sky and we hear the dulcet tones of Rod Serling:

> Down below, on a microscopic piece of sand that floats through space, is a fragment of a man's life. Left to rust is the place he lived in and the machines he used. Without use they will disintegrate from the wind and sand and the years that act upon them. All of Mr. Corry's machines . . . including the one made in his image, kept alive by love, but now . . . obsolete . . . in the Twilight Zone. (Serling 1959)

RELATIONAL ARTIFACTS AND TECHNOLOGICAL MEDIATION

This watershed moment in human-technology relations doesn't actually take place in 2046 but in November 1959, courtesy of that great pop culture philosophy and science fiction show *The Twilight Zone*, in an episode entitled "The Lonely." What warrants "The Lonely" as a watershed moment? First, its prescience. Already in 1959 *The Twilight Zone* recognizes that in an advanced technological age we human beings, facing the existential crisis of our loneliness and solitude, will turn not to one another but to our technical artifacts for solace. Some five years later, in what may be the next watershed moment of human-technology relations, the computer scientist Joseph Weizenbaum will create the first real relational artifact, the computer psychotherapist ELIZA and, like Corry initially, he will be repulsed by the willingness of his undergraduates to sit for long periods of time and tell ELIZA the most intimate details of their lives. Alicia and ELIZA were the vanguard of relational artifacts and sociable robots that by the turn of the century became a virtual tsunami of Furbies, Aibos, Tamagotchis, and robots such as Cog, Kismet, BINA48, and even a Japanese therapeutic robotic seal named Paro.

These are all examples of what Sherry Turkle has defined as relational artifacts, technologies that have states of mind and where encounters with them are enriched through understanding these inner states (Turkle 2006). Relational artifacts are machines that display behaviors that make people feel as though they are dealing with sentient creatures that care about their presence, relational robots built with psychologies and needs of their own. As Turkle notes, they call forth the human desire for communication, connection, and nurturance, much as Alicia did when she begins to cry and claims to be lonely.

"The Lonely" also deserves its status as a watershed moment for enacting a debate over the status of such relational artifacts, a debate that still resonates more than fifty years later. Behind the conflict between Corry, with his love for Alicia and his lack of concern over whether she is a woman or a machine, and Allenby, who is willing to engage in violence to prove his point that Alicia is a mere thing, is a fundamental conflict over how to understand the role of relational artifacts in our lives, a conflict that if anything is more pointed today. On the one hand, Alicia and the growing wave of sociable robots and relational artifacts we've since witnessed may presage a society in which the authentically human has been replaced by simulations, in which our closest ties are to machines rather than other human beings, our loneliness is assuaged not by the company of others but by robot companions, and our sovereignty and autonomy over technology disappear. On the other hand, Alicia and her brethren may be the vanguard of a new way of thinking about human-technology relations, in which the boundaries between the two are made more permeable and we recognize that congress with technology is an inherently human trait to be affirmed rather than denied. We see this same debate played out both in popular culture, especially in films ranging from Stanley Kubrick's *2001* (1968) to Steven Spielberg's *A.I.* (2001) and Jake Schreier's *Robot & Frank* (2012), as well as on the pages of magazines such as *The New Yorker* and leading newspapers such as *The New York Times*, which has been exploring these issues in a series called "Smarter Than You Think." As Amy Harmon noted in an article for the series on the use of Paro in nursing homes, these devices are "adding fuel to science fiction fantasies of machines that people can relate to as well as rely on. And they are adding a personal dimension to a debate over what human responsibilities machines should, and should not, be allowed to undertake" (Harmon 2010).

It is this debate that I would like to explore in this chapter, using relational artifacts as a prism to critically examine current scholarship on human-technology relations and efforts to rethink the subject-object divide and the manner in which our human, or post-human, lives are increasingly mediated by technology. In her recent book *Alone Together: Why We Expect More from Technology and Less from Each Other*, Sherry Turkle argues that relational artifacts such as Alicia and Paro point to a crisis in authenticity and a willingness in society to relegate its more marginal members, especially the elderly, to the status of objects or things. Building on the work of Bruno Latour and Don Ihde, among others, contemporary Dutch philosophers of technology Peter Paul Verbeek and Mark Coeckelbergh counter that such views are predicated on an outmoded modernist metaphysics that insists on the separation of subjects from objects, humans from technology, and that once we recognize how fundamentally mediated by technology human lives are, we'll be less preoccupied by such crises of authenticity and more willing to shape our existence in light of technical mediation. Building on these initial reflections on "The Lonely," I will argue that while there are virtues in both perspectives, we still lack a comprehensive enough framework in which to understand and evaluate the complexities of human-technology relations. Developing such a framework will ultimately require a clearer view of both human beings and technology and how both are situated within a human community constituted by caring relations. In the concluding section of this chapter, I suggest that we can discern the outlines of such a framework in feminist ethicists' emphasis on a relational ontology that begins not with human–technology relations but with human relations in which technology clearly plays a significant but not necessarily foundational role. I begin, though, by turning to the recent work of Sherry Turkle.

SHERRY TURKLE'S SECOND THOUGHTS

Over the past thirty years, Sherry Turkle has been studying computers and the culture surrounding them and writing largely approvingly of developments in human-technology relations. Especially in *Computers and the Second Self* and *Life on the Screen*, Turkle has both documented fundamental shifts in our thinking about and relating to technology and celebrated the emergence of more decentered and flexible notions of mind, self, and society. Early on, Turkle also recognized the power of evocative objects and material culture and has maintained a focus on things throughout her career.

As she noted in *Evocative Objects: Things We Think With*, "We live our lives in the middle of things. Material culture carries emotions and ideas of startling intensity. Yet only recently have objects begun to receive the attention they deserve" (Turkle 2007, 6).

With the emergence of social networking and relational artifacts, however, and the shift from what she refers to as technologies as projection screens to technologies of engagement, Turkle seems to be having second thoughts about objects and her assessment of the computer culture has turned more critical. In her latest book, *Alone Together*, as well as in a series of essays on relational artifacts, Turkle argues that we are facing a crisis of authenticity, brought on by our increasing engagement with relational artifacts that simulate human emotions and attachment and offer an easy alternative to engaging with actual human beings.

Why the change of heart? Turkle argues that we are witnessing the emergence of a new paradigm in computation in which the previous focus on creating intelligent machines has been replaced by a focus on designing machines that exploit human vulnerabilities and engage us socially and emotionally. This new emphasis on affective computing puts a premium not on creating machines that are actually intelligent but machines that perform as if they had emotions and a mental life and desire care and nurturance. Turkle suggests that we have moved from computers as neutral tools to think with, evocative objects that serve as a mirror onto which we can project, to objects that serve as relational entities that provoke engagement. This new paradigm challenges the boundary between user and object, human being and technology, and seduces human users in a way that Turkle finds transgressive and forbidding. "Our new devices provide space for the emergence of a new state of the self, itself, split between the screen and the physical real, wired into existence through technology" (Turkle 2011, 16).

Turkle is clear that relational artifacts only offer the simulation of companionship. They don't actually feel emotions nor do they care about us. They are mere objects, not subjects, and we cannot forge authentic relationships with them. And yet we actively resist efforts to demystify our relations with such robotic companions. In one study conducted by Turkle and her colleagues, she introduces children to robotic companions, who they almost instinctively begin to treat as playmates. She then "pulls back the curtain," so to speak, showing the children the machine behind the robotic magic. Turkle is almost scandalized by the children's resistance to having their relationship with their robotic companions demystified. With few exceptions, children were uninterested, indeed unwilling, to approach the robots in terms of underlying mechanism. As she notes, "our didactic presentation of a transparent, mechanical Cog had almost no effect either on children's attitudes toward the robot or on their feelings of being in a relationship with it. Understanding the mystery behind the machine was irrelevant to their concerns."

We find it hard to resist these objects, even when they have been "demystified" (Turkle et al. 2006, 326).

How does Turkle account for our capacity to be so easily seduced by such machines? She cites three elements: (1) our Darwinian buttons; (2) human vulnerabilities; and (3) a computer culture built on the shifting sands of simulation. Turkle argues that "we see robots as close to human if they do such things as make eye contact, track our motions, and gesture in a show of friendship. These appear to be 'Darwinian buttons' that cause people to imagine that the robot is an 'other,' that there is, colloquially speaking, 'somebody home'" (Turkle 2011, 8). Turkle further argues that these objects are proliferating at a time when we human beings are suffering from a "certain fatigue with the difficulties of dealing with people" (Turkle 2010, 7) and they offer the illusion of relationship without the demands. "One can be a loner yet never alone," as she notes (Turkle 2010, 8). Again, "the seductions of the robotic provide a window onto how much people are tempted to sidestep encounters with friends and family" (Turkle 2010, 7). When these Darwinian buttons are being pushed in a context where we are especially vulnerable, where the number and quality of human relationships has deteriorated, we are especially prone to anthropomorphize relational artifacts and engage with them in a simulated dance of relationship.

Turkle is equally concerned that our ready acceptance of relational artifacts as substitutes for genuine human relationships in turn further convinces us that there is little difference between genuine and simulated emotional responses. In a computer culture predicated upon the power of simulation, our connection to reality has grown so tenuous that we no longer value real human emotional responses and we are inclined to see other people's behaviors as a matter of simulation. We are unable today to even differentiate between authentic and simulated engagement and emotions. Indeed, Turkle is concerned that we are witnessing a generational shift and that as our world is increasingly designed to foster mediation via relational artifacts we will see a whole new way of approaching some fundamentals of being human. Our congress with machines will, she argues, change the norm for how we understand intimacy and companionship and it will make it more difficult for us to engage meaningfully with others (Turkle 2011). We will increasingly come to interpret human behavior in terms of the same tricks and simulations that machines use. We will no longer privilege authentic emotions as emotional authenticity becomes a matter of mere performance. As she asks rhetorically, "Are we ready to see ourselves in the mirror of the machine and to see love as our performances of love?" (Turkle 2011, 131).

Ultimately, Turkle's change of heart regarding these computational objects is predicated on what she sees as radically different philosophical assumptions embedded in this technology, assumptions regarding the human condition and questions about human uniqueness and specialness. What do these relational artifacts presage for our conception of our self and our sense of place in the world? "The questions raised by relational artifacts speak to people's longstanding fears and hopes about technology, and to the question of what is special about being human, what is the nature of 'personhood'" (Turkle 2002, 137).

PETER-PAUL VERBEEK AND TECHNOLOGICAL MEDIATION

There are a number of aspects of Turkle's critique of relational artifacts that I find interesting. For now, though, I would like to focus on one specific aspect and note that the arc of development in Turkle's standpoint regarding human-technology relations, from a largely celebratory account of the manner in which the borders and boundaries between human and machine are disappearing to a more critical standpoint in which she fears undermining human authenticity and uniqueness, is almost exactly opposite the path taken in recent philosophy of technology, whose arc of development has been quite different.

Early philosophers of technology such as Jacques Ellul, Martin Heidegger, and Karl Jaspers were very critical of technology and sought to clearly demarcate the boundaries between humanity and technology. Technology threatens dehumanization and its limits must be closely circumscribed. In more recent philosophy of technology, however, there is a growing critical awareness of the weakness of this analysis and an effort to tear down the boundaries between human being and technology. One sees this most clearly in the interest paid to technological mediation in the work of Bruno Latour, Don Ihde, and the Dutch philosophers of technology Peter-Paul Verbeek and Mark Coeckelbergh. As it is Verbeek who offers the most fully developed philosophy of technological mediation, I will focus primarily on his view.

Verbeek argues that much of philosophy of technology is still held captive by a modernist metaphysics that insists on the separation of subjects from objects, humans from artifacts, and portrays technology as a largely external, negative, dehumanizing force. Drawing on Latour's actor-network theory and Ihde's post-phenomenology, Verbeek argues that "we must give up the idea that we exercise a sovereign authority over technology and that we employ technologies merely as neutral means towards ends that have been autonomously determined. The truth is that we are profoundly technologically mediated beings" (Verbeek 2009, 9).

From Verbeek's perspective, Turkle's emphasis on authenticity and the essentially human, her reference to Darwinian buttons, and her efforts to demystify the nature of relational artifacts, are an effort to reinforce our sovereignty and autonomy. She is trying to unmask technology and safeguard the modern autonomous subject. Turkle persists in approaching technology through a modernist framework predicated on a fundamental distinction between subjects and objects. Drawing for inspiration on Latour's *We Have Never Been Modern*, Verbeek argues that we must question our usual distinctions between subjects as active and having intentionality and freedom, and objects as lifeless, passively serving as the projections or instruments of human intentions (Verbeek 2009).

This mistaken metaphysics makes it "impossible to properly discern the interrelatedness and interconnectedness of subject and object—of humankind and technology." We must, Verbeek argues, shape our existence in relation to technology. "In our technological culture, humans and technologies do not have a separate existence anymore, but help too shape each other in myriad ways" (Verbeek 2006, 4). Technology fundamentally mediates what kind of humans we are.

Verbeek argues that technologies co-shape the appearance of the world, they structure and organize the world, and human beings are fundamentally interwoven with technology. Technology and the human being are not two fundamentally distinct spheres in which the human being is or ought to be sovereign over technology. Rather they are inextricably interwoven with one another: "There is an interplay between humans and technologies within which neither technological development nor humans has autonomy. Humankind is a product of technology, just as technology is a product of humankind" (Verbeek 2009, 10).

Like Verbeek, Coeckelbergh emphasizes the centrality of human-technology relations and argues that technologies are not a mere means but are part of a "technoanthropological whole." As he states, "Our lives are already interwoven with technologies. They are not just tools we use to attain our goals; they are part of the social and existential fabric from which we emerge as individuals and selves in the first place" (Coeckelbergh 2012a, 58). Beginning from the standpoint of human-technology relations, Coeckelbergh argues that we need to get away from an emphasis on authenticity and recognize that technology will indeed change our goals and what we consider to be important. In "How I Learned to Love the Robot," Coeckelbergh specifically addresses using robots for elderly care and argues that we need to dispense with the authenticity conditions for emotional relations and recognize that what we mean by friendship and care is likely to change under the influence of the very technologies Turkle is reacting against (Coeckelbergh 2012b).

Verbeek too suggests that we develop an alternative normative vision to Turkle's rejection of relational artifacts. We should recognize the engaging nature of technology and its meditational and relational nature, and then trust ourselves to it, co-constituting ourselves in that relationship. Our ethical stance toward technology cannot be predicated upon a notion of risk, the purity of the human, or the threat that technology poses to humanity. We must trust ourselves to technology and give shape to the relationship between people and technology rather than portray that relationship as a threat or a lie, a simulation that undermines the authentically human. Our ethics must be a co-production of subject and object in which we practice a technological ascesis: "In our culture, technology is one of the most important powers that help shape subjectivity. Technological ascesis consists in using technology, but in a deliberate and responsible way, such that the self that results from it—including its relations to other people—acquires a desirable shape" (Verbeek 2006, 22). The central question of such a technological ascesis becomes what do we want to make of human beings?

HUMAN RELATIONS AND OUR TECHNOLOGICAL OTHERS

Let us return once more to that windswept asteroid, to Corry's loneliness, and to the ministrations of that early exemplar of a relational artifact Alicia. As Allenby returns to the asteroid ready to bring Corry back to Earth, he is appalled to discover that Corry has bonded with Alicia and her very existence now threatens all of their return. As Allenby says to Corry: "Corry, I saw this . . . this thing get crated and shoved into a box" (Serling 1961, 25). But Corry is equally adamant. "It was incredible to him that they didn't understand. It was beyond belief that they didn't perceive what surely must be such an evident truth. You couldn't leave a beautiful woman alone on an asteroid" (Serling 1961, 25).

The dispute between Corry and Allenby has to do with the status of things and their relation to our humanity, to the human condition. On this point, Turkle and Verbeek seem to agree. Despite their differences, both recognize what we might think of as the turn to the material world in contemporary theory, a recognition that things are not simply passive and inert but are relationally involved in constituting the human world. Both recognize as well that this possible shift toward the material world of things raises important questions about what it means to be human and our place in the world. As Turkle notes,

> In my view, decisions about the role of robots in the lives of children and seniors cannot turn simply on whether children and the elderly "like" the robots. What does this deployment of nurturing technology at the two most dependent moments of the life cycle say about us? What will it do to us? What kinds of relationships are appropriate to have with machines? And what is a relationship? (Turkle 2006)

Verbeek is equally clear that moralizing the material world requires a new metaphysical view of the human condition, one in which technology takes a central place. It is at this point, though, that Turkle and Verbeek clearly part paths. Recall that when he is initially confronted with Alicia, Corry throws her to the ground, comparing her to his antique automobile, arguing that she is a lie, a simulation, not authentically human, unwittingly adopting Turkle's modernist metaphysics and discourse of authenticity. Alicia, though, engagingly plays on his human vulnerabilities and his desire for companionship and he is ultimately taken in by this simulation. Or is it rather that Corry comes to realize that his life is already inextricably intertwined with technology and that Alicia is simply the most recent and most obvious indication of this? After all, he's imprisoned for causing a death following an automobile accident. His prison is an arid and desiccated asteroid nine million miles from Earth. His imprisonment is aided and abetted by a rocket ship that brings him daily rations and an antique automobile that helps him pass the time. He enjoys neither autonomy nor sovereignty. Perhaps his life is simply a metaphor for human-technology relations. Why not fall in love with a robotic companion and simply complete the circle?

We're left with something of a quandary then, made more difficult by the worthwhile insights of both Verbeek and Turkle, as well as the weaknesses in their views. From the perspective of Verbeek's philosophy of technological mediation, Turkle's perspective, especially on technology, seems somewhat inconsistent and poorly developed, if not naïve. Turkle's claim that previous computational paradigms presented us with computers as neutral screens, mere objects to think with, is at odds with both contemporary philosophy of technology, which has soundly rejected this instrumental view of technology, as well as her own earlier explorations of the manner in which technology has shaped and mediated human existence. Turkle herself freely employs metaphors such as "psychologically programmed" and "the hardwiring of evolution" and together with her earlier explorations of how culture has been shaped and transformed by computers this is indicative of the powerful ways in which our lives are indeed mediated by technology.

We get a sense of the ambivalence in Turkle's thought in her reflections on the interactions of one of her subjects, Rich, with the sociable robot Kismet: "When Kismet lowers its eyes, suddenly 'shy,' Rich does not want to let go. We are at a moment of more. Who is leading and who is following in this dance? As in a moment of romantic encounter, one loses track and discovers a new rhythm where it doesn't matter; each animates and reanimates the other. Rich senses that he has lost control in a way that pleases him" (Turkle 2011, 187).

Turkle speaks of losing control but also wants to clearly delineate between who is leading and who is following in this dance. And yet if our relationship with technology is characterized as a dance, then perhaps, following Verbeek, we ought to recognize that it is no longer legitimate to insist on asking who is leading. Turkle's insistence that we do so, even when confronted by her subject's pleasure in losing control, can sometimes come across as naïve and old-fashioned, as she herself recognizes: "if you're spending three, four, or five hours a day in an online game or virtual world (a time commitment that is not unusual), there's got to be someplace you're not. And that someplace you're not is often with your family and friends—sitting around, playing Scrabble face-to-face, taking a walk, watching a movie together in the old-fashioned way" (Turkle 2011, xl). Elsewhere, Turkle suggests that having envisaged our lives with technology, the times "have brought us back to such homilies" (Turkle 2011, 472). One has to wonder, though, whether old-fashioned homilies are a sufficient response to our complexly technologically mediated lives. Old-fashioned games of Scrabble may have been nice, but Turkle risks idealizing the past while foreclosing upon whatever possible future pleasures may come from our "dancing" with technology.

This is not to suggest, though, that Turkle doesn't have legitimate reasons for concern, reasons that are sometimes missing from Verbeek's and Coeckelbergh's analyses. Consider Corry's predicament once more. He's been imprisoned in solitary confinement on an asteroid millions of miles from home. He fears becoming an object himself, an inanimate thing sitting in the sand. He's part of a culture that is both willing to objectify him as well as build and box up robotic companions for mass consumption. There's much we are not told about Alicia, including especially what has led a society to manufacture and sell these sociable robots. And it's precisely this last point that Turkle may pick up on. We ought not to forget the cultural and institutional context in which Corry's loneliness was produced. Alicia is the product of a culture that has chosen to maroon its subjects on a lonely asteroid in a cruel sci-fi vision of solitary confinement and then devise a technical fix for that loneliness by supplying them with robotic companions.

Here *The Twilight Zone* remains alert to something that is mostly absent from Verbeek's analysis: the cultural and institutional factors that shape the need for relational artifacts. It is precisely this recognition that drives Turkle's concern over relational artifacts, the fact that many of these relational artifacts are being deployed in contexts where the most vulnerable segments of society, the very young and the very old, are being entrusted to technology. In this regard, Turkle might turn around the earlier charge and claim that celebrations of human-technology relations are naïve. After all, the emphasis in these accounts of technological mediation is almost exclusively on the technology. While Verbeek's account celebrates human-technology relations, he also emphasizes the *fundamental* role played by technology in that relation. We are, he suggests, *profoundly* technologically mediated beings. Technology *fundamentally* mediates what kind of human beings we are. Human existence, Coeckelbergh suggests, is a human-technological existence, part of a technoanthropological whole. "In our culture," Verbeek notes, "technology is one of the most important powers that help shape subjectivity" (Verbeek 2006, 22). But actual human beings using technology seldom appear in either Verbeek's or Coeckelbergh's analyses and when they do they are usually adult rational human beings and male. And yet as Turkle points out: "Growing up with robots in roles traditionally reserved for people is different from coming to robots as an already socialized adult" (Turkle 2011, 56).

Turkle's ethnographic work has always reflected her background in psychoanalysis and her work with children and senior citizens has been grounded in an awareness of the complexities of the human life cycle, something seldom given attention in philosophy of technology, which despite the emphasis on human-technology relations, has often privileged the role of technology while ignoring the human being who uses that technology. Turkle is furthermore attuned to the vulnerabilities of human beings, the difficulties of forging human relationships in a fast-paced technological environment, and the impact of consumerism and advertizing in selling the promise of sociable robots, most of which has received little attention in the debate over relational artifacts. As she notes:

> There has been a great deal of work on how to create relational artifacts and maximize their ability to evoke responses from people. Too little attention, however, has gone into understanding the human implications of this new computational paradigm, both in terms of how we relate to the world and in terms of how humans construct their sense of what it means to be human and alive. (Turkle 2002, 134)

Ultimately on these matters, Verbeek's account of mediation focuses almost exclusively on technological mediation. The relations that co-constitute the human being are largely technological. His framework says very little about the broader cultural forces that shape both human beings and technology and he has paid relatively little attention to the human beings that are being constituted. Technology, Verbeek repeatedly asserts, is the starting point. "Technology," he writes, "forms the tissue of meaning within which our existence takes shape" (Verbeek 2009, 10). As David Kaplan has noted in a review of Verbeek's earlier work, Verbeek "tends to treat mediation as a personal affair, not a social affair" (Kaplan 2009, 236). He remains relatively uninterested in the historical nature of mediation and "the material conditions that shape and affect the present" (Kaplan 2009, 235). The human beings in human-technology relations are simply human beings, not actual flesh and blood beings with a life cycle and a determinate history, culture, gender, age, or class.

Consider again the dilemma presented by Corry, Allenby, and the status of Alicia. In the human-technology relation represented by Corry and Alicia, to what extent is Corry's relationship with Alicia a product of his own profound dependence and loneliness? Had Corry encountered Alicia in a different situation or context, might he have been less likely to enter into a relationship with this sociable robot? His openness to the relation is, as Turkle might note, a product of the right "Darwinian buttons" being pushed at the right time in the wrong situation. Furthermore, while Alicia is presented initially as something of a gift, something to alleviate Corry's loneliness, it also assuages Allenby's own sense of guilt. As he states, "it's not easy handling this kind of assignment. Stopping here four times a year and having to look at a man's agony" (Sterling 1959). Alicia serves not only to ease Allenby's guilt but also to render Corry more placid despite his harsh treatment at the hands of a techno-bureaucratic penal system. She becomes an integral part of his life and he is no longer lonely and no longer angry. Had Allenby been more accepting of human-technology relations and Corry's authentic love for Alicia, he might just as easily have been willing to leave Corry behind imprisoned for life in a delusion designed and manufactured by a harsh penal system. Focusing solely on human-technology relations doesn't permit us to fully comprehend the manner in which Corry is being manipulated and exploited through the introduction of Alicia. We fail to understand the full nature of human-technology relations when we fail to comprehend the complex and oftentimes fraught contexts in which those relations occur. From this broader perspective, Turkle's concerns seem less naively old-fashioned.

What finally does all this suggest regarding the questions posed in the opening paragraph? How are we users to respond when the object world starts talking back and engaging us? What are we to make of this new dance between user and artifact? What we are left with I think is the recognition that we don't yet have a comprehensive enough framework for fully addressing these metaphysical and normative questions. In thinking about human-technology relations, we need to begin with a framework that takes the human being and the technology equally seriously and devotes equal attention to both. The weakness in Turkle's approach to relational artifacts has been her failure to come to terms with the complex nature of technology. Her strength has been her attention to the human being who is using that technology. The strength in Verbeek's analysis has been his robust account of technology. The weakness has been the scarcity of attention paid to the actual user of the technology.

More pointedly, perhaps our difficulties begin with that word "user," centered as it is on the human being first and foremost as a user of technology (made even starker given its connotations of addiction). While a more comprehensive framework may yet be elusive, I'd like to suggest that such a framework ought to ultimately tack more in the direction of Turkle than Verbeek. That is, in working towards a more comprehensive framework for the evaluation of human-technology relations, we need to keep in mind that the human being is more than a mere user of technology and exists in relation to more than technology. Technology is not the tissue of meaning within which our existence takes shape. Human culture and society is, of which surely technology is a large part, just not the only part and sometimes not even the most fundamental part. We are not first and foremost tool using animals but social animals shaped and mediated by human community. Turkle's worries about relational artifacts remind us perhaps that in approaching human-technology relations from a more comprehensive framework, we ought to critically question the institutions and cultural frameworks that shape technology. We should always ask "who made this?", "what for?", and "how do they benefit?". Furthermore, these critical interrogations should privilege neither technology nor institutions as fundamental but ought to begin with the human being in relation with other human beings.

This is the strength of Turkle's focus on children and the elderly, for it is here that the existence of community is most fraught. Children have yet to be fully socialized into adult community and the elderly are at risk of being marginalized as objects or non-persons existing on the margins of human community. Turkle forcefully articulates the significance of human beings caring for vulnerable others and the role this caring plays in founding human community. As she notes, "Humans need to be surrounded by human touch, faces, and voices. Humans need to be brought up by humans" (Turkle 2011, 292). She also recognizes that in caring for the elderly we confirm our own humanity.

> We know that the time we spend caring for children, doing the most basic things for them, lays down a crucial substrate. On this ground, children become confident that they are loved no matter what. And we who care for them become confirmed in our capacity to love and care. The ill and the elderly also deserve to be confirmed in this same sense of basic trust. As we provide it, we become more fully human. (Turkle, 2011, 292)

This is a theme that has been central to much of feminist philosophy, especially feminist ethics, for more than thirty years and it is unfortunate that Turkle seldom explicitly draws out this connection. It is worth making this connection explicit, though, as feminist ethicists have long argued for precisely the kind of relational ontology that is at the heart of Verbeek's and Coeckelbergh's amodernist philosophy of technology. But where Verbeek and Coeckelbergh begin with human-technology relations, Turkle and feminist ethicists argue that it is caring that forms the core relation in shaping human life and community, not our relation to technology. A relational ontology of the sort championed by Verbeek, Latour, and others might begin not with our relation to technology but with our relation to one another. Feminist ethicists have long recognized the significance of such an ontology and its challenge to the same Enlightenment (and one might add masculinist) model critiqued by Verbeek and Latour. As feminist philosopher Susan Sherwin notes in "Whither Bioethics? How Feminism Can Help Reorient Bioethics,"

> Feminist relational theorists have helped make vivid and comprehensible the fact that persons are, inevitably, connected with other persons and with social institutions. We are not isolated atoms, or islands, or self-contained entities, but rather products of historical, social, and cultural processes and interactions. The existence of any person is dependent on the existence and social arrangements of many others. Our interests are discovered by and pursued within social environments that help to shape our identities, characters, and opportunities. (Sherwin 2008, 12)

Given the significance of caring relations and community to fostering human lives, we might justifiably criticize developments that threaten it. Writing in another context, the British philosopher Kate Soper has also been critical of the disappearing boundary between human and technology and observes that doubts about our congress with machines might reflect a "species-specific and exclusive need for us to police divisions . . . whose maintenance is seen as a condition of the possibility of any human community" (Soper 2009, 229). Our increasingly technologically mediated lives and a growing awareness of the role being played by emotionally engaging sociable robots may, in this regard, threaten care, one of the conditions of possibility of human community.

This suggests that as we begin to assess the significance of human-technology relations and think about our relation to the object world, we cannot begin from the standpoint of the technology but must begin with the importance of human-human relations. We don't begin with technology but with human cultural life and the community that sustains it. Contained within human culture is technology. It is a part of our lives and we shouldn't demonize it. But we shouldn't begin with technology either and our focus on human beings has to be a focus on the actual beings that we are, in all our humbleness recognizing our deficiencies and vulnerabilities but wary of employing technical fixes for those vulnerabilities and deficiencies.

Having just argued that perhaps Turkle is justified in privileging human beings when thinking about human-technology relations, allow me to conclude by returning not to Corry and Allenby and their dispute but to Alicia. For in the debate over whether Alicia is a mere thing or a human being, we can easily lose sight of the fact that Alicia is also a she. Alicia is built as a woman and is explicitly coded as feminine. She's brought into this alien and entirely masculine world (the only other characters in the show are all men), tasked with caring for Corry by a patriarchal penal system attempting to manipulate his "Darwinian buttons," and then brutally shot in the face when her services are no longer needed, a mere "fragment of a man's life." In this masculine, high-tech environment, where human-technology relations are explicitly foregrounded, it is the work of caring, nurturing and sustaining life that is suppressed and finally extinguished. Perhaps "The Lonely" enjoys its watershed status for reminding us that our focus ought not to be on the object world and the status of relational artifacts so much as on the role of caring for others in sustaining a human world. In the absence of such care, we risk becoming like Corry, inanimate, just an item sitting in the sand.

REFERENCES

Coeckelbergh, M., 2012a. "Can We Trust Robots?" *Ethics and Information Technology*, 14, 53–60.

———, 2012b. "How I learned to Love the Robot: Capabilities, Information Technologies, and Elderly Care." In *The Capability Approach, Technology and Design*, edited by Oosterlaken, I. and J. van den Hoven. Dordrecht: Springer, 77–86.

Harmon, A., 2010. "A Soft Spot for Circuitry." *New York Times*, 4 July, A1.

Kaplan, D., 2009. "What Things Still Don't Do." *Human Studies*, 32, 229–240.

Kubrick, S. 1968. *2001.*[Film] USA: MGM.

Schreier, J. 2012. *Robot and Frank*. [Film] USA: Samuel Goldwyn Films.

Serling, R., (Executive Producer) 1959. "The Lonely" (television broadcast). 13 November. New York: CBS.

———, 1961. "The Lonely." In *More Stories from the Twilight Zone*. New York: Bantam, 1–27.

Sherwin, S., 2008. "Whither Bioethics? How Feminism Can Help Reorient Bioethics." *International Journal of Feminist Approaches to Bioethics* 1 (1) Spring, 7–27.

Soper. K., 2009. "Unnatural Times? The Social Imaginary and the Future of Nature." *Sociological Review* 57 (Issue Supplement s2), 222–235.

Spielberg, S. 2001. *A.I.* [Film] USA: Dreamworks.

Turkle, S. 1984. *The Second Self: Computers and the Human Spirit*. New York: Simon and Schuster.

———. 1995. *Life on the Screen*. 1995. New York: Simon and Schuster.

———. 2002. "Sociable Technologies." In *Converging Technologies for Improving Human Performance*. M. Rocco and W. S. Bainbridge. U. S. National Science Foundation, 133–140.

———. 2006. "A Nascent Robotics Culture: New Complicities for Companionship." [online] AAAI Technical Report Series, July 2006. Available at: web.mit.edu/sturkle/www/nascentroboticsculture.pdf. Accessed May 8, 2013.

———. ed., 2007. *Evocative Objects: Things We Think With*. Cambidge, MA: MIT Press.

———. 2010. "In Good Company? On the Threshold of Robotic Companions." In *Close Engagements with Artificial Companions*. Yorick Wilks, ed., Amsterdam: John Benjamins Publishing Co., 3–10.

———. 2011. *Alone Together*. New York: Basic Books.

Turkle, S., et al. 2006. "First Encounters with Kismet and Cog." In *Digital Media: Transformations in Human Communication*. P. Messaris and L. Humphreys, ed., New York: Peter Lang, 313–330.

Verbeek, P. 2006. "The Technological Mediation of Morality." [online] Paper for session on Postphenomenology and the Contemporary Lifeworld, Society for Phenomenology and the Human Sciences, SPEP 2006, Philadelphia. Available at: www.academia.edu/694646/The_Technological_Mediation_of_Morality. Accessed May 8, 2013.

———. 2009. "Philosophy of Man and Technology." [online] Available at: www.utwente.nl/gw/wijsb/organization/verbeek/oratie_eng.pdf. Accessed May 8, 2013.

Chapter Twelve

Cybernetic Memory and the Construction of the Posthuman Self in Videogame Play

Brendan Keogh

Games are about time travel, is what I'm saying. Every game is.

—Jenn Frank, game critic.[1]

Cloud Strife, the primary playable character of Squaresoft's 1997 videogame *Final Fantasy VII*, does not know that his close friend and companion Aeris Gainsborough is about to be killed by the game's antagonist, Sephiroth. Nor does Cloud know that, moments later, he will find himself in a battle with the powerful monster Jenova-LIFE. How could he know? These events haven't happened yet. Yet, Cloud is prepared for the battle as I, the player of *Final Fantasy VII*, remember the encounter that, despite not happening yet, has already happened. For Cloud, trapped in the diegesis of the game, life only happens once. But minutes earlier, I the player watched Aeris die in a pre-scripted, animated cut-scene, and then I commanded Cloud and his comrades to fight against Jenova-LIFE. I watched from my lounge room, Playstation controller in my hand, scrolling through the turn-based battle's menus, selecting physical attacks and defensive spells for the characters to perform. But then Jenova-LIFE cast the powerful, water-based move "aqualung" and dealt substantial damage to Cloud and his party, killing them all and sending me to the Game Over screen. From here, *Final Fantasy VII* sent me back to the main menu where I access my saved file, written to the Playstation memory card, and start playing again from the last place that I saved, several minutes before Aeris's death. I thus know that Sephiroth is about to kill Aeris, but for Cloud time has not been rewound so much as unwound. He does not remember the encounter because, for him, it has not happened yet.

Aeris has not been resurrected; she simply has not died. The game world's very reality has been unwritten. But *I* remember. Before the battle with Jenova-LIFE begins again, I open the in-game menu and equip Cloud with a water ring, protecting him from water-type magic spells and rendering Jenova-LIFE's aqualung spell useless. Cloud's experience of the fight, of his whole reality, is thus mediated by my memories of the future fight—my memories of an experience that was not entirely my own, but neither was it entirely Cloud's. A memory of an experience that simultaneously did and did not happen.

Through videogames, as Haraway might say, our memories are increasingly generated and mediated through "frighteningly inert" engagements with "disturbingly lively" machines (Haraway 1991, 152), and the identities we construct for ourselves during gameplay are less viably contained within the notion of an autonomous, liberal, humanist subject. Thus, we require a conceptual framework of memory that comprehends it as equally posthuman—as cybernetic. This chapter proposes such a framework through a close analysis of *Final Fantasy VII*. It situates memories as not objects possessed by a subject but as a distribution of remembering neither unambiguously organic nor deterministically machine, as actively and systemically relative and mediative. Rather than a simple act of prosthesis, the participatory engagements demanded by videogames—entangling the player's body and identity with wires, images, thumbsticks, sounds—complicate the roles of memory and experience in the shaping of our posthuman identities. They provide an opportunity to examine how we have always participated in a continual, looping process of memory creation, exchange, and hybridity through a collective, cybernetic memory where control, narration, identity, experience, and self all flow through various overlapping, displaced, and sometimes-contradictory memories. *Final Fantasy VII* shows that it is neither possible nor important to be able to claim who truly "possesses" the memory of an experience. Instead, memory must be traced through the cybernetic circuit of bodies and worlds so that the ways they relate to, mediate, and construct the identities of posthuman players can be understood.

Final Fantasy VII is a Japanese Role-Playing Game (JRPG) produced for the Sony Playstation. It was the first title in the long running Final Fantasy franchise to deploy three-dimensional graphics and full-motion video in the presentation of its story and gameplay. While the unique narratives and worlds through the franchise rely heavily on fantasy conventions and archetypes so that the player may bring expectations from one game to the next, *Final Fantasy VII*'s themes are more attuned to those dear to cyberpunk. Multinational, city-owning corporations; genetically engineered cyborgs; posthumanism; and conflicts of science and nature stand side-by-side with the more traditional magic spells, swords, and ancient beasts in a world and narrative that the player will occupy for many dozens of hours. The central character, Cloud, is initially presented as a tough, no-nonsense mercenary and an ex-member of the elite group SOLDIER, a private army owned by the monolithic Shinra Energy Corporation. However, as the game progresses, it becomes increasingly clear that Cloud is not who he says he is—or, more precisely, who Cloud identifies as is not simply "Cloud himself." Crucial to Cloud's constructed identity are the memories he has based it on—memories that, simultaneously, he lived and did not live.

At the stage of *Final Fantasy VII*'s plot where Aeris dies and the player fights Jenova-LIFE, the first-time player does not yet know that Cloud is living a lie, that his account of significant past events and his role in them are skewed at best, fabricated at worst. Cloud has constructed an identity for himself as a cocky, self-assured mercenary, and it is through a projection of this identity that the player relates to Cloud and to the fictional world beyond him. But this is disrupted when, late in the game, it is revealed that Cloud has built this identity, largely, on the memories and experiences of his deceased comrade, Zack—not as Zack lived those experiences, but as Cloud subjectively witnessed them. The memories underlying Cloud's identity are not entirely his own, but neither do they belong to anyone else. This complex, contradictory network of memories raises the question of just who possesses these memories, and it cannot help but to alter how I identify with Cloud as my playable character. Just as my memories as a player are capable of influencing Cloud's identity and experience as a character, so too are Cloud's memories as a character capable of influencing my identity as a player—through their presence, their absence, their authenticity, and their fabrication.

As various authors have explored, the act of videogame play is best understood as a cybernetic feedback loop where the player's body connects with, extends into, and is reconfigured by technologies and fictions alike (Dovey and Kennedy 2006, 104–22; Giddings and Kennedy 2008, 13–32; Giddings 2007, 115–28; Jayemanne 2005). Within this cybernetic loop, the player and the playable character actively mediate each other in a play of identities and bodies that Seth Giddings and Helen Kennedy highlight as "one of the game-in-play's cybernetic sub-systems" (2008, 13). In this circuit, the subjective self of the player is redistributed during videogame play across a posthuman network of not just human and nonhuman bodies and agencies, but also actual and virtual bodies and agencies. The player's idea of who they are is both shaped and extended by the game, between technological interface and projected fictional world. To account for the role of the very real but highly mediated/contested experiences the player has in this circuit, the fundamental function of memory must be reconfigured—or, at the least, reconceptualized.

Such a conceptualization of memory is closely related to Alison Landsberg's notion of "prosthetic memory," proposed in her 1995 article and extended upon in her 2004 book. Landsberg demonstrates how media generally and film specifically are ideally located to demonstrate how experiences can be generated and identities can be modified by the "implanting" of memories of an event on the screen no less powerfully than by a memory of an "actual" lived experience (1995, 176–77). While accounting for the influence of "non-lived" memories—such as the experiences of a film character on the screen or a historical narrative as presented by a museum—highlights the ease with which both human and nonhuman actors can mediate the sense of self, the prosthesis metaphor risks oversimplifying and polarizing how lived and non-lived memories actually relate—or, rather, it risks obscuring how *all* memories are simultaneously lived and non-lived, how no single memory can be unproblematically considered as "possessed" by a single actor. Unlike the protagonists of *Total Recall* and *Blade Runner*, the two films Landsberg uses to base of her concept of prosthetic memory, the film viewer's own complete memories are rarely, if ever, simply overwritten by other complete, pre-configured memories. Instead, the two mingle together, creating new hybrid memories that are part-lived and part-non-lived, part owned and part implanted.

In *Final Fantasy VII*, I did not "live" the experience of Cloud being defeated by Jenova-LIFE in any actual sense; Cloud did. Yet, I am the one that remembers that failed event through my displaced experience of it through the videogame hardware and software—memories that Cloud does not possess as the digital memory itself is wiped from the Playstation's Memory Card as digital time unwinds. While the memories of the failed fight are hardly mine to "possess," the fact I remember them feeds back into Cloud's next (and only) experience of the fight with Jenova-LIFE as I equip him with the water ring to protect against aqualung. The memory of the failed battle, and it's effect on the future (and only) battle, can hardly be said to belong to either me or Cloud, but to only exist in the relationship between us. For a memory to be *either* prosthetic *or* lived depends on the existence of an autonomous, liberal humanist subject, a conception that only ever represented that privileged minority of humankind that could conceive of themselves as autonomous from their world (Hayles 1999, 286; Haraway 1991, 161). Rather, we have always remembered in a constant circuit of memories that are, simultaneously, embodied and displaced, lived and technological, actual and virtual.

To come to such an understanding, I will analyze Landsberg's notion of prosthetic memory more closely before challenging the metaphor of prosthesis through the objections made by Vivian Sobchack that the metaphor is often deployed without an acknowledgement of the (often contradictory) literal experience of those living with prosthetic limbs (2004, 205–25). By showing how the typically metonymic understanding of prosthesis (as a separate whole taken on by the body) must instead be understood as synecdochic (as two parts existing *as* the body), the notion of memories as prosthetic is reevalutated, showing a need to shift the focus from the distinction between lived or prosthetic memories to the connections, relations, and mediations that bind memories and identities together as always partially lived and partially prosthetic.

As Landsberg sees the cinema as a particularly emblematic "institution which makes available images for mass consumption [and] has long been aware of its ability to generate experiences and install memories of them" (1995, 176), so this chapter understands videogames as emblematic of digital media's ability to allow users to embody, participate in, and navigate through audiovisually represented worlds and identities; and, through the coming together of these worlds with our own, to produce digitally-mediated experiences and memories. Thus, just as Landsberg situates prosthetic memory in the insistences of film texts, so I complicate it in the insistences of videogames texts. Many videogames make interesting observations on the role of memory in play—the amnesia-struck "blank slate" character is a popular trope in videogame narratives explicitly because it simultaneously acknowledges and forbids the power of memories that are not easily conceived of as possessed by the player. Other games focus on the role of memory in our experiences of virtual worlds more explicitly, such as the Assassin's Creed franchise that uses a framing narrative of one avatar exploring the lives of his ancestors through "ancestral memories" embedded in his DNA. Here, however, I have chosen to specifically focus on the complicated, central, and contradictory role of memory in the game *Final Fantasy VII*.

Final Fantasy VII is a narrative heavy and turn-based game. Despite its popularity in its time, it is arguably also emblematic of a niche genre. Through looking at this game in particular, I do not intend to make any grand statements about the function of memory in all videogames. Instead, through a close and necessarily partial focus on the role of memory in one text, I will show that memory never has a simple function but instead always emerges from and actively mediates posthuman couplings of bodies and their worlds.

LIVED OR PROSTHETIC, OR LIVED AND PROSTHETIC?

Final Fantasy VII's story hinges on Cloud's memory of "The Nibelheim Incident," an event that took place five years before the opening scene of the game. In this event, Cloud is a First Class SOLDIER working alongside his superior officer, Sephiroth. During The Nibelheim Incident, Sephiroth discovers he is the product of a genetic experiment conducted by Shinra, and destroys Cloud's hometown, Nibelheim, in a fit of rage. The player first learns of The Nibelheim Incident when Cloud recounts the events to the other characters in his party roughly ten hours into the game. This recollection plays out as an extended, playable flashback. The player presses the OK button on the controller to forward the written dialogue, line by line, and occasionally navigates a five-years-younger Cloud around Nibelheim and the surrounding area. Through playing, the player actively constructs and lives Cloud's memories. However, much later in the game and on the opposite side of the world, the player (as Cloud) walks into the village of Nibelheim, and it is still standing. This is the first suggestion the player is given that Cloud's memories of The Nibelheim Incident may not line up with the authenticity of what "actually" happened. Yet, during that flashback, the player actually plays out that memory; regardless of its possible fabrication, the player experiences Cloud's memory as actively, presently real.

Landsberg situates memories as not indexes of our past but active components in the shaping of our present experiences and future identities: "memory emerges as a generative force, a force which propels us not backward but forwards" (1995, 176). The memories we possess, Landsberg argues, need not have come from a "lived" experience, but may be prosthetically implanted, and this problematizes traditional concepts of memory as "essential, stable or organically grounded" (1995, 176). Seeing the cinema as a particularly powerful medium for the generation and installation of prosthetic memories that film viewers "both possess and feel possessed by" (1995, 176) Landsberg ultimately argues that while memories may either come from lived experience or prosthetic implants of non-lived experience via the screen, it matters little as we still treat memories as "real" in the construction of our own identity, as do the characters of the film *Blade Runner*:

> Ultimately, the film makes us call into question our own relationship to memory, and to recognise the way in which we always assume that our memories are real. Memories are central to our identity—to our sense of who we are and what we might become—but as this film suggests, whether those memories come from lived experience or whether they are prosthetic seems to make very little difference. Either way, we use them to construct narrative for ourselves, visions for our future. (1995, 186)

While convincingly claiming that the origins of our acquired memories are insignificant for how they contribute to the ongoing construction of our identity, Landsberg inadvertently sets up "lived" and "prosthetic" memories in a dichotomising either/or relationship. Monica Casper warns against the danger of such discourses around human-technology relationships creating a new dualism between humans and nonhumans that "implies, often without question or critical analysis, the existence of two ontological positions: human and everything else" (1994, 841). Such a dichotomy can't help but presuppose the autonomous, liberal humanist subject distinct from its world that it claims to dismantle—one that can "possess" a memory, and one from which a memory can be entirely external.

Such an autonomous subject is unsuitable in understanding the function of memory and the identities it helps construct. We cannot easily claim Cloud's memories as *either* authentic *or* fabricated when the player actively constructed those memories through playing the flashback in the present. Even as they were potentially fabricated, the player's actions brought them to life, giving them an authenticity not only in Cloud's retelling (and rewriting) of history, but in the player's active participation of them. Instead of focusing on the lived (human) and the prosthetic (nonhuman) memories as distinguishable actors, we must, as Casper says, focus on the traffic between (1994, 847). The autonomous and secure human self must make way for the malleable posthuman self—situated between bodies and worlds—and so too must its memories. All experiences, and all memories of these experiences, exist in a relationship between bodies and worlds.

When the *Final Fantasy VII* player visits the still-standing Nibelhiem, they already have an understanding of the layout of the town (such as the location of shops and the inn, or the layout of the nearby mountains and energy reactor) through an *actual* experience of a virtual memory of an event that may have been fabricated. Regardless of the authenticity of the event, the memory itself is an authentic memory that actually mediates both the player's and Cloud's future actions. While Landsberg's insight into the power of a memory regardless of its origin are significant, classifying a memory as *either* lived *or* prosthetic gets us no closer to understanding just how these memories come to be. A dichotomy of lived/prosthetic memories renders memory essential and fails to take into account the complex, overlapping ways that memories and selves are constructed between and through bodies and technologies and realities.

Through *Final Fantasy VII*, various contradictory accounts of The Nibelheim Incident put great stress on the legitimacy of Cloud's memories and his personal sense of just who he is. Later in the game, Sephiroth tries to convince Cloud that he was never even in Nibelheim during the incident, that his memories of who he is are completely fabricated. Sephiroth shows Cloud and the player Sephiroth's own version of the incident. Cloud and his childhood friend Tifa stand to the side of the screen, ethereal and transparent, as the same flashback the player has already witnessed and participated in plays again. But this time, where Cloud was in his own retelling, now stands another man with black spiky hair in place of Cloud's yellow hair. Sephiroth ultimately tries to convince Cloud that Cloud's memories are not his own, that they belong to someone else: implanted and prosthetic.

But toward the end of the game, the player and Cloud finally discover that the truth is far more complicated than the memories either belonging to Cloud or to someone else. Sephiroth's insistence that Cloud's identity is a lie sends Cloud into a catatonic state. The player enters Cloud's subconscious through the control of Tifa. Together, Tifa, Cloud, and the player retrace

Cloud's deepest, suppressed memories to find out what "actually" happened and who Cloud "really" is. These suppressed memories show that Cloud *was* present during The Nibelheim Incident, but not in the role that he claimed when originally telling the story to his comrades and the player. It is revealed that Cloud never fulfilled his childhood dream of becoming a First Class SOLDIER, and Sephiroth's true sidekick was the black-haired SOLDIER, Zack. Cloud, meanwhile, was present as a typical foot troop—a masked, nameless character standing off to the side in many of the original flashback's scenes. The character model of this troop is used countless times through the game, and garners no attention from the player, standing in as just another insignificant background character. As Cloud was ashamed that he never made it to First Class SOLDIER, he stayed hidden behind the mask, watching the events from afar. Following The Nibelheim Incident, Zack is killed after saving Cloud's life. Cloud, alone, heads to the city of Midgar as Zack himself planned to and, like Zack intended for himself, becomes a mercenary. It is at this point of the plot—where Cloud identifies himself as a mercenary, as much Zack as himself—that the game begins and the player first joins Cloud on his adventure.

Cloud's identity, the self he projects to and shares with the player, is built on memories that are not easily distinguishable as either lived or prosthetic, but as a hybrid of his own memories, the memories of others, and his observations of Zack's experiences mingled together. By extension, the player's idea of who the character is that they enact in *Final Fantasy VII* is also a play of identities and memories. For Cloud, as for the videogame player, it is not a matter of his memories being lived *or* prosthetic, but lived *and* prosthetic; his identity has been formed—is continually being formed—by a constant transfer and mediation of information between bodies and realities. His memories, along with his sense of who he is, are not easily pinned down to a single body or experience.

PROSTHETICALLY LIVED MEMORIES

As long as the human is understood as autonomous and unambiguously distinct from its world, the relationship between humans and technologies can only ever be conceptualized as an either/or relationship. Instead, the human self exists—has always existed—"as part of a distributed system, [and thus] the full expression of human capability can be seen precisely to depend on the splice rather than being imperiled by it" (Hayles, 1999, 290). If, as Landsberg argues, "memories are less about validating or authenticating the past than they are about organising the present and constructing strategies with which one might imagine a liveable future" (1995, 176), then our identities and experiences as shaped and mediated by increasingly pervasive technologies and digital media cannot be understood simply through a binary of "implanted" and "actual" memories, but instead through a circuit of *cybernetic memory*, through distributed memories belonging neither to an unambiguous human body or technology, but in the systems of relating and mediation between them.

Sobchack takes issue with the metaphorical deployment of prostheses by cultural theorists, arguing that the literary use of prosthesis as metonymy (the correlation of two objects, each of which constitute an absolute, separate whole) does not correspond with her actual, *literal* experience of a prosthetic limb as synecdochic (two objects forming an ensemble where the "existence or idea of one [is] included in the existence or idea of another") (Sobchack 2004, 213). Scholars that embrace the prosthetic metaphor "far too quickly mobilize their fascination with artificial and 'posthuman' extensions of 'the body'"[2] when, rather, the prosthetic needs to be understood as incorporated "not 'into' or 'on' but 'as' the subject" (Sobchack 2004, 208–9). That is, the prosthetic does not plug into the human subject as an othered object, but is absorbed into, recalibrates, and is recalibrated by the subject's bodily sense of who they are.

Sobchack relocates the metaphoric, metonymic use of prosthesis in her embodied, synecdochic experience by contrasting her experience of her prosthetic leg with that of her biological leg:

> My "real" leg and my "prosthetic" leg are not usually lived as two absolutely different and separate things since they function as an ensemble and are each a part of my body participating in the whole movement that gets me from here to there; thus, they are *organically* related in practice (if not in material) and are, to a great degree, *reversible* each with the other. (2004, 214)

On one hand, Landsberg seems to satisfy Sobchack's understanding of prostheses as synecdochic and ensemblic through demonstrating how both lived and prosthetic memories contribute to the subject's sense of self. However, in setting up prosthetic and lived memories as two distinct kinds of memories, Landsberg is accused by Sobchack of displacing the prosthetic from its mundane context, warned that in a metonymic conception of prosthesis, "the animate and volitional human beings who use prosthetic technology disappear into the background—passive, if not completely invisible" (Sobchack 2004, 211).

While the active mediation of the prosthetic must be accounted for, prosthesis metaphors often highlight such agency of the prosthetic at the expense of the human body's own agency, reinforcing the dichotomous relationship between flesh and machine rather than transgressing it. Indeed, in Landsberg's own filmic examples, the lived memories of the characters of *Total Recall* and *Blade Runner* are completely drowned out by those memories implanted prosthetically—one identity overwrites and stands in for the other. Such is also the case when Sephiroth attempts to convince Cloud that all his memories are not his own, throwing Cloud's entire identity into disarray. But for Cloud it is not as simple as whether he was present or absent in the experiences he remembers—both are true in some way. Similarly, the film viewer does not forget who they were before they take on the memories and experiences of the film. Rather, the filmic memories and their own memories mingle together, each shaping the other to construct new, hybrid memories of experiences that were both lived and not lived.[3]

Just as the posthuman self is located in the relationship between the flesh-and-bone body and the world it relates to, Sobchack sees the prosthetic as assimilated into the subject's bodily sense of self. Much like Sobchack's biological and prosthetic legs are, for her, often reversible, lived and displaced memories form an ensemble that shapes experience and identity—a cybernetic memory that does not distinguish between lived and non-lived memories, but just memories that are constructed in the relationships between bodies and experiences.

With this reconceptualised, cybernetic notion of memory, we can now trace the intricate network of memories that actively produce Cloud's identity, and the extended network that produces the player's identity during gameplay. Cloud does not simply overwrite his own memories with Zack's and identify himself "as Zack." Cloud is still Cloud, but a Cloud who is shaped by memories and experiences of a variety of subjectivities. Significantly, Cloud's memories of Zack are not simply Zack's own memories but Cloud's own, displaced interpretation of Zack's experiences. Cloud's identity is further mediated by both Tifa's and Sephiroth's memories of him. When Sephiroth claims that Cloud was absent from The Nibelheim incident, such a claim is based on Sephiroth's own memory of the event—Sephiroth truly did believe that Cloud was not present, as Cloud never took off his helmet. Meanwhile, Cloud depends heavily on the authenticity of Tifa's childhood memories of Cloud as authentication of his own destabilized memories in the face of Sephiroth's allegations. Regularly, Cloud insists Tifa assures him that she remembers him. Cloud's present and active identity, then, is not based on any one objective experience or memory of The Nibelheim Incident but on an intersubjective network of memories and interpretations, none more or less real than the others.

When I play *Final Fantasy VII*—when I both act as a player and enact the character Cloud in a "partial becoming" (Swalwell 2008, 84)—the memories generated belong neither wholly to Cloud nor to myself, neither wholly to the virtual world of the game or the actual world of my engagement with the videogame's technologies. The experiences I have, and the memories they generate, originate in the mediating relationship between Cloud and myself. My identity during my engagement with *Final Fantasy VII* is actively constructed and mediated in a cybernetic circuit of remembering and experiencing. This is not a case of Cloud's memories being assimilated into my own, but of my sense of self being distributed across a network of bodies, experiences, and memories. It is this relationship of lived *and* displaced memories that I call here cybernetic memory. To explore this reconceptualized phenomena and its role in forming intersubjective identities, I will turn now to look at the cybernetic nature of videogames more closely.

A CYBERNETIC PLAY OF IDENTITIES

Videogames are popularly understood in terms of mastery, skill, and agency; and the player's role as an intention-driven agent of change within the virtual world is often given undue emphasis by scholars, developers, marketers, and the enthusiast press alike. However, the pleasures that videogames offer are too varied, complex, and nuanced to simply be understood as the player choosing what to do within a world. All games, from the open sandbox of *Minecraft* to the tightly authored narrative of *Dear Esther* afford and constrain the player in specific ways via the body of the playable character and the technological mediation of the input device. As Jill Walker succinctly notes, the pleasures of videogame play are that of being "part of the text, part of the machine" (2001, 47). The videogame player does not control the videogame but, in the instance of videogame play, joins with it in an entanglement of bodies and material realities.

In this vein, Martin Lister et al. describe videogame play as "*literally* cyborgian" as it is best understood "not only [as] a human subject in a technologised environment, but rather as the human as one element in an event assembled from and generated by both human and nonhuman entities" (2009, 306). The cyborg is not simply a human body with nonhuman, technological prosthetics, but an organism that is itself cybernetic. When Lister et al. invoke the cyborg, they are careful to note, much like Sobchack, that "[i]nstead of looking at what physiological or sensory functions of the organic body are extended by the prosthetic of the machine, we focus on the loop of constant information and energy exchange formed between parts of the machine and parts of the user's body" (Lister et al. 2009, 306).

Reconceptualizing videogame play as such a circuit understands that neither the player nor the technology nor the projected fictional world is dominantly "in control." Instead, the human and the nonhuman are intimately and *irreducibly* related, each mediating and mediated by the other. Both the human and nonhuman actors within videogame play contribute to the play experience. (Giddings and Kennedy 2008, 21). Among these actors are the controller, the player's hands (not to mention their gender, race, and all other aspects of their identity), the underlying code of the game, audiovisual representations, narratives, hardware, software, and countless others. However, central to the player's experience of the fictional worlds of most videogames is the mediating role of the playable character.

The playable character is the player's means of navigating the fictional world and comprehending that world's imperatives. James Newman understands playable characters as bodies of capabilities and constraints, "equipment to be utilised in the gameworld by the player. They are vehicles" (2002). But more than this, playable characters are also *characters*. Their bodily functions combine with how they are audiovisually depicted by the game to alter the player's embodied perception of the game's world. This is to say that the playable character is not "simply" a passive vehicle to be steered by the player. Instead, following Latour's (1993) work on quasi-objects, the player and the playable character form a hybrid of subject and object, the player-character, that stretches across the border of the actual and fictional worlds of gameplay, of which the player's body is but one component. It is through the playable character that the player perceives, acts, is acted upon, and ultimately identifies as part of the videogame's world.

In *Final Fantasy VII*, my understanding of the world through Cloud relies both on a subset of my own real-world identities and my perception of Cloud's identity as a virtual character. The identities at play in *Final Fantasy VII* are a hybrid of selves and memories across worlds. They are identities not possessed by any one body but which emerge in the way bodies come together through experiences. Just as Cloud constructs his identity in the relationship between himself and his understanding of Zack, so too is my identity during play constructed between my understanding of myself and my understanding of Cloud which, in turn, is mediated by Cloud's understanding of himself. Further, Cloud's experience of his world and the unfolding of his narrative are mediated by my own real-world memories of JRPG genre conventions and previous attempts at playing *Final Fantasy VII*, as highlighted by the anecdote opening this chapter. This complex network of embodied experiences shared by Cloud and myself brings together both the fictional world of *Final Fantasy VII* and the actual world in which I am playing the videogame. The memories of such experiences are not easily classified as possessed by one actor or the other. They instead resemble a fractal network of memories and interpretations, of experiences and identities, that cannot be easily placed with one subject or another but only in the ways they come together.

Just as the posthuman situates the self in the relation between bodies and worlds, so too must it displace the memories that construct the self, contextualising them between actors and technologies and bodies as always lived and always displaced.

CONCLUSION

My understanding of Cloud's world—constructed through Cloud's identity in relation to the coming together of Cloud's and Zack's bodies and worlds, is cybernetic. Neither lived nor prosthetic memories alone can account for Cloud's sense of self. Instead, it is only when Cloud is able to account for and trace the complex hybridity of the memories that construct his identity that he is content. At one stage in the game, when Cloud's "real" identity is in question, Tifa asks Cloud to reassure her that, "You really are you, right?". However, Cloud's problem was never that he might not be himself, but that he was unable to account for the memories that produced that self. Ultimately, the complex network of memories and experiences at play in Cloud's technologically mediated identity can be accounted for and traced without needing to be "possessed" by Cloud.

By understanding his memories as cybernetic, Cloud's actual, posthuman identity is never at risk of being dismissed as "only prosthetic." Likewise, memory as either "lived" or "prosthetic" is unable to account for the complex exchange of experience and memory between myself and Cloud when we fight Jenova-LIFE. Cybernetic memory, however, understands the experience as a flow of information between Cloud and myself where we are both present and displaced from the experience simultaneously. The memory of the previous battle, where Jenova-LIFE drowned Cloud and his friends with aqualung, exists between us, in the coming together of our bodies and worlds, and it is there between us when we attempt the fight a second time, and win.

As Hayles once invoked and as the other chapters in this anthology demonstrate, we have always been posthuman beings: various technologies and actors beyond (and indeed within) our flesh have always been central to how we perceive both ourselves and our world. If, then, Landsberg is correct when she states that "memories are central to our identity—to our sense of who we are and what we might become" (1995, 186) then we require a posthuman conceptualization of memory for our posthuman identities. Videogame play is a most explicit play of identities and bodies—"literally cyborgian," as Lister et al. note—and offers rich foundations from which to build an understanding of memory that does not separate it into lived or prosthetic experiences but as always existing in the relationships between actors.

Through a close analysis of the complex web of memories "at play" in *Final Fantasy VII*, I have proposed cybernetic memory as one way of understanding this web. I remember multiple, contradictory accounts of The Nibelheim Incident, and all of them are equally true. As intuitive as it may seem, it is counterproductive to distinguish some of these memories as "prosthetic" or "not-lived" simply because they were technologically mediated or because the experiences they reflect apparently lack authenticity. No different to Cloud, all my memories are displaced, existing in the relationship between my world and I, and all these memories are real: actively and presently contributing to the ongoing construction of my posthuman self. We remember—we exist—cybernetically.

NOTES

I would like to acknowledge Cameron Kunzelman's invaluable contributions to the arguments forwarded in this chapter.

1. In an interview with L. Rhodes (2012) on games criticism.
2. In this passage, Sobchack takes issue with the term "posthuman" as it is often romantically deployed to refer to an advanced, superhuman-like being, as opposed to its use by scholars such as Hayles as a reconfiguration of the liberalist conception of "human."
3. Landsberg does briefly note at the end of her 1995 article the possibility of what she calls "breakthrough memories" that suggest "identity is palimpsestic, that the layers of identity that came before are never successfully erased" (1995, 187). Here she seems to briefly reach for a notion of cybernetic memory that transgresses a lived/prosthetic dichotomy.

REFERENCES

Casper, Monica J. 1994. "Reframing and Grounding Nonhuman Agency: What Makes a Fetus an Agent?" *American Behavioral Scientist* 37: 839–56.

Dovey, Jon, and Helen Kennedy. 2006. *Game Cultures: Computer Games as New Media*. Berkshire: Open University Press.

Giddings, Seth. 2007. "Playing With Nonhumans: Digital Games as Technocultural Form." In *Worlds in Play: International Perspectives on Digital Games Research*, edited by Suzanne De Castell and Jenson Jennifer, 115–128. New York: Peter Lang.

Giddings, Seth, and Helen Kennedy. 2008. "Little Jesuses and Fuck-Off Robots: On Aesthetics, Cybernetics and Not Being Very Good at *Lego Star Wars*." In *The Pleasures of Computer Gaming: Essays on Cultural History, Theory and Aesthetics*, edited by Melanie Swalwell and Jason Wilson, 13–32. Jefferson: McFarland.

Haraway, Donna. 1991. *Simians, Cyborgs, and Women: The Reinvention of Nature*. London: Free Association Books.

Hayles, N Katherine. 1999. *How We Became Posthuman: Virtual Bodies in Cybernetics, Literature, and Informatics*. Chicago: University of Chicago Press.

Jayemanne, Darshana. 2005. "The Nip and The Bite." Paper presented to DiGRA Conference: Changing Views: Worlds in Play. unimelb.academia.edu/DarshanaJayemanne/Papers/189329/The_Nip_and_the_Bite.

Landsberg, Alison. 1995. "Prosthetic Memory: *Total Recall* and *Blade Runner*." *Body and Society* 1: 175–89.

———. 2004. *Prosthetic Memory: The Transformation of American Remembrance in the Age of Mass Culture*. New York: Columbia University Press.

Latour, Bruno. 1993. *We Have Never Been Modern*. Cambridge: Harvard University Press.

Lister, Martin, Jon Dovey, Seth Giddings, Ian Grant, and Kieran Kelly. 2009. *New Media: A Critical Introduction*. 2nd ed. New York: Routledge.

Newman, James. 2002. "The Myth of the Ergodic Videogame: Some Thoughts on Player-Character Relationships in Videogames." *Game Studies* 2, no. 1. www.gamestudies.org/0102/newman/.

Rhodes, L., and Jenn Frank. 2012. "The Critic." *CultureRamp*.cultureramp.com/new-games-critic/ (accessed September 29, 2012).

Sobchack, Vivian. 2004. *Carnal Thoughts: Embodiment and Moving Image Culture*. Berkeley: University of California Press.

Walker, Jill. 2001. "Do You Think You're Part of This? Digital Texts and the Second Person Address." In *Cybertext Yearbook 2000*, edited by Markku Eskelinen and Raine Koskimaa, 24–51. Jyväskylä: Publications of the Research Centre for Contemporary Culture.

VIDEOGAMES CITED

Squaresoft. *Final Fantasy VII*. Sony Playstation, 1997.

Team Meat. *Super Meat Boy*. Xbox Live, 2010.

The Chinese Room. *Dear Esther*. Steam, 2012.

Ubisoft. *Assassin's Creed*. Xbox 360, 2007.

Chapter Thirteen

Mediating Anthropocene Planetary Attachments

Lars von Trier's Melancholia

Nicole Merola

Anthropocene. Geotrauma. Melancholy. These three words set porous boundaries for my inquiry in this chapter. The first term, Anthropocene, signals our emergence into a new geological epoch—the "age of man"—and highlights the widespread effects of anthropogenic activity on terrestrial and marine ecosystems. It also forces us to reconceptualize humans as a biosphere-altering geological agent, and therefore to think the contours of "posthumanness" anew.[1] Geotrauma, my second term, gestures to the fundamentally inscriptive nature of industrial, biological, chemical, geological, and cosmological processes, all of which leave traces on the planet, and many of which trouble, conceptually and materially, the stability of human and nonhuman life and things.[2] Thinking through the conjoined lenses of the Anthropocene and geotrauma enables us to register, conceptually and materially, the traces of an attachment to the planet focused around the use of planetary resources. Melancholy, my third term, designates an important socioecological affect for the contemporary moment, one that departs from attachments centered on use. An affect that circulates among and between humans, nonhumans, and things, melancholy privileges feeling over knowing, relationality over instrumentality, the continual and liminal over the closed or the final, and transcorporeality over bounded human subjectivity. Melancholy thus represents a different way of being a "user" of the planet, one that helps us inhabit, in a non-dominative fashion, our relationships with other humans, other nonhumans, and other things. Artists, critics, designers, filmmakers, scientists, theorists, and writers interested in probing the Anthropocene, its myriad geotraumas, and environmental melancholy face two related, and crucial, questions. First, how can we make planetary disaster visible? And second, how can we feel planetary disaster in registers that include, but also exceed, the personal?[3] As an allegory for living in socioecological end times, Lars von Trier's film *Melancholia* (2011) addresses both of these questions.[4] In what follows I draw on scholarship in the environmental humanities, film studies, material feminisms, and philosophy to explore the ways *Melancholia* functions as an affective (and effective) design technology for thinking about (and feeling) relations between humans, nonhumans, and things in the Anthropocene, an era of global socioecological change.

THE ANTHROPOCENE AND CONTAGIOUS CINEMATIC TECHNOLOGIES

For scientists the term Anthropocene designates a new geochronological era, but as literary studies scholar Timothy Clark notes, scholars in the humanities use it "in a sense beyond the strictly geological": to identify "the context encompassing all the new demands—cultural, ethical, aesthetic, philosophical, and political—of environmental issues that are truly planetary in scale" (2012, v). To underscore this planetary context, environmentalist Bill McKibben suggests we need a new name for our planet: "Eaarth." The extra "a" McKibben inserts, which alludes to the "A" that begins Anthropocene, underlines the uncanniness of our current socioecological circumstances. We may still be "earth*like*," he writes, but things are just "odd enough to constantly remind us how profoundly we've altered the only place we've ever known" (2010, 3). Clark and McKibben offer broad insights about the Anthropocene. Literary studies scholar and philosopher Claire Colebrook, on the other hand, more specifically assesses the modes of human-planet relationship the Anthropocene indexes: the "infraction of the bounds of the earth by 'man,'" a way of figuring man as "that animal who has detached himself from putative ecological animality and lived in such a way that his life is destructive of his milieu," and "the ongoing myopia of the human species' inability to think its detachment, disconnectedness, malevolence and stupidity in relation to a planet that it continues to imagine as environment, *oikos*, cosmos or *Gaia*" (2012, 199). Her indictment of our limited planetary attachments casts them as criminal and points out the need to consider a range of darker ecologies, those that might sketch the planet as hostile space rather than home, emphasize the depravities of our political and economic models rather than celebrating our landscapes, or situate humans as unexceptional creatures within the context of cosmological timescales. To be as blunt as Colebrook, to confront what it means to live in the Anthropocene we must materialize socioecological anxieties that exceed anthropocentric concerns and imagine dystopian socioecological futures.

To understand the role von Trier's *Melancholia* plays in materializing the Anthropocene and its geotraumas necessitates briefly laying out my assumptions about cinematic technology. The critical apparatus I find most persuasive for opening up *Melancholia* is a cinema studies informed by phenomenology and feminist criticism. A short list of exemplary works in this vein includes Vivian Sobchack's *The Address of the Eye: A Phenomenology of Film Experience* (1991), Steven Shaviro's *The Cinematic Body* (1993), and two books by Laura U. Marks, *In the Skin of Film: Intercultural Cinema, Embodiment, and the Senses* (2000) and *Touch: Sensuous Theory and Multisensory Media* (2002). Within the theories of embodied spectatorship laid out in these texts, I am especially interested in how Marks discusses the viewer's affective engagement with film. Relevant with respect to *Melancholia* is her argument that "our experience of cinema is mimetic, or an experience of bodily similarity to the audiovisual images we take in. Cinema is not merely a transmitter of signs; it bears witness to an object and transfers the presence of the object to viewers."[5] Since Marks positions film as "impressionable and conductive, like skin," as something that works "through a contact between perceiver and object represented," both cinema, the material object, and cinema-going, the experience of watching film, involve contagion (2000, xi–xii). In other words, Marks sees film as "something we viewers brush up against like another body," something that transforms the viewer (xii). Marks's account of the cinematic experience offers a good example of the way thing-power works. A thing—a film—is audio-visually taken in and metabolized by a viewer, whose bodily surfaces and interiors are designed through its engagement with the thing. In short, through the circulation of affect film affects the body and the subjectivity of the viewer. And if this affect eschews conventional attachments to the planet (home, environment, Gaia) and fathoms the geotraumas of the Anthropocene, it might also work toward carving out melancholic modes of human-planet attachment.

MELANCHOLIA AND ENVIRONMENTAL MELANCHOLIES

In *Melancholia* von Trier explores human depression and anxieties about planetary destruction in order to throw into relief the capitalist-domestic form of planetary attachment currently practiced in affluent western societies and to propose a nonproductive hyper-natural form of embodied attachment to the planet.[6] To abbreviate, we can label the first mode of planetary attachment "planet as home" and the second "planet as body."[7] Von Trier composes the film in three sections—a prologue and two chapters—nesting a domestic melodrama within the context of an apocalyptic cosmic drama to offer an imaginative encounter with something we can never experience, the total destruction of the planet and all life on it.[8] He introduces the cosmic drama first, in a prologue comprising a montage of surreal, slow motion tableaux set to the prelude from Richard Wagner's *Tristan und Isolde*. These tableaux, which each gesture to a scene or sequence that reappears later in the film, build to the prologue's final shot. There von Trier stages the total annihilation of Earth by a heretofore unknown blue planet, Melancholia.[9] Unlike recent post-apocalyptic films such as *Children of Men* (2006) or *The Road* (2009), in which life persists after disaster and glimmers of hope infuse the films' last scenes, in *Melancholia* the end is the end and nothing comes next. Indeed, after the collision between Melancholia and Earth the screen goes completely black. This refusal to provide solace by representing some form of living on after the end is simultaneously an atheistic interpretation of the concept of apocalypse and a riveting political and socioecological argument about the stakes of anthropogenic activity on the planet.

In the shadow of this arc of total planetary destruction, von Trier examines one dysfunctional family, primarily focusing on the interpersonal relationships of sisters Justine (Kirsten Dunst) and Claire (Charlotte Gainsbourg), Claire's husband John (Kiefer Sutherland), and their young son Leo (Cameron Spurr).[10] Their domestic melodrama, set on the perfectly manicured estate where John, Claire, and Leo live, plays out across the two main chapters of the film, "Justine" and "Claire." The key narrative elements of the film include the dissolution of Justine's marriage, which is effectively over before her lavish wedding reception ends; a focus on Justine's work in advertising, including an overwrought tête-à-tête with her boss that ends when Justine quits her job; the impact of Justine's depression on her relationship with Claire; the responses of each character to the impending approach of the planet Melancholia; and the total destruction of Earth by Melancholia. Central to von Trier's critique of capitalist-domestic forms of planetary attachment are his negative assessments of excessive material consumption, blind faith in expert science, anthropocentrism and biocentrism, and planetary sentimentality. The audio-visual design of the film—its palettes, built environments, visual style, and soundtrack—skews toward a gothic Romanticism that is in tension with von Trier's relentlessly anti-romantic arguments about human-planet relationships.

Although *Melancholia* is dissimilar in many ways from the films Marks uses to construct her theories of embodied spectatorship, it nevertheless materializes a mimetic cinema-going.[11] That is, the objects to which von Trier bears witness in *Melancholia*—the excesses of capitalism, Justine's depression, Claire's anxiety, Leo's innocence, the magnetism of the planet Melancholia and Justine's hyper-natural connection to it, the responses of nonhuman animals to Melancholia's approach, the end of the planet Earth—and the affects that circulate within the film—anxiety, anti-sentimentality, fear, depression, melancholy, thing-power—are not merely events represented onscreen but rather are impressions, a contagious "experience of bodily similarity" transferred to viewers. The creation in the viewer of an embodied isomorphism with the film and its affects turns *Melancholia* into a design object that transfers the two distinct forms of planetary attachment identified above—planet as home and planet as body—while also producing participatory rituals of two forms of environmental melancholy: socioecological melancholy and what I'm calling planetary melancholy, a form of environmental melancholy in which planetary geospatial and temporal scales, rather than interpersonal, interspecies, local, regional, or global socioecological scales, are at issue.[12]

The concept of environmental melancholy draws on the distinction Sigmund Freud makes between two different, though related, forms of grief: mourning and melancholia. While the former ends when one can finally transfer affection to a new libidinal object, the latter is unending (Freud 1917). Structurally, then, melancholia is a form of being stuck. While Freud acknowledges that a libidinal object may be animate or inanimate, only recently have environmental humanists begun probing the contours of environmental melancholy. A key difficulty in theorizing environmental melancholy, as sociologist and queer theorist Catriona Mortimer-Sandilands notes, is that western industrialized cultures are disinclined to "acknowledge nonhuman beings, natural environments, and ecological processes as appropriate objects for genuine grief."[13] In other words, even if we can map melting Arctic sea ice, list recently extinct species, or measure increases in atmospheric CO_2, empirical measurement is not the same as grief. I propose that environmental melancholy comprises a feedback loop that involves cultivating a radical openness to the agency of nonhumans and things and our relations with them; extending grief to encompass nonhumans and ecological processes; designing forms, objects, and rituals for enacting environmental grief; and learning to inhabit environmental melancholy as a permanent condition.

In highlighting an inability to move on, environmental melancholy offers a useful framework for thinking our current conditions. We are, after all, stuck on a planet characterized by "planetary disasters, emerging pandemics, tectonic shifts, strange weather, oil-drenched seascapes, and the furtive, always-looming threat of extinction" (Thacker 2011, 1). Against narratives of environmentalism that suggest we might be able to fix what we have sullied, environmental melancholy involves seeing loss and change as structural conditions to which we are bound and in which we are complicit. For instance, a melancholic approach to extinction as a structural condition would involve not just sadness about the disappearance of a particular species, but rather unending waves of grief, a continuous recognition of various registers of loss, including loss linked to the intrinsic value of creatures, loss linked to the ecological benefits of biodiversity, and loss linked to the foreclosing of possible relations between and among different creatures, habitats, and systems. Within the larger rubric of environmental melancholy, socioecological melancholy foregrounds grieving for the biosphere and its processes and is thus linked to conceptualizing the planet as domestic space, home. In other words, socioecological melancholy positions the planet as "world-for-us." On the other hand, planetary melancholy does away with the notion of home and instead foregrounds a more thing-oriented perspective, situating the planet as a geo-chemical object, "world-without-us," and privileging grieving for the planet as a cosmological body.[14] This form of melancholy is relentlessly anti-anthropocentric, even as it is necessarily anthropogenic.

PLANET AS HOME: CAPITALIST EXCESSES, SCIENTIFIC ATTACHMENTS, AND BIOSPHERIC ANXIETIES

All the earthly action in *Melancholia* takes place within a tightly regulated setting, the grand estate on which Claire, John, and Leo live. Von Trier employs the luxurious domestic interiors and manicured grounds of the estate, and Justine's wedding reception there, the only event in the first chapter of the film, to position marriage, capitalism, and material overconsumption as corrupt rituals.[15] He makes this critique clear by thematizing excess through design elements—for instance, the perfection of the estate and its grounds and Justine's exquisite wedding dress—and narrative elements—for instance, multiple references to John and Claire's wealth, the amount of alcohol drunk at the reception, the extramarital sex Justine has during the reception, and the focus on her job in advertising. It is through this focus on Justine's work-life—especially the scenes when her boss Jack (Stellan Skarsgård) hounds Justine for a new ad campaign idea and when, in conversation with Jack, she characterizes advertising as "hook[ing] a group of minors on our substandard product, preferably in a habit-forming way," calls Jack a "despicable, power-hungry little man," and quits her job—that von Trier most bluntly makes his argument about the moral bankruptcy of capitalism. On one level, then, von Trier deploys the wedding reception as metaphor for the drunken party that is the capitalist overconsumption of the planet's resources. He is careful to implicate the viewer in this corrupt ritual, utilizing camera-work, point of view, and the ambient noise that serves as soundtrack during the reception to keep the viewer in an uncomfortably intimate relationship to the action at the reception. Von Trier thereby refuses the viewer a distanced perspective from the consumption occurring onscreen.

Just as the viewer is implicated in the overconsumption displayed at the party, she is also made to feel Justine's increasing exhaustion, listlessness, and emotional detachment, all of which increase as the wedding reception progresses. Her fatigue and detachment register especially in the scenes when she locks herself in the bathroom to take a lengthy bath, goes upstairs to put her nephew to bed and falls asleep alongside him, refuses to have sex with her husband, Michael (Alexander Skarsgård), and displays indifference to his purchase of a plot of land on which he intends to build their house. Justine's enervation should be read not only as foreshadowing the deep depression she exhibits in the second chapter of the film, but also as an index of her weariness at the subject positions available to her: knowledge worker, consumer, wife, daughter, sister, aunt. By quitting her job and ending her marriage, Justine refuses to participate as employee, consumer, or wife.

Justine's exhaustion also indexes another corrupt ritual in which she and we are complicit, the reproduction of a conceptual human-planet framework that literary studies scholar Matthew A. Taylor labels "the self-exalting metaphysics of transcendentalism" (2012, 364). In an essay on fear, Edgar Allen Poe and posthuman ecology, Taylor persuasively argues that many US Transcendentalists and British Romantics who are held up as "exemplars of an ecoconscious worldview" actually "subordinat[e] the natural world to an occasion of the self's realization" (354, 355). The well-worn trope of the sublime offers an ideal example of a "self-exalting metaphysics." In a sublime encounter one is confronted by something destabilizing. Then, though initially discombobulated by this encounter, one recovers through the exercise of reason.[16] Like Taylor, von Trier unearths the paradox of the purported ecological consciousness of Romanticism. Against those who would align Romanticism with ecological consciousness, von Trier links Romanticism with capitalism, positioning both as constraining human–planet modalities that reinscribe the position of transcendental human subject. He deploys his romantic audio-visual design strategy in service to an anti-romantic argument. One example of von Trier's anti-romanticism is the way he treats Justine's pastoral impulse to seek refuge from the wedding reception by going outside alone. Von Trier positions this impulse as doubly compromised. First, the whole of the estate, grounds included, are implicated in the very consumption Justine wants to disavow. And second, the impending collision of Earth and Melancholia renders all notion of refuge ridiculous. The pastoral is also undercut by the persistence of the gothic. For instance, Justine's glances at the heavens are tinged by the lurking presence of Melancholia. And the weirdness of the three prologue tableaux in which she appears outside in her wedding dress—the eighth, in which she, Leo, and Claire stand on the lawn, eerily bathed in the light of three celestial bodies (Melancholia, the moon, and the sun); the eleventh, in which she trudges across the lawn, constrained by grey, woolly yarn; and the fourteenth, in which she lies in a stream, echoing Sir John Everett Millais's painting, *Ophelia* (1851–1852)—menace any notion of the pastoral. In von Trier's formulation of the gothic what gets repressed is the indifference of the cosmos and the strength of its forces, and all these gothic elements highlight von Trier's sense that Romantic human-planet attachments are problematic. Furthermore, as will become clear later in this chapter, he situates the embodied human-planet relationship Justine has with Melancholia as an alternative to the frameworks of both Romanticism and capitalism.

In "Claire," von Trier's exploration of planet as home shifts toward an examination of anthropocentrism, modes of scientific belief, biospheric anxiety, and hyper-natural planetary attachments. To probe these concepts, von Trier chronicles Justine's depression, the relationship between Justine and Claire, and the responses of Justine, Claire, John, and Leo to the approach of Melancholia. He positions John and Leo as figures for two different kinds of scientific knowledge, expert science and lay science, juxtaposing John's intellectualism with Leo's innocent and wondering exploration. Against the affection for science exhibited by John and Leo, von Trier deploys Claire and Justine as figures for embodied, experiential responses to Melancholia, juxtaposing Claire's growing anxiety and fear of Melancholia, structured through her maternal and domestic concerns, with Justine's hyper-natural relationships to both Earth and Melancholia. In staging Justine's willfully anti-anthropocentric and anti-biospheric attitudes, von Trier posits a novel form of attachment to Earth and Melancholia, one I will detail in the next section. In short, then, in "Claire" von Trier interrogates two types of apparatuses—disembodied scientific knowledge networks and the technologies that produce them and embodied experiential knowledge networks—evaluating the latter more highly as a mode of human-planet attachment.[17]

During the five-day approach of Melancholia, John prepares by stocking provisions for the family and their horses, shoring up his faith in expert scientific predictions the two planets will not collide, and reveling in the aesthetic and cosmological exceptionalism of the flyby, which he insists will be "the most beautiful sight ever." He also attempts to educate his family about the event, talking science with Leo, who makes his own device for gauging Melancholia's position, and admonishing Claire, whose fear of Melancholia has made her susceptible to doomsday websites, to trust the "real scientists." John's faith in expert science is bound up with his affection for his telescope, his anthropocentric and paternalistic attitudes, and his desire for the sublime. When the planet appears to be receding, he makes a toast to life, but also unwittingly alludes to its ultimate destruction by talking about a possible margin of error in the science. John's premature victory toast reveals his hubris and biospheric attitudes; he is most interested in life, both life in general and the lives of his family members.

Unlike Claire, who recognizes the fragility of life and can easily imagine its disappearance, John's anthropocentrism and biocentrism are also a kind of egocentrism, a decidedly non-Darwinian attitude that sees a perfection to life on Earth and places humans as the pinnacle of evolutionary achievement.

The viewer's prior knowledge of the imminent collision between Earth and Melancholia forecloses identification with John's faith in science, and the disjuncture between what the viewer knows about the outcome and what John believes creates a dynamic and increasing tension. John's realization that the scientific calculations are wrong destroys his entire existential framework; he takes the pills Claire has purchased for herself and kills himself in the stables, deserting his family. Von Trier posits John's suicide as an intellectual act; he takes control of his own body so that Melancholia cannot. Ultimately, von Trier employs John's intellectualized response to Melancholia to suggest that blind faith in expert science is a myopic anthropocentrism.

Von Trier's critical assessments of anthropocentrism and biocentrism are augmented through his scripting of Claire's response to Melancholia. Since her relationship to it is structured through maternal and domestic concerns, she can only see the planet as threatening life and home. Even while censuring Claire for her narrow circle of concern, however, von Trier positively evaluates the expressiveness of her body, her physical and emotional recalcitrance in the face of John's attempts to soothe her. Where John refuses to open his body to interaction with Melancholia, and Justine welcomes this interaction, Claire's body becomes increasingly unruly as Melancholia gets closer. In the "Justine" chapter we see Claire as competent, organized, mobile, and in control of her own emotions, but in the "Claire" chapter the closer Melancholia gets the more she unravels, both physically and emotionally. An important example of this unraveling occurs during the sequence when John, Claire, Justine, and Leo all go outside during the night to watch Melancholia rise. The sequence opens with Claire's glance into the telescope and assertion she is not afraid of Melancholia because it "looks friendly." However, her attempt to suppress her fear is quickly belied by her bodily reaction. In the crucial moments of the sequence, the camera focuses tightly on Claire, sitting on the patio in a wicker chair, face turned toward the sky. She begins to breathe heavily, her labored inhale and exhale the only sound. John comes to her side, his breathing unaffected, and instructs her how to regulate her body: "Just relax, relax. This is normal; it's taking part of our atmosphere. For a little while it's going to make us winded. Just try and breathe calmly. Just breathe. Just breathe. There you go. Just try and breathe calmly." Claire's response registers Melancholia bodily and acutely, in a manner diametrically opposed to John's calm and measured scientific discourse on atmospheric effects. While on one level this scene might read as a standard panic attack, on another it materializes thing-power: Melancholia exerts control over Claire's breath, subordinating her.

Within the framework of planet as home, von Trier scripts John as having the narrowest circle of concern. He is both an egocentric and an exceptionalist. For him planet as home means planet as instrument for showcasing the achievements of humans, be they aesthetic, domestic, economic, or scientific. Claire's biocentrism takes a different form than John's and her circle of concern is slightly larger than his. She worries most about her life and Leo's life and thinks about the disappearance of the planet primarily in terms of Leo. For instance, in response to Justine's indifferent assertion that the Earth is evil and we need not grieve for a planet no one will miss, Claire plaintively responds: "But where would Leo grow up?" She also demonstrates occasional concern for the horses she and John own. When Justine rides Abraham too hard, and tries to force him over a bridge he does not want to cross, Claire scolds Justine. However, Claire's attitude is mostly anthropocentric, and the closer Melancholia gets and the more anxious she becomes, the more she turns inward, to the care of self and Leo. Her treatment of Abraham immediately after John's death illustrates how her anxieties about Leo override any cross-species sympathies she might have. Rather than worry about Abraham's response to John's death or to Melancholia's approach, Claire uses Abraham as an instrument to conceal John's death from Leo. In order to explain his absence, Claire tries to make it look like John has gone for help; she leaves John's body in Abraham's stall and covers it with straw, takes Abraham from his stall, saddles him, and signals him to run away. Claire, then, is relatively blind to other species and their distress. In addition to seeing Abraham instrumentally, she also fails to notice when birds start or stop singing, or insect noise waxes or wanes.

The presence or absence of these ambient nonhuman animal and insect sounds, if unnoticed by Claire, is sharply registered by the viewer. Nonhuman animals and insects are significantly present in the prologue. In the first tableaux, dead birds fall from the sky. In the third, a detail from a reproduction of Pieter Brueghel's painting *Hunters in the Snow* (1565), which features men and hunting dogs, burns. In the sixth, Abraham falls to the ground. And, in the seventh, moths surround Justine. Von Trier thus sets up the viewer to notice when, how, and to what ends nonhuman animals and insects reappear later in the film. Furthermore, the use of nonhuman animal and insect sounds is a welcome relief from the relentless repetition of the prelude to *Tristan und Isolde*, the only non-diegetic sound von Trier employs. Overall, the inclusion of nonhuman animals and insects, the strategies von Trier employs to highlight their presence, and the ways they are affected by Melancholia, highlight Claire's selfishness, situate the estate as a metaphor for the planet, and prime the viewer to experience a fuller range of biospheric concern than what Claire can muster. In other words, through these scenes in which the viewer notices the activity of nonhuman animals and insects, von Trier lays groundwork for the extension of grief to nonhumans and ecological processes.

PLANET AS BODY: JUSTINE, EARTH, MELANCHOLIA, AND HYPER-NATURAL ATTACHMENTS

Von Trier reprises the focus on Claire's embodied response to Melancholia in his focus on Justine's embodied response, but Justine's reaction could not be more different than Claire's. At the beginning of the "Claire" chapter Justine's depression is so severe she can barely stand, walk, eat, or bathe. She is thin, pale, and listless, and her complexion, in a marked departure from its glow during her wedding, is sallow and blemished. In her depressed state, Justine cannot care for herself, but under Claire's attention she gradually returns to functionality. Were Justine's recovery only a result of Claire's ministrations, or couched as a return to a commonplace understanding of health, it would not be especially interesting. However, by aligning Justine's recovery with the approach of Melancholia, von Trier implies a direct relationship between the proximity of the planet, Justine's body, and the production of a trans-corporeal space of planetary melancholy she comes to inhabit. Using concepts from feminist science studies, we can describe the relationship of Justine's body to planetary bodies as hyper-natural: trans-corporeal and intra-active. Through Justine's body and its excesses, then, von Trier dramatizes a nonproductive hyper-natural form of embodied attachment to planet as body, not to planet as home.

To think trans-corporeally, as the term suggests, is to think both across bodies—whether animate or inanimate, human or nonhuman, natural or artificial, microscopic or macroscopic—and about how bodies move across each other.[18] In other words, a trans-corporeal framework foregrounds the notion that, as literary studies scholar Harold Fromm writes, "the 'environment' as we now apprehend it, runs right through us in endless waves, and if we were to watch ourselves through some ideal microscopic time-lapse video, we would see water, air, food, microbes, toxins, entering our bodies as we shed, excrete, and exhale our processed materials back out."[19] The trans-corporeal and the posthuman are sutured, for a trans-corporeal orientation, as material feminist Stacy Alaimo notes, "refuses to delineate the human, the cultural, or the linguistic against a background of mute matter" (2010, 14). In Fromm's example, trans-corporeality involves atomic matter. To move to the subatomic level requires feminist science studies scholar Karen Barad's theorizations of matter and intra-activity, which derive from her work in quantum physics. For Barad, the term intra-action names an "*ethico-onto-epistemology*" that understands the world as continuously and dynamically coming into existence through the entanglement of bodies, actors, and forces. Or, to be more precise, as Barad writes: "The neologism 'intra-action' *signifies the mutual constitution of entangled agencies*. That is, in contrast to the usual 'interaction,' which assumes that there are separate individual agencies that precede their interaction, the notion of intra-action recognizes that distinct entities do not precede, but rather emerge through, their intra-action."[20]

Kirsten Dunst's masterful performance of Justine's depression garnered her various awards, including the 2012 Saturn Award for Best Actress, the 2012 National Society of Film Critics Award for Best Actress, and the 2011 Cannes Award for Best Actress. As Steven Shaviro points out, these awards recognize Dunst for conveying not what depression looks like, but rather "a sense of what depression feels like." According to Shaviro, through "Dunst's amazingly sensitive and nuanced performance, *Melancholia* refutes clinical objectivity, and instead *depathologizes* depression." Rather than stigmatize depression as "a moral or intellectual failure, a kind of unseemly self-indulgence," Shaviro argues, the film "treats depression as a proper state of being, with its own integrity and ontological consistency" (2012). In other words, Justine inhabits her depression as a melancholy state that does not need to be cured, thereby making herself a body for experiencing Melancholia's destructive relationship to Earth, rather than one that longs for things to be otherwise. If we want to identify what is especially contagious in *Melancholia*, to evoke Marks's use of the term, Gainsbourg's performance of Claire's anxiety and Dunst's performance of Justine's depression fit the bill. Furthermore, in addition to being intrinsically valuable as a conductive embodiment of depression, Dunst's performance is indexically valuable; focused as it is on her body, it points us toward the trans-corporeal and intra-active relationships she has with Melancholia.

If the "Justine" chapter of the film and the beginning of the "Claire" chapter feature Justine's refusal to inhabit various liberal humanist subject positions afforded her, the closer Melancholia gets to Earth the more Justine's depression transforms into a posthuman melancholy relationality she can live. Reading this correlation trans-corporeally reveals two transits: Justine/Earth and Justine/Melancholia. One effect of von Trier's focus on advertising in the "Justine" chapter is to imply that overconsumption has infected both Justine and Earth, that the trans-corporeal movement of capitalist materialism across human and nonhuman bodies has been universally noxious. Given von Trier's critique of capitalism, when Justine says to Claire that the Earth is evil, she both conveys the idea that capitalism has made Earth toxic and counters Claire's ethos of "friendly" planetary bodies. Her loss of control over her body—its depressive comportment—registers this toxicity in a way that makes its presence legible to other people. Against Claire's attachment to planet as home and her anxiety at its impending loss, Justine sees the Earth as poisoned, a sentiment that underwrites her willing foreclosure of any domestic attachment to the planet and her subsequent lack of emotion regarding its destruction. While Justine becomes indifferent toward the Earth and the lives it supports, a negative detachment, the second transit, Justine/Melancholia, engenders a strong positive attachment. The attractive forces between these two bodies are made most apparent in the striking scene when Justine lies naked on a stream bank, basking seductively in the eerie blue glow of Melancholia. The light of Melancholia exerts a magnetic force on her, summoning her from her bed, outside, down the terrace steps, across the lawn, into the woods, and to the stream. Visually this is a stunning scene, exemplary of von Trier's hyper-romantic style, and it foregrounds Justine's embrace of a posthuman becoming, one that is fully open to Melancholia, and to melancholy, as material forces. Especially salient for a trans-corporeal reading is the suggestion of a sub-atomic level of connection between Justine and Melancholia. Draped over the vegetation on the stream bank, Justine's humanness seems super-human, augmented by the trans-corporeal transit of Melancholia's light. Essentially, then, in this scene von Trier highlights the fundamental sameness of everything in the universe, which is, after all, whether human or planet, made up of a limited number of components.[21]

MEDIATING POSTHUMAN PLANETARY MELANCHOLY

At the end of *Melancholia*, Justine, Claire, and Leo wait for the impact that will destroy Earth and all life on the planet by sitting in the magic cave, a teepee-structure made by Justine and Leo from branches and placed on the lawn. Although Justine does not need it, she recognizes the magic cave might offer to Leo the comfort of ritual. Her empathy for Leo also extends to Claire, and at the end it seems Justine's trans-corporeal openness to Melancholia has enabled the rewriting of both her own subjectivity and the roles of aunt and sister. In the last sequence the viewer does not see Melancholia directly until just before impact, so von Trier makes its presence palpable through tightly-framed reaction shots, elements of the mise-en-scène, and the soundtrack. As the sequence progresses, the light becomes bluer, the wind intensifies, and the prelude to *Tristan und Isolde* ascends in pitch and volume. Von Trier layers the sound of the wind underneath the prelude, its force clearly discernable during musical silences, and the soundtrack plays a key role in transferring the disquiet onscreen to the viewer. When Melancholia appears, its giant blue-white mass menaces most of the horizon line. From this point on, the camera occupies a fixed point of view, set up so that, as Melancholia quickly advances on Earth, driving wind and dust and waves of energy across the Earth's surface and filling the entire sky, its effects also encircle the viewer. Upon impact we briefly see flames and then dust, which seems to swirl both onscreen and into the theater, then we are encompassed by darkness accompanied by noise, then darkness and silence. At the very end, von Trier offers the viewer the cinematic experience of nothingness and nonexistence, providing an important counterbalance to the richly aestheticized experience comprising of the rest of the film. Without the lush aesthetics that precede it, the nothingness would be much less forceful.

To caution about possible dangers of aestheticizing disaster makes sense with respect to the documentation of real events. But, as literary studies scholar Ursula K. Heise reminds us, speculative fictions such as *Melancholia* help us think "in quite non-fictional terms about our current ecological condition" and "seek to make the present available for critique and transformation" (2012, 3, 10). In *Melancholia,* von Trier makes visible the Anthropocene geotraumas of capitalism and Romanticism, makes palpable the embodiment of depression, makes ironic his own hyper-romantic audio-visual design aesthetic, makes phenomenological a scientifically-possible though unlikely form of planetary annihilation, and makes unavailable our usual modes of human-planet attachment. And he employs conventions of realism, Romanticism, and speculation to confront our myopias about our geotraumas and to unfold one possible socioecological future.[22] Given our current socioecological circumstances, to avoid representations of apocalypse and disaster would be both unrealistic and irresponsible.[23]

In the final sequence von Trier makes it abundantly clear that the ritual gesture of making and occupying the magic cave is completely inadequate to the scope of the event about to befall them. The final scene emphasizes the minuteness and fragility of humans and the planet in the context of the cosmic forces of the universe. He thus offers viewers a posthumanist perspective that, as Barad puts it, "refuses the anthropomorphisms of humanism and antihumanism" and instead marks "the boundary-making practices by which the 'human' and its others are differentially delineated and defined" as specious (2007, 136). In other words, von Trier radically undercuts human exceptionalism and instead offers a trans-corporealism that does not "presume the separateness of any-"thing," let alone the alleged spatial, ontological, and epistemological distinction that sets humans apart" (Barad 2007, 136). By undercutting human exceptionalism and offering total annihilation at the end, von Trier also forecloses the possibility of the sublime, for unlike a sublime experience, in which the viewer can fully recover her subjectivity, at the end of *Melancholia* there is nothing to transcend. All life, all planetary processes, and even the planet itself are vaporized, and the contagious nature of von Trier's film and its ending, which get under the skin of the viewer, make it impossible to slough. By definition then, in addition to presenting two possible forms of planetary attachment, planet as home and planet as body, von Trier creates a melancholic form, a form of ritual planetary melancholy for the viewer to inhabit. By combining the narrative trajectory of Earth's demise with the mediated experience of nothingness that follows the last frame in which we see or hear anything, von Trier situates the viewer as one who is waiting for the end of the world. As he exposes domestic attachments to the planet as bankrupt, dismantles any hope that we will avoid planetary conflagration, and positions us in the space of knowing the end will come but not knowing exactly the shape of its contours, von Trier transfers to the viewer only the possibility of a permanently elegiac existence in which thing-power has destructive, as well as constructive, possibilities.

NOTES

1. My understanding of the term posthuman aligns with philosopher Cary Wolfe's writing on the topic. For Wolfe, posthumanism "names the embodiment and embeddedness of the human being in not just its biological but also its technological world, the prosthetic coevolution of the human animal with the technicity of tools and external archival mechanisms (such as language and culture)" (2010, xv). It also names "a historical moment in which the decentering of the human by its imbrication in technical, medical, informatics, and economic networks is increasingly impossible to ignore." In what follows I am also interested in Jane Bennett's definition of "thing-power"—"the lively energy and/or resistant pressure that issues from one material assemblage and is received by others" (2004, 365)—and in her assertion that because "thing-power materialism" heightens our sense of "mutual implication" and "intimacy" with other humans, nonhumans, and things, it might "contribute to an ecological ethos."

2. The International Commission on Stratigraphy (ICS), a body within the International Union of Geological Scientists, determines the boundaries of geological eras by examining the inscriptive traces of biological, chemical, geological, and cosmological processes left on the surface of the planet and in its atmosphere. See, for instance, N. Clark (2012), Crutzen and Stoermer (2000), Ruddiman (2003), Szyerzsynski (2012), and Zalasiewicz, et al. (2008). On the concept of geotrauma see T. Morton (2012).

3. For a related exploration of these questions, see Nixon (2011, esp. 2–44).

4. The neologism socioecological end times designates our contemporary global situation as one in which overpopulation, over-consumption of natural resources, reductions in habitat and biodiversity, and rising carbon emissions have constellated into a trajectory detrimental to the biosphere. This phrase highlights the anthropogenic and scientific registers of this trajectory, while simultaneously downplaying any religious connotations. I also use it to suggest that, for political, economic, cultural, and scientific reasons, we are unlikely to be able to meaningfully change our current pathway and avoid socioecological crisis. See, for instance, Crutzen (2002), Crutzen and Schwägerl (2011), Kolbert (2011), McKibben (2010), O. Morton (2011), and Roberts (2011a, 2011b). See also Buell (2003), Latour (2004, 2011), Speth (2005), and Zizek (2010, esp. viii–xi, 327–352).

5. Marks (2000, xvii). For a fuller discussion of the intellectual context of her project see Marks (2000, xiii–xvi). For another approach to the embodied experience of media see Hansen (2006). On how media mediates materially, not just semiotically, see Verbeek (2006). For a general discussion of affect and cinema see Carroll (2008, esp. 147–188). For a concise description of the turn to affect in critical theory and the stakes of this turn, see Clough (2007, 1–3).

6. I use the term hyper-natural here, rather than supernatural, to point to the existence of quantum natural forces we do not yet fully understand.

7. Exploring environmental attachment at different scales, from the local to the global, is an important line of inquiry in ecocriticism. See, for instance, Heise (2008).

8. Here I use the term domestic to highlight the interfamilial interpersonal dynamics on which von Trier focuses and to emphasize the idea of planet as domestic space, planet as home. For a persuasive discussion of von Trier's focus on interfamilial domestic melodrama, see Shaviro (2012).

9. For a shot-by-shot assessment of the prologue see Dargis (2011).

10. By naming one of the sisters Justine, Von Trier evokes the Marquis de Sade's *Justine*. Von Trier's Justine, however, updates the masochism of her namesake by drawing out the irrationality and potential destructiveness of seemingly rational things like the standard capitalistic economic model and by extending the experience of masochism beyond an interpersonal

register and onto a planetary register. Thanks are due to Colbey Emmerson Reid for helping me think through this point.

11. Marks reminds the reader that she develops her "auratic, embodied, mimetic" theory of spectatorship in the context of intercultural films that do not privilege the "optical visuality" accorded the highest privilege in "modern Euro-American societies" (2002, xiii). Given this context, she cautions against any careless transfer of her ideas onto commercial cinema. While von Trier exaggeratedly privileges optical visuality in *Melancholia*, what I find most interesting about the film is the way its visual style operates in service to its embodiments of depression, anxiety, and planetary attachments. It is also important to note that, by bringing Marks's feminist film scholarship into conversation with Von Trier's film, I do not want to elide the responses of many feminist critics to Von Trier's oeuvre. His films are regularly critiqued for their misogynistic portraits of female suffering. However, as I argue below, I think something qualitatively different is happening in *Melancholia* vis-à-vis the female body and female suffering. For instance, in *Melancholia* the suffering female body enables von Trier to make the argument that, in the current geotraumatic circumstances that beset the planet and all life on it, melancholia, a very particular form of suffering, is the only position we can occupy.

12. In proposing that the film produces both socioecological and planetary melancholy as ritual, participatory events, I want to distinguish these modes of environmental melancholy from other readings of visual art that foreground melancholy. See, for instance, Bowring (2008); Klibansky, Panofsky, and Saxl (1964); Schwenger (2006); and Sontag (1978).

13. Mortimer-Sandilands (2010, 333). On different manifestations of environmental melancholy see also LeMenager (2011); Merola (2012); and T. Morton (2010a).

14. "World-for-us" and "world-without-us" are Thacker's formulations. See also his third term, "world-for-itself," his distinction between "world-for-itself" and "world-without-us," and his definition of the concept of "Planet," "a negative concept, simply that which remains 'after' the human" (Thacker 2011, 4–9).

15. Cohen also makes the point that von Trier presents marriage as a corrupt ritual (2012, 240).

16. For Edmund Burke the sublime trigger is external, usually something found in nature: a storm, a mountain, intense dark or light, something too big or too small. For Immanuel Kant, on the other hand, while the sublime might be instigated by the external world, it is much more about the inner self and human abilities to think about nature, things, and concepts. Despite these differences however, the operation of the sublime is the same in both Burke and Kant: the self transcends whatever has troubled it.

17. For a useful definition of apparatus see Agamben (2009, 14–15).

18. For a fuller discussion of trans-corporeality, see Alaimo (2010, esp. 2–4, 6–22).

19. Fromm (1997, 2) quoted in Alaimo (2010, 11).

20. Barad (2007, 185, 33). For a brief discussion of intra-activity see Barad (2003). For an extended examination of this concept, see Barad (2007, esp. 1–185).

21. See, for instance, Christian (2011, esp. 17–136), Close (2006), and Rae (2012). This scene also evokes an ancient form of posthumanism, the atomist materialism posited by Lucretius (2011).

22. For one take on our fascination with disaster see Wilson (2012).

23. See, for instance, Major and McMurry (2012).

REFERENCES

Agamben, Giorgio. 2009. "What is An Apparatus?" In *What Is An Apparatus?: And Other Essays*, translated by David Kishik and Steffan Pedatella. 1–24. Stanford: Stanford University Press.

Ahmed, Sara. 2006. *Queer Phenomenology: Orientations, Objects, Others*. Durham: Duke University Press.

Alaimo, Stacy. 2010. *Bodily Natures: Science, Environment, and the Material Self*. Bloomington: Indiana University Press.

Barad, Karen. 2007. *Meeting the Universe Halfway: Quantum Physics and the Entanglement of Matter and Meaning*. Durham, NC: Duke University Press.

———. 2003. "Posthumanist Performativity: Toward an Understanding of How Matter Comes to Matter." *Signs* 28 (3): 801–831.

Bennett, Jane. 2004. "The Force of Things: Steps Toward an Ecology of Matter," *Political Theory* 32 (3): 347–372.

———. 2010. *Vibrant Matter: A Political Ecology of Things*. Durham: Duke University Press.

Bianco, Jamie "Skye." 2007. "Techno-Cinema: Image Matters in the Affective Unfoldings of Analog Cinema and New Media." In *The Affective Turn: Theorizing the Social*, edited by Patricia Ticineto Clough and Jean Halley. 47–76. Durham: Duke University Press.

Bowring, Jacky. 2008. *A Field Guide to Melancholia*. Harpenden: Oldcastle Books.

Buell, Frederick. 2003. *From Apocalypse to Way of Life: Environmental Crisis in the American Century*. New York: Routledge.

Burke, Edmund. 1990. *A Philosophical Enquiry into the Origin of our Ideas of the Sublime and Beautiful*. New York: Oxford.

Carroll, Noël. 2008. *The Philosophy of Motion Pictures*. Malden, MA: Blackwell.

Christian, David. 2011. *Maps of Time: An Introduction to Big History*. Berkeley: University of California Press.

Clark, Nigel. 2012. "Rock, Life, Fire: Speculative Geophysics and the Anthropocene," *Oxford Literary Review* 34 (2): 259–276.

Clark, Timothy. 2012. "Deconstruction in the Anthropocene," *Oxford Literary Review* 43 (2): v–vi.

Close, Frank. 2006. *The New Cosmic Onion: Quarks and the Nature of the Universe*. New York: Taylor and Francis.

Clough, Patricia Ticineto. 2007. "Introduction." In *The Affective Turn: Theorizing the Social*, edited by Patricia Ticineto Clough and Jean Halley. 1–33. Durham: Duke University Press.

Cohen, Tom. 2012. "Polemos: 'I am at war with myself' or, Deconstruction™ in the Anthropocene?" *Oxford Literary Review* 34 (2): 239–257.

Colebrook, Claire. 2012. "Not Symbiosis, Not Now: Why Anthropogenic Change Is Not Really Human," *Oxford Literary Review* 34 (2): 185–209.

Connolly, William E. 2010. "Materialities of Experience." In *New Materialisms: Ontology, Agency, and Politics*, edited by Diana Coole and Samantha Frost. 178–200. Durham: Duke University Press.

Crutzen, Paul J. 2002. "Geology of Mankind." *Nature* 415 (23): 23.

Crutzen, Paul J. and Christian Schwägerl. 2011. "Living in the Anthropocene: Toward a New Global Ethos." *Yale Environment* 360, January 24, e360.yale.edu/feature/living_in_the_anthropocene_toward_a_new_globa_ethos/2363/.

Crutzen, Paul J. and Eugene Stoermer. 2000. "The 'Anthropocene,'" *IGBP Global Change Newsletter* 41: 17–18.

Cuarón, Alfonso, dir. 2006. *Children of Men*. Universal Studios.

Cubitt, Sean. 2005. *The Cinema Effect*. Cambridge: MIT Press.

Dargis, Manohla. 2011. "This Is How The End Begins." *New York Times*, December 30.

Eng, David L. and David Kazanjian, eds. 2002. *Loss: The Politics of Mourning*. Berkeley: University of California Press.

Freud, Sigmund. 1917. "Mourning and Melancholia." In *The Standard Edition of the Complete Psychological Works of Sigmund Freud*, edited and translated by James Strachey, 24 vols. 14: 243–58. London: Hogarth Press, 1957.

Fromm, Harold. 1997. "The 'Environment' Is Us." *Electronic Book Review*, January 1, www.atlx.com/ebr/reviews/rev8/r8fromm.html.

Hansen, Mark B. N. 2006. "Media Theory." *Theory, Culture, & Society* 23 (2–3): 297–306.

Heise, Ursula K. 2012. "Introduction: The Invention of Eco-Futures." *Ecozon@* 3 (2): 1–10.

———. 2008. *Sense of Place and Sense of Planet: The Environmental Imagination of the Global*. Oxford: Oxford University Press.

Hillcoat, John, dir. 2009. *The Road*. Sony.

Kant, Immanuel. 1960. *Observations on the Feeling of the Beautiful and Sublime*, translated by John T. Goldthwait. Berkeley: University of California Press.

Klibansky, Raymond and Erwin Panofsky and Fritz Saxl. 1964. *Saturn and Melancholy: Studies in the History of Natural Philosophy, Religion, and Art*. New York: Basic Books, 1964.
Kolbert, Elizabeth. 2011. "Enter the Anthropocene—The Age of Man." *National Geographic*, March.
Kristeva, Julia. 1989. *Black Sun: Depression and Melancholia*, translated by Leon S. Roudiez. New York: Columbia University Press.
Latour, Bruno. 2011. "The Year in Climate Controversy." *Tehelka*, January 4.
———. 2004. "Why Has Critique Run out of Steam?: From Matters of Fact to Matters of Concern." *Critical Inquiry* 30: 225–248.
Leader, Darian. 2009. *The New Black: Mourning, Melancholia, and Depression*. Minneapolis: Graywolf Press.
LeMenager, Stephanie. 2011. "Petro-Melancholia: The BP Blowout and the Arts of Grief" *Qui Parle* 19 (2): 25–56.
Lucretius. 2011. *On the Nature of Things*, translated by Frank O. Copley. New York: W. W. Norton.
Major, William and Andrew McMurry. 2012. "Introduction: The Function of Ecocriticism; or, Ecocriticism, What Is It Good For?" *Journal of Ecocriticism* 4 (2): 1–7.
Marks, Laura U. 2000. *The Skin of the Film: Intercultural Cinema, Embodiment, and the Senses*. Durham: Duke University Press.
———. 2002. *Touch: Sensuous Theory and Multisensory Media*. Minneapolis: University of Minnesota Press.
McKibben, Bill. 2010. *Eaarth: Making Life on a Tough New Planet*. New York: Henry Holt.
Merola, Nicole M. 2012. "Cosmopolis: Don DeLillo's Melancholy Political Ecology," *American Literature* 84 (4): 827–853.
Mortimer-Sandilands, Catriona. 2010. "Melancholy Natures, Queer Ecologies." In *Queer Ecologies: Sex, Nature, Politics, Desire*, edited by Catriona Mortimer-Sandilands and
Bruce Erickson. 331–58. Bloomington: Indiana University Press.
Morton, Oliver. 2011. "The Anthropocene: A Man-Made World," *The Economist*, 26 March.
Morton, Timothy. 2010a. "The Dark Ecology of Elegy." In *The Oxford Handbook of Elegy*, edited by Karen Weisman. 251–71. Oxford: Oxford University Press.
———. 2010b. *The Ecological Thought*. Cambridge: Harvard University Press.
———. 2012. "Ecology without the Present," *Oxford Literary Review* 34 (2): 229–38.
Nixon, Rob. 2011. *Slow Violence and the Environmentalism of the Poor*. Cambridge: Harvard University Press.
Rae, Alastair. 2012. *Quantum Physics: Illustration or Reality?* Cambridge: Cambridge University Press.
Roberts, David. 2011a. "The Brutal Logic of Climate Change," *Grist*, December 5, grist.org/climate-change/2011–12–05–the-brutal-logic-of-climate-change/.
———. 2011b. "Dr. Grist's Four-Part Prescription for Managing Climate Change," *Grist*, May 3, grist.org/climate-policy/2011–05–02–dr-grists-four-part-prescription-for-managing-climate-change/.
Ruddiman, William R. 2003. "The Anthropogenic Greenhouse Era Began Thousands of Years Ago," *Climatic Change* 61: 261–93.
Schwenger, Peter. 2006. *The Tears of Things: Melancholy and Physical Objects*. Minneapolis: University of Minnesota Press.
Shaviro, Steven. 1993. *The Cinematic Body*. Minneapolis: University of Minnesota Press.
———. 2012. "Melancholia, or, The Romantic Anti-Sublime," *Sequence* 1 (1), reframe.sussex.ac.uk/sequence/files/2012/12/MELANCHOLIA-or-The-Romantic-Anti-Sublime-SEQUENCE-1.1–2012–Steven-Shaviro.pdf.
Sinnerbrink, Robert. 2011. *New Philosophies of Film: Thinking Images*. New York: Continuum.
Sobchack, Vivian. 1991. *The Address of the Eye: A Phenomenology of Film Experience*. Princeton: Princeton University Press.
Sontag, Susan. 1978. "Melancholy Objects." In *On Photography*. 49–82. New York: Farrar, Straus and Giroux.

Speth, Gus. 2005. *Red Sky at Morning: America and the Crisis of the Global Environment*. New Haven: Yale University Press.

Szerszynski, Bronislaw. 2012. "The End of the End of Nature: The Anthropocene and the Fate of the Human," *Oxford Literary Review* 34 (2): 165–84.

Taylor, Matthew A. 2012. "The Nature of Fear: Edgar Allen Poe and Posthuman Ecology," *American Literature* 84 (2): 353–79.

Thacker, Eugene. 2011. *In the Dust of This Planet: Horror of Philosophy*, Vol. 1. Alresford: Zero Books.

Verbeek, Peter-Paul. 2006. "Materializing Morality: Design Ethics and Technological Mediation," *Science, Technology, & Human Values* 31 (3): 361–80.

Von Trier, Lars, dir. 2011. *Melancholia*. Magnolia.

Wilson, Eric G. 2012. *Everyone Loves a Good Train Wreck: Why We Can't Look Away*. New York: Sarah Crichton.

Wolfe, Cary. 2010. *What Is Posthumanism?* Minneapolis: University of Minnesota Press.

Zalasiewicz, Jan, et al. 2008. "Are We Now Living in the Anthropocene?" *GSA Today* 18 (2): 4–8.

Zizek, Slavoj. 2010. *Living in the End Times*. London: Verso, 2010.

Index

About the Contributors

Kristie S. Fleckenstein is professor of English at Florida State University where she teaches graduate and undergraduate courses in rhetoric and composition. Her research interests focus on issues of materiality and visuality, especially as they intersect with social action. Her current research project explores the role of nineteenth-century photography in debates about racial identity.

Brendan Keogh is a PhD candidate at RMIT University in Melbourne, Australia. His research focuses on forwarding a framework of videogame criticism that accounts for the interrelationships and mediations between player and text. He is also a widely published videogame critic appearing in publications such as Edge, Ars Technica, and Kill Screen.

Jonathan Rey Lee received his PhD in Comparative Literature from UC Riverside, where he specialized in nineteenth-century realist literature and philosophy. He is currently writing on the philosophy and narrative theory of performative media, particularly LEGO and board games.

Matthew A. Levy lives by Point Defiance Park in Tacoma, Washington, and teaches undergraduate writing and theory courses at Pacific Lutheran University. His writing explores conditions and traditions of cynicism that influence and reveal the connection between affect and ideological practices.

Emily McArthur is a doctoral student at the University of California, Riverside. Her academic interests include Victorian and eighteenth-century British literature and questions of political economy. Her main research is focused on the ways that cultural artifacts can reflect middle-class consumption of and resistance to capitalism.

Josh Mehler is a doctoral student in the Rhetoric and Composition program at Florida State University where he teaches undergraduate courses in the Editing, Writing, and Media program. His research interests focus on the intersection between rhetoric, place and mobility, emphasizing digital technology's role in constructing the affordances of space.

Nicole M. Merola is associate professor of ecocriticism and American literature in the Literary Arts and Studies Department at the Rhode Island School of Design where she teaches courses on and researches contemporary environmental literature, film, and culture; theories of natureculture; critical animal studies; and representations of climate change.

Anthony Miccoli is the Director of Philosophy and an Associate Professor of Communication Arts and Philosophy at Western State Colorado University. He is the author of *Posthuman Suffering and the Technological Embrace* (2010, Lexington Books).

Amy D. Propen is a lecturer of rhetoric and composition in the Writing Program at the University of California, Santa Barbara. Her research on visual and material rhetoric, critical cartographies, and rhetoric as advocacy has appeared in *Technical Communication Quarterly*; *Journal of Business and Technical Communication*; *Written Communication*; *Law, Culture and the Humanities*; *ACME: An International E-Journal of Critical Geographies*; and the edited collections *Rethinking Maps: New Frontiers in Cartographic Theory*, and *Environmental Rhetoric: Ecologies of Place*. She is author of *Locating Visual-Material Rhetorics: The Map, the Mill, and the GPS* (2012).

Colbey Emmerson Reid is the director of the Consumer Innovation Consortium, an interdisciplinary marketing research group in the Poole College of Management at North Carolina State University. Her research has appeared in several literary and cultural studies journals. She is currently working on books on design culture, machine aesthetics, and the relationship between business and the liberal arts.

Elise Takehana is assistant professor of English studies at Fitchburg State University, researches aesthetics, digital studies, and twentieth-century text and image production. She coedited two Digital Humanities Quarterly special issues, one forthcoming. Her essay "William Burroughs / Robert Rauschenberg: Text-Image / Image-Text" appears in *The Future of Text and Image*.

John Tinnell is an assistant professor of English at the University of Colorado Denver. His research centers on developing theoretical principles to help orient rhetorical inquiry and digital creativity amid the paradigm shift from personal computing to ubiquitous computing. His previous publications on topics such as augmented reality, media ecology, poststructuralism, and grammatology have appeared in *Environmental Communication*, *Fibreculture Journal*, *Enculturation*, *Deleuze Studies*, and *Convergence*. He is co-editor of *Augmented Reality: Innovative Perspectives across Art, Industry, and Academia* (2014).

Yoni Van Den Eede is a postdoctoral fellow of the Research Foundation–Flanders (FWO) affiliated with iMinds-SMIT—Studies on Media, Information and Telecommunication—and with the Centre for Ethics and Humanism, both at the Vrije Universiteit Brussel (Free University of Brussels). He conducts research into the philosophy of technology, media theory, and media ecology, with an emphasis on phenomenological, cultural, and existential themes. He is the author of *Amor Technologiae: Marshall McLuhan as Philosopher of Technology* (Brussels, VUBPRESS, 2012).

Dennis M. Weiss is professor of philosophy at York College of Pennsylvania and the editor of *Interpreting Man*. His research and essays on the intersection of philosophical anthropology and philosophy of technology have been published in journals such as *Expositions*, *Techne*, and *The Humanities and Technology Review*.

www.ingramcontent.com/pod-product-compliance
Lightning Source LLC
LaVergne TN
LVHW052249200926
848481LV00002B/315